# Fundamental and Research Frontier of Atmospheric Corrosion

## Special Issue Editor

Manuel Morcillo

MDPI • Basel • Beijing • Wuhan • Barcelona • Belgrade

**MDPI**

*Special Issue Editor*
Manuel Morcillo
National Centre for Metallurgical Research (CENIM-CSIC)
Spain

*Editorial Office*
MDPI AG
St. Alban-Anlage 66
Basel, Switzerland

This edition is a reprint of the Special Issue published online in the open access journal *Materials* (ISSN 1996-1944) in 2017 (available at: http://www.mdpi.com/journal/materials/special_issues/atmospheric_corrosion).

For citation purposes, cite each article independently as indicated on the article page online and as indicated below:

Author 1; Author 2. Article title. *Journal Name*. **Year**. Article number/page range.

**First Edition 2017**

**ISBN 978-3-03842-643-1 (Pbk)**
**ISBN 978-3-03842-642-4 (PDF)**

Image courtesy of Manuel Morcillo Linares

# Table of Contents

# About the Special Issue Editor

**Manuel Morcillo** is PhD in Chemistry from the Complutense University of Madrid (UCM). He is currently a Research Professor in the Department of Surface Engineering, Corrosion and Durability at the National Centre for Metallurgical Research (CENIM-CSIC). He has been the Director of CENIM, President of the International Corrosion Council (ICC) and President of the Ibero-American Corrosion and Protection Association (AICOP). Author of more than 250 scientific publications in Spanish and international journals and author or editor of several books on this speciality. He has received awards from a number of different scientific journals and institutions, including FSCT (1990), AICOP (1992, 2003, 2006), SSPC (1994), AETEPA (1994), NACE (1995), PCE (1997), JPCL (1998, 2003), ICC (2003, 2017), ACMM (2004), ITPTS (2013). His main lines of research are concerned with corrosion in natural environments, mainly atmospheric corrosion, and anticorrosive protection by coatings, especially anticorrosive paints.

# Preface to "Fundamental and Research Frontier of Atmospheric Corrosion"

Metallic corrosion is an expensive problem in industrialised countries, creating annual costs that are equivalent to around 3.5% of GDP. More than half of this amount is attributable to atmospheric corrosion.

Atmospheric corrosion has been extensively researched over the last hundred years and a large amount of scientific and technical literature, numerous books and many general treatises have been published on this topic. In the period between 1970-2000 ASTM-STP made great efforts to disseminate progress in this field through the publication of monographies and the organisation of symposia on the atmospheric corrosion of metals and alloys.

This special issue (SI) of Materials (MDPI) brings together the state of the art in atmospheric corrosion, highlighting the advances made in the last fifteen years and identifying the areas where future research will need to concentrate.

The effects of meteorological and pollution variables on atmospheric corrosion are now quite well known. Even so, our knowledge of this issue still holds many gaps, such as how to accurately estimate the total time of wetness of metallic surfaces and the real effects of climate change and acid rain. Average climatic variables and regression methods have traditionally been used in atmospheric corrosion studies. However, as is remarked in the first chapter of this SI, the processes controlling corrosion are highly dependent on the characteristics of wet/dry cycles occurring on the metal surface exposed to the atmosphere, which cannot be properly identified by average data, and factors such as aerosols content, wind, global radiation, rain, etc. interacting with each other. In this respect, laboratory studies under droplets may provide a new insight into the scientific knowledge of atmospheric corrosion of metals and alloys.

For engineers and political policy-makers it is fundamental to be able to predict atmospheric corrosion well into the future (25, 50, 100 years). Thus, in some highly developed countries efforts are now being made to design civil structures such as bridges and other load-bearing structures for 50-100 years of service without any maintenance. Data mining and modelling tools can help to improve atmospheric corrosion forecasts and anti-corrosive designs. Thus it is important to underline the efforts made in wide-scale international cooperative research programmes such as ICP/UNECE, ISOCORRAG and MICAT, as well as the studies undertaken by the Russian Academy of Sciences, SVUOM in the Czech Republic, etc. The SI presents four chapters dedicated to this matter, where it can be seen that despite great progress in the development of damage (dose-respose) functions there is still a way to go for such long-term modelling of atmospheric corrosion processes.

Until relatively recently surprisingly little attention has been paid to the action of airborne marine salts in chloride-rich atmospheres. Corrosion in coastal regions is a particularly relevant issue due to the latter's great importance to human society. About half of the world's population lives in coastal regions and the industrialisation of developing countries tends to concentrate production plants close to the sea. The corrosion mechanisms that act in chloride-rich and marine-industrial atmospheres have not yet been sufficiently clarified. This matter is addressed in this SI considering three types of materials: carbon steel, low alloy steels and concrete.

The effects of atmospheric corrosion on old buildings and structures, statues and monuments, etc. have resulted in substantial degradation of artistic and historic objects. Mankind is concerned about the degradation, restoration and conservation of its historic heritage. Two chapters on this topic appear in the SI, one on the long-term atmospheric corrosion mechanisms of low alloy steel reinforcements and another on ancient historic copper patinas.

Finally, over the past few decades the new analytical techniques developed to study properties of solid surfaces have continued to increase and improve in terms of resolution and sensitivity. The more recent analytical techniques are both surface-sensitive and able to provide information under in-situ conditions. In this respect, the SI presents a review article on conventional and advanced vibrational spectroscopic techniques.

**Manuel Morcillo**
*Special Issue Editor*

*materials*     MDPI

*Review*

# Recent Progress and Required Developments in Atmospheric Corrosion of Galvanised Steel and Zinc

## Ivan S. Cole

School of Engineering, RMIT University, Melbourne, Victoria 3001, Australia; ivan.cole@rmit.edu.au

Received: 16 October 2017; Accepted: 9 November 2017; Published: 9 November 2017

**Abstract:** This paper reviews the progress in atmospheric corrosion of zinc since 2009. It firstly summarises the state of the art in 2009, then outlines progress since 2009, and then looks at the significance of this progress and the areas the need more research. Within this framework, it looks at climate effects, oxide formation, oxide properties, pitting, laboratory duplication of atmospheric corrosion, and modelling. The major findings are that there have been major advances in the fields understanding of the structure of corrosion patina, in particular their layered structure and the presence of compact layers, local corrosion attacks have been found to be a significant process in atmospheric corrosion and experiments under droplets are leading to new understanding of the criticality of drop size in regulating atmospheric corrosion processes. Further research is indicating that zinc oxide within corrosion products may promote the oxygen reduction reaction (ORR) and that, in porous oxides, the ORR would control pore chemistry and may promote oxide densification. There is a strong need for more research to understand more deeply the formation and properties of these layered oxides as well as additional research to refine and quantify our emerging understanding of corrosion under droplets.

**Keywords:** zinc; corrosion; oxides; droplets

---

## 1. Introduction

There has been significant progress in our understanding of atmospheric corrosion of metals over the last decade. A range of studies into the interaction of aerosols with surfaces, oxide development, and the role of oxides in controlling the corrosion process have deepened the field's understanding and together provide a fuller and more profound picture of atmospheric corrosion. To tackle this subject for all metals is too large a challenge for this modest paper so the developments will be highlighted by discussing the atmospheric corrosion of zinc. Wider developments will be incorporated when required to illustrate a common point. The current author was able to contribute a review on this subject in 2009 [1] and thus there is no need to repeat the material in that review, although it does form a useful definition of our knowledge a decade ago. In this paper for several critical processes that control atmospheric corrosion I will summarise the state of the art in 2009, highlight developments in the past decade, and discuss where greater research is required and what problems need to be resolved.

## 2. Climate Effects

### 2.1. State of the Art in 2009

Cole et al.'s [1] paper of 2009 extensively maps out aerosol formation, chemistry and transportation, and deposition while other papers [2] have outlined how an aerosol wets and other environmental effects on corrosion including solar radiation, rain, wind, etc. and how this is related to local and global climate so that, in 2009, a strong understanding of the effect of climate on corrosion existed.

In this, it was appreciated that there was a variety of pollutant deposition mechanisms that could trigger corrosion, dry deposition, wet deposition through rain or fog, or deposition of aerosols.

## 2.2. Developments Since 2009

In two papers, Cole et al. [3,4] developed a corrosion map of Abu Dhabi and collected site climatic and corrosion data. Despite being very dry, significant corrosion was observed, higher than equivalent sites in Australia. The model indicated that this could be attributed in the low rainfall resulting in higher salt retention rates on the exposed metals combined and facilitating significant night time condensation. Chico et al. [5] have developed a corrosion map of zinc for Spain based on a dose function approach. Panchenko et al. [6] developed dose functions for predicting the corrosion of steel and corrosion for continental Russia based on data from 12 exposure sites. Tidblad et al. [7] looked at data for corrosion from 19 sights in Europe of a range of materials including steel, zinc, aluminium, and copper from 1987–2014. They found that the corrosion rates and $SO_2$ levels significantly decreased after 1997.

Cole et al. [8] and Trivedi et al. [9] looked at the effect that climate change may have on metals in infrastructure in Australia. They used the effect of the latest climate scenarios on actual climate data to give predicted temporal climate sequences (rather than adjust average parameters). In general, they found that in coastal zones overall lower and less frequent rainfall would result in a higher level of retained salinity and thus a higher corrosion rate while in inland areas decreased RH would result in decreased night-time condensation and so decreased corrosion.

## 2.3. Significance of Developments and Need for More Research

The work up to 2009 and the recent work all indicated that the effect of climate on the atmospheric corrosion of zinc is best understood by studying how integrated temporal climate data affects local processes of corrosion. This is in contrast to the use of average climatic variables and parametric or regression methods. The processes controlling corrosion are highly dependent of cycles (wet/dry, etc.) which cannot be properly captured by average data and factor—such as wind, global radiation, and rain—all interact and so need to be studied by methods that facilitate this interaction.

# 3. Oxide Formation

## 3.1. State of the Art in 2009

It had been well established that the development of oxides can provide a barrier to reduce the corrosion rate of zinc or galvanised steel and that the effectiveness of this barrier depends on the nature of the exposure environment and of the oxide [10]. Most studies of oxides had looked at surface properties using such techniques as XRD, SEM-EDAX, or FTIR and thus a good knowledge of the nature of near surface oxides in short-term exposures and long-term field exposures had been obtained [11–15]. However, there were some minor differences in observation or theories of oxide formation based on the different nature of exposures. While a number of measurements had been made of oxide properties (surface wettability, IEP, electrochemical parameters) there was no consensus on which oxides were the most effective in reducing corrosion.

## 3.2. Developments Since 2009

A number of workers [16–18] have continued to undertake laboratory or field studies of oxide developing on the surface of zinc plates, though for some such as Odnevall et al. [16], the motive has widened to look at the quantity of zinc run off into the environment. Heldberg et al. [17] from Odnevall's group studied zinc exposed for four years in marine and urban environments in Sweden. Using Raman spectroscopy, they observed that the corrosion products consisted mainly of hydrozincite and zinc oxide for both exposure conditions. However, while the ZnO was predominately crystalline in the marine location, it was predominately amorphous at the urban location. Cui et al. [18] examined the oxides developed when zinc was exposed for four years in tropical marine locations in China.

They observed a three-layered structure with simonkolleite ($Zn_5 Cl_2[OH]_8 \cdot H_2O$) on the inner-most layer sodium zinc chlorohydroxy-sulfate ($NaZn_4Cl[OH]_6SO_4 \cdot 6H_2O$) in the middle layer and zinc hydroxyl carbonates and zincite ($ZnO$) on the surface. They observed a decrease in corrosion rate with time which they associated with the protective properties of simonkolleite. Liu et al. [19] exposed zinc in a Chinese industrial marine site and noted that the corrosion rate decreased and the polarization resistance increased with time. The noted that, initially, the rust layer consisted of zinc hydroxyl carbonates while sodium zinc chlorohydroxy-sulfate and $Zn_2(OH)_{15} Cl_3(SO4)_3 \cdot 5H_2O$ compact layers developed with longer exposures. Interestingly, Liu et al. noted that the corrosion was initially localised but developed uniform coverage with time. Persson et al. [20] undertook a worldwide (Europe, East Asia, and USA) study of the corrosion of galvanised steel over periods from 0.5 to 2 years. Corrosion pits were observed for all exposure conditions and contained zinc hydroxyl-sulfates, sodium zinc chlorohydroxy-sulfate, and zinc hydroxyl chloride. Zinc hydroxyl carbonates were found outside the pits while, overall, the sulphate content was higher for industrial sites.

Li et al. [21] looked at salt deposition and corrosion for very short exposure times (30 min) in a marine Hawaiian locations on low carbon steel and zinc. In this, they duplicated the earlier test of Cole et al. [22] in Australia with very similar results. Li et al. [21] found that the corroded region or the original droplet was from 10–30 µm. Corrosion occurred at all sizes and secondary spreading was observed for the larger droplets. The most common corrosion product was simonkolleite with some zinc oxide and zinc hydroxyl carbonate with the latter two compounds being most common in the secondary spread zones.

One of the significant developments since 2009 has been the substantial use of focused ion beam SEM and TEM to study the full depth of oxides. Cole et al. [23] carried out one of the first studies of oxide development on zinc using FIB-SEM. They exposed zinc to fine seawater droplets under controlled humidity for 15 min to 6 h and observed the oxide that formed by Raman, XRD, SEM, and FIB-SEM. The SEM observations are consistent with their previous work but what was novel was their observation on the build-up of the oxide layer with time. There key observations were that:

(a) A three layer structure built up over the 6 h with consisting of a zone below the original metal surface where oxide has filled in previous areas of metal attack which appear to be along microstructural features, a zone above the original metal surface that is relatively compact, and a third zone above the first two where a highly porous crystalline structure is observed. The elemental composition of the first layer was predominately zinc and oxygen with some areas of zinc, oxygen, and chloride; the second layer contained zinc, oxygen, and carbon; and the third zinc, oxygen, carbon, and chloride.

(b) After the build-up of an initial oxide, the underlying metal is subject to localized attack that appears to be along grain boundaries.

(c) The oxide in the second and third layers appears to decrease in void content with time until it is relatively compact (except the upper surface) after 6 h.

Subsequently, Thomas et al. [24] in similar studies of zinc under a saline drop refined this layered structure. By both FIB-SEM and TEM analysis, he found a very fine (20 to 70 µm) crystalline compact layer (similar to that observed by McDonald [25], although this observation was after anodic polarization into the passive region in alkaline solution). TEM analysis indicated that the passive layer was ZnO. Above this compact layer was a precipitated layer of low void content and over this was a porous layer of zinc based chloride sulphate compounds. EIS studies indicated that the summation of these barrier layers was protective. In zinc substrates electrochemically oxidized in NaCl solution, Prestat et al. [26] observed similar compact layers to Thomas et al. The zinc was oxidized at a potential of −0.64 V in 0.1 M NaCl solution for 2 h. They observed a three-layered structure, the upper most layer consisted off flakes of simonkolleite (with pores between) underlayed by a compact layer of ZnO and an "interface layer" which was zinc-rich with some oxygen. Electrochemical studies indicated that the layer structure was partially protective. On the basis of their electrochemical measurements (cathodic polarization), Prestat et al. [26] postulate that the ZnO layer is inactive to cathodic dissolution and that only a fraction of the Zn sites are active for oxygen reduction.

In a recent paper, Thomas et al. [27] looked at the electrochemical behaviour and oxides that formed at high pH values. They conducted tests on zinc from pH = 12 to 13 in NaOH solution. They noticed a marked decrease in the anodic currents in potentio-dynamic scans and an even more marked decrease in corrosion current in potential hold tests as the pH was increased towards 13. FEB-SEM studies were undertaken of potential hold specimens and a parallel set of tests where droplets at pH = 12 were placed on zinc for times from 30 to 120 min. In all cases, compact oxides and precipitated oxides were seen. At short times, extensive voiding was apparent above or below the compact layer. For the droplet tests, the integrity of the compact layer was markedly higher after 120 min, voiding was substantially decreased, and the precipitation layer thickened.

Fattah-alhosseini and Mirshekari [28] also looked at zinc in alkaline solutions (0.01 M NaOH). Their study showed that Mott–Schottky analysis revealed that the passive films displayed n-type semi-conductive characteristics with the calculated donor density increasing with increases in the applied formation potential while EIS indicated that increasing this potential leads to the growth of a thicker but more defective film.

### 3.3. Significance of Developments and Need for More Research

Continued field studies are broadening or knowledge of the relation of locality and microclimate to oxide formation while the patterns observed prior to 2009 still holding in recent literature. Some finer details are emerging, such as observed climatic or geographic difference between where ZnO is found as a crystalline or amorphous phase. This could be quite important as crystalline ZnO can support the ORR (as discussed in a later section) and the barrier properties and defect transport mechanisms (particularly relevant to compact oxides) are likely to be different for the two phases.

The deepening understanding of the nature of zinc corrosion patina as a layered structures, where in some circumstances the bottom layer may be a compact oxide is critical. This indicates that surfaces methods of identifying corrosion product are not a reliable indicator of the full structure and may bear little relationship to the corrosion performance of this layered structure. A great deal more work is required in this area to generate a much greater understanding of how the layered structures form and how common and under what conditions the various elements form (particularly the compact oxide). Lastly, the role of each layer in forming a corrosion barrier needs to be elucidated.

## 4. Pitting

### 4.1. State of the Art in 2009

In 2009, it was generally held that zinc corrosion was a general process (rather than localised) although it was well established that oxide dissolution [15] was a critical element in on-going corrosion and that this could be local and that corrosion could initiate from local features [29] (e.g., grain boundaries or triple points) on the zinc surface.

### 4.2. Developments Since 2009

Recently, scanning vibration electrode technique (SVET) studies by Mena et al. [30] provided direct electrochemical evidence of pitting in zinc in laboratory tests. They used SVET to study the potential distribution on zinc in 1:1000 diluted ocean water. They observed localised anodic areas and more distributed cathodic areas after 2 h. Over the period of the test (5 h), the location of the anodic zones shifted. They attributed this to the development of oxides blocking the initial pit development and allowing the anodic (pitting) areas to redistribute to new locations.

Some workers have looked at pitting of zinc in specialized solutions, such as Abd El-Rehim et al. [31] in $Na_2S_2O_3$ solution. These studies observed deep pitting and are primarily relevant to cases of high industrial pollutants.

Cole et al. [32] conducted an extensive study of zinc corrosion morphology and form across sites in Southeast Asia and Australia with severe marine, marine, severe industrial, industrial, marine/industrial, urban, and highland environments over a period of one year. Localized corrosion was observed at all sites and Cole et al. classified this localised corrosion in three modes depending on form and aspect ratio ($\alpha$): very shallow $\alpha < 0.7$, sharper pits $\alpha > 0.7$ and nearly occluded pits. Interestingly, the results indicate that shallow pits occurred in the high $SO_2$ environments and that sharp or occluded pits occurred in the moderate or low $SO_2$ environments or marine locations. In the industrial environments where shallow pits occurred, S was found throughout the oxide, but where sharp pits occurred it was found almost exclusively at the bottom of the pits. Cole et al. observed strong oxide dissolution in industrial areas with high $SO_2$ levels and they concluded that deposition of acid aerosols or rain locally dissolved the oxide, promoting relatively vigorous corrosion in these relatively large patches, leading to the formation of shallow pits. In contrast, in lower $SO_2$ environments, the oxides remained intact and corrosion was facilitated by defects and voids in the oxide layer that lead to a highly localised Galvelle [33] type mechanisms and sharp or occluded pits.

*4.3. Significance of Developments and Need for More Research*

Recent research has demonstrated that localised pitting is very important to zinc corrosion. The question of whether this is to be regarded as 'true pitting' is somewhat of a definition issue as the aspect ratios of the local attack are much lower than the pits observed in stainless steel or aluminium. In field observations, pitting appears to be associated with local breakdown in oxide layers and thus the nature of oxide dissolution is critical to its form. Pitting-like phenomena have also been observed in the early stages of corrosion in laboratory experiments but this may have the added dimension of crystallographic orientation.

More work is required to assess how widespread such local corrosion is and what form it takes and how it is related to oxide breakdown and microstructural features. Importantly, the relationship between pitting-like and general corrosion in zinc needs to be elucidated.

## 5. Laboratory Duplication of Atmospheric Corrosion

*5.1. State of the Art in 2009*

In the 20th century, most laboratory studies of atmospheric corrosion where undertaken either in climate chambers or in solution. In the late 90s, it was realised that, it in order to closely duplicate actual field atmospheric corrosion, testing either under droplets or thin films was required.

*5.2. Developments Since 2009*

Valuable work has continued in studying corrosion of zinc under thin electrolyte layers (TELs) such as that of Cheng et al. [34]. In their study, they exposed zinc electrodes to 0.1 M NaCl with an initial thickness of 467 μm at RH of 75%, 85%, and 97% for 72 h. The solution will gradually evaporate, with evaporation being greatest at 75%. They found that the cathodic process was dominated by oxygen reduction and that initial corrosion rates were highest for the 75% RH exposure. However, this reversed with time and at 72 h the corrosion rates were highest for 97% RH exposure. The development of surface oxides was postulated to affect the corrosion rates.

Song et al. [35] used a three electrode system to investigate the electrochemical properties under a drop formed by the wetting of a NaCl crystal. The electrode system was a zinc working electrode (10 mm by 2 mm) and a reference and a counter electrode embedded in epoxy resin. This system was placed in a chamber at 97% RH, 25 °C for 2, 8, and 24 h. Polarization curves and EIS were undertaken, demonstrating that the corrosion rate decreased with time while the polarization resistance increased with time. EIS indicated that the system moved from one of charge transport control to one of mixed charge transport and mass transport control. The authors attributed this to the development of corrosion product which hindered mass transfer.

Azmat et al. [36] introduced a new technique to undertake high throughput conditions under droplets in order to better simulate corrosion under aerosols or rain. Multiple 0.5 μL (diameter approximately 1.6 mm) droplets of saline solution were deposited on a single zinc plate in a humidified environment for 6 h and then the volume loss (after etching) was measured with a profilometer. The observed mass loss (converted from volume loss) was very close to that calculated from multi-channel microelectrode analyser (MMA) measurement using same drop size and chemistry. They demonstrated the technique with $MgCL_2$ and NaCl droplets of different concentrations (0.05 to 3.42 M), and acidified to a pH of 1 or not acidified. Interestingly, the 0.3 M $MgCl_2$ acidified droplet showed the highest volume losses. In order to reproduce aerosol corrosion in a more accurate manner, Azmat et al. [37] inkjet printed very fine aerosol droplets with diameters ranging from 0.2 to 1 μm. Drop chemistry was seawater and NaCl solution at concentrations of 0.42, 0.6, and 3.42 M at its natural pH or acidified to a pH of 1 using a range of acids (HCl, $H_2SO_4$, $HNO_3$, $CF_3SO_3H$) and held at 90% RH for 6 h. Corrosion rates at the lower salinities were relatively constant and not dramatically affected by acidification. With the 3.42 M solution, corrosion rates rose dramatically and varied considerable depending on acidified or not and the acidification method. Thus, the order of corrosion rates for the 3.42 M solutions were, in terms of acidifying agent: HCl, $HNO_3$, unacidified, $H_2SO_4$, and $CF_3SO_3H$. Seawater demonstrated comparable but slightly lower corrosion rates with the order, again in terms of acidifying agent being: $HNO_3$, $H_2SO_4$, natural, HCl, and $CF_3SO_3H$. SEM and FIB-SEM analysis of corrosion under the fine natural pH seawater droplets (1 μm) showed a porous oxide of 0.5 to 0.75 μm in thickness having elemental composition of Zn, O, and C. This compares with the more complex oxides containing Zn, O, C, Cl, S, and Mg that occur with larger (5 μm in diameter) droplets. Nazmat et al. attributed this difference to the high oxygen and thus high hydroxyl concentration likely throughout the small droplets that would promote significant $CO_2$ absorption and precipitation of zinc hydroxyl carbonates. In larger drops, anode/cathode separation would occur and the rate of $O_2$ diffusion would be lower and thus, at the anode, lower $OH^-$ and carbonate concentrations would occur (these being higher at the cathode), promoting the precipitation of zinc hydroxychlorides and other more complex species.

Azmat et al. [38] undertook an extensive study of the nature of droplet, pH, stability, and corrosion product formation with droplets of moderate size (1 μL) that represent larger aerosols or raindrops. Droplets were placed on zinc held for 6 h in humidified conditions (RH = 90%). Droplet chemistry was 3.5% NaCl, 3.5%NaCl acidified to pH = 1 with either HCl or $H_2SO_4$, or alternatively natural seawater. Secondary spreading occurs for all droplets with the pH of the original drop showing significant spatial variation prior to secondary spreading. The edges of the drops all become less acidic or more alkaline while the central regions maintained or increased in acidity. Mass loss values were very different, being lowest for the sulphate-containing droplets (3.5% NaCl acidified with $H_2SO_4$ or seawater) and the highest for the unacidified NaCl solution or NaCl acidified with HCl. Within these two groups, the acidification increased corrosion rates only slightly. A combination of XRD, SEM, and Raman spectroscopy identified a wide range of corrosion products, with the main constituents being $Zn_5(OH)_8(Cl)_2 \cdot H_2O$ (simonkolleite) in the central region of the non-sulphate containing solutions and $NaZn_4Cl(OH)_6SO_4 \cdot 6H_2O$ (gordiate) and simonkolleite being the major and minor components respectively in the sulphate containing solutions. ZnO, $ZnCO_3$, or $Zn_5(CO_3)_2(OH)_6$ were the major constituents in the secondary spread region. The study has three notable results: that secondary spreading can occur even in acidic droplets, that there can be significant precipitation in such acidic droplets, and that corrosion rates under conditions that produce gordiate appear to be lower than those where the main product is simonkolleite. Risteen [39] refined the ink-jet printer method to deposit drops of a controlled size on low-carbon steels.

## 5.3. Significance of Developments and Need for More Research

This work has demonstrated that studies under thin electrolyte layers and droplets are useful and necessary for atmospheric corrosion and that standardized methods can be developed. The corrosion products formed under TELs or relatively large droplets appear similar to those observed in the field.

Similarly, processes such as secondary spreading are observed for large droplets and moderate droplets both in laboratory experiments and in the field. However, when the droplets become small, corrosion products and reactions change. The exact size when this change occurs has not been determined, but it has certainly occurred by 1 μm. For these droplets, the Evans effect with anode and cathode separation does not occur, nor is the ORR likely to be limited by oxygen diffusion. One result is the predominance of zinc oxide and zinc hydroxide as corrosion products.

Under many circumstances, atmospheric corrosion will just be the net effect of a large number of droplet surface interactions. Thus, it is important to determine the rate and forms of this type of corrosion. These studies are a start in this direction, but more work is required to define the corrosion rate as a function of drop size, the effect of enhanced oxygen levels (relative to corrosion in solution) on droplet corrosion, and the change and nature of the transition from Evans controlled to non-Evans processes. Importantly, droplet studies should be undertaken in cases where an oxide film has already built up—this is required to reproduce service effects as most exposed metals will have significant patinas.

## 6. Oxide Properties

### 6.1. State of the Art in 2009

There had been little direct study of the properties of oxides on zinc up until 2009. Most properties were inferred from electrochemical measurements of laboratory experiments or by combining mass loss data with surface determination of oxide properties. Because of the layered structure of the zinc patina, relating mass loss to nature of the surface oxide is unreliable. One exception to the lack of basic studies on oxides are the works of Muster et al. [40,41] who looked at surface charge, surface energy, and wettability on zinc corrosion products.

### 6.2. Developments Since 2009

Both theoretical and experimental researchers [42,43] have been concerned with how the development of crystalline ZnO may affect the oxygen reduction reaction. One school of thought arguing that, since ZnO is a semiconductor, it should support the ORR [44].

Thomas et al. [42] looked at this issue using a scanning electrochemical microscope (SECM) with a Pt microelectrode to measure the oxygen consumption in solution above a zinc sample. Here, a Pt tip is held at specific potentials where it undergoes the ORR to measure the local oxygen concentration. The experiments were conducted at pH of 7 in a 0.001M sodium carbonate salt solution and at pH of 13 in a 0.1 M NaOH solution. In the former condition, the zinc will be active electrochemically while in the latter it will be passive and covered by a zinc oxide film. At a pH of 7, the oxygen concentration measured at a distance of 50 μm from the zinc surface, using the Pt microelectrode, is 15–60% of the bulk oxygen concentration. Therefore, correspondingly, the oxygen consumption by zinc (due to ORR) is 40–85% of the bulk oxygen concentration. At a pH of 13, where zinc undergoes passivation, the oxygen consumption by zinc is 70–80% of the bulk oxygen concentration. Thus, the oxygen consumed and the ORR reaction in the passive state is considerably higher than in the active state. Thomas et al. argue that this is because significant oxygen reduction is occurring on the zinc oxide in the passive state. Nazarov et al. [45] looked at local corrosion of zinc and zinc alloys contaminated with NaCl crystals using a Kelvin probe. They found that the rate of ORR was high on oxide films that formed on zinc (which they attributed to the semiconducting properties of ZnO) but dramatically decreased if alloy agents such as magnesium were included in the zinc as these altered the composition of the resulting oxide films.

Prestat et al. [43] electrodeposited ZnO onto a copper and then investigated the ORR in $10^{-3}$ M KOH solution using a rotating disk electrode. They found that ZnO has a relatively poor electrocatalytic activity towards oxygen reduction and that oxygen reduction occurs via a direct pathway without a hydrogen peroxide intermediary. They also found that the ORR kinetics depend on film thickness and pH.

*6.3. Significance of Developments and Need for More Research*

The work on the role of zinc oxides in supporting the ORR reaction is vitally important for, as discussed in the next section, it has major implications the for the corrosion rate (when the corrosion rate is limited or controlled by the ORR) and the growth and densification of oxide, and thus their ability to develop barrier protection. In general, the properties of oxides on zinc are not well-defined and substantial work is required if our understanding of how the multi-layered structures occur is to be advanced.

## 7. Modelling

The aim of this section is not to look at the overall state of corrosion modelling but rather to consider any developments relative to zinc corrosion. The reader can refer to Gunnesagaran et al. [46] or Simillion et al. [47] for a recent detailed review of modelling.

*7.1. State of the Art in 2009*

A number of modelling schemes such as Graedel's [48,49] GILDES or gas, interface, liquid, electrodic and surface regime approach or Cole et al's [50] holistic model had been applied to zinc, while others such as Spence and Haynie [51] had modelled particular processes (oxide dissolution) and McDonald [25] had modelled void transport and oxide stability for passive films on zinc. These works provided a strong framework. However, how porous oxide and layered structures on zinc form and function had not been extensively studied.

*7.2. Developments Since 2009*

Venkatraman et al. [44] analysed electrochemical and diffusional processes occurring in a porous oxide. The model can thus provide guidance as to the processes that may occur as oxide layers build up and densify on zinc. They assumed that, in such a porous oxide, ORR could occur both at the bottom of pore at the metal surface and on the walls of the oxide surface (assuming the oxide was a semi-conductor such as zinc oxide). They analysed the percentage of the ORR that occurred in these two positions and found that these percentages were a function of oxide thickness, oxide conductivity, potential difference between the zinc surface and the oxide, and specific contact resistivity of the metal–oxide interface. Under appropriate circumstances, a significant-to-dominant fraction of the ORR can occur at the oxide pore surface. This in turn can change the chemistry in the pores, leading to depletion of oxygen in the pores and build-up of metal ions. The depletion of oxygen means that the ORR will move up the pore wall to the oxide solution interface. Although not proven by Venkatraman et al. [44], this change in pore chemistry could provide the conditions for precipitation or oxide growth in the pores, leading to the densification of oxides as highlighted by Cole et al. [23]. Sherwood et al. [52,53] used Kelvin's approach to look at condensation and moisture retention in fine porous structures. They looked at under what conditions moisture would connect from the outer boundary of the oxide to the metal surface and defined a time of percolation which is the fraction of time (in a year) that that interconnected moisture exists. They found that, for certain oxide conditions, the time of percolation could be substantially higher than the time of wetness on an uncorroded surface.

Venkatraman et al. [54] modelled corrosion under films and derived the corrosion current density and corrosion potential as a function of kinetic, thermodynamic, and mass transport parameters. The model can be used in refining multi-scale models of atmospheric corrosion of zinc. Simillion et al. [55] developed a multi-ion transport and reaction model under thin NaCl containing films. Two important conclusions for this model were that the effect of macro-level geometric factors decreased as the thickness of the films decreased while chloride accumulation had a major role in controlling corrosion rates for thin films.

In very recent work of critical importance in understanding the resistance of compact oxides in zinc, Todorova et al. [56] undertook electronic structure calculations to look at the dominant defects that control the growth and dissolution of these oxide layers. Their calculations indicated that, in ZnO films, the doubly negatively charged oxygen vacancy may not be present and rather oxygen interstitial or unexpected charge states, such as the neutral oxygen vacancy, are found.

*7.3. Significance of Developments and Need for More Research*

These studies highlight some of the possible roles of porous oxides in the zinc system in terms of promoting longer wetting times, facilitating the ORR, and how the ORR on the pores of zinc oxide could promote densification of the oxide. All these works are preliminary and need further development. In particular, the porous oxide model of Venkatraman et al. needs to be expanded to include the tri-layer structures observed experimentally and in particular the compact oxide. It is notable that compact oxides are effectively grown in alkaline conditions in bulk electrochemical experiments yet form under droplet exposure in neutral conditions. It may be that the development of the porous oxide promotes sufficient alkalinity at the oxide–metal interface to promote the formation of a compact layer. Greater theoretical development is also required around how compact oxides function in zinc and how defects cross the oxide

## 8. Overall Significance of Developments and Need for More Research

This survey of the last decade's literature on the atmospheric corrosion of zinc is by no means complete, but it is clear that there has been significant progress, particularly in:

(a)   The development of corrosion patinas both in field and laboratory experiments and in particular the formation of multi-layered structures and compact layers.
(b)   The identification of local attack as a significant phenomenon in zinc corrosion.
(c)   Formulating methods to undertake corrosion under droplets and the critical role of drop size in determining the dominant corrosion processes and the oxides that form.
(d)   Role of oxide layers, particularly ZnO, in promoting the ORR.
(e)   The modelling or porous oxide and how the occurrence of ORR on oxide pore walls could have strong implications for pore chemistry and oxide densification.

Many of these elements are coming together and a new and more refined understanding of how atmospheric corrosion of zinc occurs is emerging. For instance, the modelling of porous oxides and growth studies of compact oxides in alkaline conditions may if combined help explain how compact oxides may occur in corrosion under neutral droplets (the local pH at the zinc/oxide boundary may be made alkaline due to ORR on the oxide pore surface). However, this new synthesis is very incomplete and substantial work is required to understand how the layered corrosion patinas form and what their properties are, as well as to grasp a detailed understanding of corrosion under droplets and other processes that reflect the actual corrosion events in service. A combined modelling and experimental approach would lead to the most profound advance.

**Conflicts of Interest:** The authors declare no conflict of interest.

## References

1.   Cole, I.S.; Azmat, N.S.; Kanta, A.; Venkatraman, M. What really controls the atmospheric corrosion of zinc? Effect of marine aerosols on atmospheric corrosion of zinc. *Int. Mater. Rev.* **2009**, *54*, 117–133. [CrossRef]
2.   Cole, I.S.; Muster, T.H.; Paterson, D.A.; Furman, S.A.; Trinidad, G.S.; Wright, N. Multi-scale modeling of the corrosion of metals under atmospheric corrosion. In *Pricm 6: Sixth Pacific Rim International Conference on Advanced Materials and Processing, pts 1–3*; Chang, Y.W., Kim, N.J., Lee, C.S., Eds.; Trans Tech Publications Ltd.: Zürich, Switzerland, 2007; Volume 561–565, pp. 2209–2212.
3.   Ganther, W.D.; Cole, I.S.; Helal, A.M.; Chan, W.; Paterson, D.A.; Trinidad, G.; Corrigan, P.; Mohamed, R.; Sabah, N.; Al-Mazrouei, A. Towards the development of a corrosion map for Abu Dhabi. *Mater. Corros.-Werkstoffe Und Korrosion* **2011**, *62*, 1066–1073. [CrossRef]

4.    Cole, I.S.; Ganther, W.D.; Helal, A.M.; Chan, W.; Paterson, D.; Trinidad, G.; Corrigan, P.; Mohamed, R.; Sabah, N.; Al-Mazrouei, A. A corrosion map of Abu Dhabi. *Mater. Corros.-Werkstoffe Und Korrosion* **2013**, *64*, 247–255. [CrossRef]

5.    Chico, B.; de la Fuente, D.; Vega, J.M.; Morcillo, M. Corrosivity maps of Spain for zinc in rural atmospheres. *Rev. Met.* **2010**, *46*, 485–492. [CrossRef]

6.    Panchenko, Y.M.; Marshakov, A.I.; Nikolaeva, L.A.; Kovtanyuk, V.V.; Igonin, T.N.; Andryushchenko, T.A. Comparative estimation of long-term predictions of corrosion losses for carbon steel and zinc using various models for the Russian territory. *Corros. Eng. Sci. Technol.* **2017**, *52*, 149–157. [CrossRef]

7.    Tidblad, J.; Kreislova, K.; Faller, M.; de la Fuente, D.; Yates, T.; Verney-Carron, A.; Grontoft, T.; Gordon, A.; Hans, U. ICP materials trends in corrosion, soiling and air pollution (1987–2014). *Materials* **2017**, *10*, 969. [CrossRef] [PubMed]

8.    Cole, I.S.; Paterson, D.A. Possible effects of climate change on atmospheric corrosion in Australia. *Corros. Eng. Sci. Technol.* **2010**, *45*, 19–26. [CrossRef]

9.    Trivedi, N.S.; Venkatraman, M.S.; Chu, C.; Cole, I.S. Effect of climate change on corrosion rates of structures in Australia. *Clim. Chang.* **2014**, *124*, 133–146. [CrossRef]

10.   Leygraf, C.; Graedel, T.E. *Atmospheric Corrosion*; Wiley-Interscience: Hoboken, NJ, USA, 2000.

11.   Odnevall, I.; Leygraf, C. Formation of $NaZn_4Cl(OH)_6SO_4 \cdot 6H_2O$ in a marine atmosphere. *Corros. Sci.* **1993**, *34*, 1213–1229. [CrossRef]

12.   Odnevall, I.; Leygraf, C. The formation of $Zn_4Cl_2(OH)_4SO_4 \cdot 5H_2O$ in an urban and an industrial atmosphere. *Corros. Sci.* **1994**, *36*, 1551–1559. [CrossRef]

13.   Friel, J.J. Atmospheric corrosion products on Al, Zn, and AIZn metallic coatings. *Corrosion* **1986**, *42*, 422–426. [CrossRef]

14.   Cole, I.S.; Muster, T.H.; Furman, S.A.; Wright, N.; Bradbury, A. Products formed during the interaction of seawater droplets with zinc surfaces: I. Results from 1-and 2.5-day exposures. *J. Electrochem. Soc.* **2008**, *155*, C244–C255. [CrossRef]

15.   Bernard, M.C.; Hugotlegoff, A.; Phillips, N. In-situ raman-study of the corrosion of zinc-coated steel in the presence of chloride .1. Characterization and stability of zinc corrosion products. *J. Electrochem. Soc.* **1995**, *142*, 2162–2167. [CrossRef]

16.   Wallinder, O.; Leygraf, C. A critical review on corrosion and runoff from zinc and zinc-based alloys in atmospheric environments. *Corrosion* **2017**, *73*, 1060–1077. [CrossRef]

17.   Hedberg, J.; Le Bozec, N.; Wallinder, I.O. Spatial distribution and formation of corrosion products in relation to zinc release for zinc sheet and coated pre-weathered zinc at an urban and a marine atmospheric condition. *Mater. Corros.-Werkstoffe Und Korrosion* **2013**, *64*, 300–308. [CrossRef]

18.   Cui, Z.Y.; Li, X.G.; Xiao, K.; Dong, C.F.; Liu, Z.Y.; Wang, L.W. Corrosion behavior of field-exposed zinc in a tropical marine atmosphere. *Corrosion* **2014**, *70*, 731–748. [CrossRef]

19.   Liu, Y.W.; Wang, Z.Y.; Cao, G.W.; Cao, Y.; Huo, Y. Study on corrosion behavior of zinc exposed in coastal-industrial atmospheric environment. *Mater. Chem. Phys.* **2017**, *198*, 243–249. [CrossRef]

20.   Persson, D.; Thierry, D.; Karlsson. Corrosion and corrosion products of hot dipped galvanized steel during long term atmospheric exposure at different sites world-wide. *Corros. Sci.* **2017**, *126*, 152–165. [CrossRef]

21.   Li, S.X.; Hihara, L.H. Aerosol salt particle deposition on metals exposed to marine environments: A study related to marine atmospheric corrosion. *J. Electrochem. Soc.* **2014**, *161*, C268–C275. [CrossRef]

22.   Cole, I.S.; Lau, D.; Paterson, D.A. Holistic model for atmospheric corrosion—Part 6—From wet aerosol to salt deposit. *Corros. Eng. Sci. Technol.* **2004**, *39*, 209–218. [CrossRef]

23.   Cole, I.S.; Muster, T.H.; Lau, D.; Wright, N.; Azmat, N.S. Products formed during the interaction of seawater droplets with zinc surfaces II. Results from short exposures. *J. Electrochem. Soc.* **2010**, *157*, C213–C222. [CrossRef]

24.   Thomas, S.; Cole, I.S.; Birbilis, N. Compact oxides formed on zinc during exposure to a single sea-water droplet. *J. Electrochem. Soc.* **2013**, *160*, C59–C63. [CrossRef]

25.   Macdonald, D.D.; Ismail, K.M.; Sikora, E. Characterization of the passive state on zinc. *J. Electrochem. Soc.* **1998**, *145*, 3141–3149. [CrossRef]

26.   Prestat, M.; Holzer, L.; Lescop, B.; Rioual, S.; Zaubitzer, C.; Diler, E.; Thierry, D. Microstructure and spatial distribution of corrosion products anodically grown on zinc in chloride solutions. *Electrochem. Commun.* **2017**, *81*, 56–60. [CrossRef]

27. Thomas, S.; Cole, I.S.; Sridhar, M.; Birbilis, N. Revisiting zinc passivation in alkaline solutions. *Electrochim. Acta* **2013**, *97*, 192–201. [CrossRef]
28. Fattah-alhosseini, A.; Mirshekari, M. Effect of film formation potential on the electrochemical behavior of the passive films formed on zinc in 0.01 m NaOH. *Trans. Indian Inst. Met.* **2015**, *68*, 851–857. [CrossRef]
29. Muster, T.H.; Ganther, W.D.; Cole, I.S. The influence of microstructure on surface phenomena: Rolled zinc. *Corros. Sci.* **2007**, *49*, 2037–2058. [CrossRef]
30. Mena, E.; Veleva, L.; Souto, R.M. Mapping of local corrosion behavior of zinc in substitute ocean water at its initial stages by SVET. *Int. J. Electrochem. Sci.* **2016**, *11*, 5256–5266. [CrossRef]
31. Abd El-Rehim, S.S.; Hamed, E.; Shaltot, A.M.; Amin, M.A. Pitting corrosion of zinc in $Na_2S_2O3$ solutions. Part I. Polarization studies and morphology of pitting. *Int. J. Res. Phys. Chem. Chem. Phys.* **2012**, *226*, 59–85.
32. Cole, I.S.; Ganther, W.D.; Furman, S.A.; Muster, T.H.; Neufeld, A.K. Pitting of zinc: Observations on atmospheric corrosion in tropical countries. *Corros. Sci.* **2010**, *52*, 848–858. [CrossRef]
33. Keitelman, A.; Alvarez, M.G. 40 years of JR galvele's localized acidification pitting model: Past, present, and future. *Corrosion* **2017**, *73*, 8–17. [CrossRef]
34. Cheng, Q.L.; Song, S.H.; Song, L.Y.; Hou, B.R. Effect of relative humidity on the initial atmospheric corrosion behavior of zinc during drying. *J. Electrochem. Soc.* **2013**, *160*, C380–C389. [CrossRef]
35. Song, S.H.; Chen, Z.Y. Initial corrosion of pure zinc under NaCl electrolyte droplet using a Zn-Pt-Pt three-electrode system. *Int. J. Electrochem. Sci.* **2013**, *8*, 6851–6863.
36. Azmat, N.S.; Ralston, K.D.; Muster, T.H.; Muddle, B.C.; Cole, I.S. A high-throughput test methodology for atmospheric corrosion studies. *Electrochem. Solid State Lett.* **2011**, *14*, C9–C11. [CrossRef]
37. Azmat, N.S.; Ralston, K.D.; Muddle, B.C.; Cole, I.S. Corrosion of Zn under fine size aerosols and droplets using inkjet printer deposition and optical profilometry quantification. *Corros. Sci.* **2011**, *53*, 3534–3541. [CrossRef]
38. Azmat, N.S.; Ralston, K.D.; Muddle, B.C.; Cole, I.S. Corrosion of Zn under acidified marine droplets. *Corros. Sci.* **2011**, *53*, 1604–1615. [CrossRef]
39. Risteen, B.E.; Schindelholz, E.; Kelly, R.G. Marine aerosol drop size effects on the corrosion behavior of low carbon steel and high purity iron. *J. Electrochem. Soc.* **2014**, *161*, C580–C586. [CrossRef]
40. Muster, T.H.; Cole, I.S. The protective nature of passivation films on zinc: Surface charge. *Corros. Sci.* **2004**, *46*, 2319–2335. [CrossRef]
41. Muster, T.H.; Neufeld, A.K.; Cole, I.S. The protective nature of passivation films on zinc: Wetting and surface energy. *Corros. Sci.* **2004**, *46*, 2337–2354. [CrossRef]
42. Thomas, S.; Cole, I.S.; Gonzalez-Garcia, Y.; Chen, M.; Musameh, M.; Mol, J.M.C.; Terryn, H.; Birbilis, N. Oxygen consumption upon electrochemically polarised zinc. *J. Appl. Electrochem.* **2014**, *44*, 747–757. [CrossRef]
43. Prestat, M.; Vucko, F.; Lescop, B.; Rioual, S.; Peltier, F.; Thierry, D. Oxygen reduction at electrodeposited zno layers in alkaline solution. *Electrochim. Acta* **2016**, *218*, 228–236. [CrossRef]
44. Venkatraman, M.S.; Cole, I.S.; Emmanuel, B. Corrosion under a porous layer: A porous electrode model and its implications for self-repair. *Electrochim. Acta* **2011**, *56*, 8192–8203. [CrossRef]
45. Nazarov, A.; Thierry, D.; Prosek, T. Formation of galvanic cells and localized corrosion of zinc and zinc alloys under atmospheric conditions. *Corrosion* **2017**, *73*, 77–86. [CrossRef]
46. Gunasegaram, D.R.; Venkatraman, M.S.; Cole, I.S. Towards multiscale modelling of localised corrosion. *Int. Mater. Rev.* **2014**, *59*, 84–114. [CrossRef]
47. Simillion, H.; Dolgikh, O.; Terryn, H.; Deconinck, J. Atmospheric corrosion modeling. *Corros. Rev.* **2014**, *32*, 73–100. [CrossRef]
48. Graedel, T.E. Gildes model studies of aqueous chemistry. 1. Formulation and potential applications of the multi-regime model. *Corros. Sci.* **1996**, *38*, 2153–2180. [CrossRef]
49. Graedel, T.E. Corrosion mechanisms for zinc exposed to the atmosphere. *J. Electrochem. Soc.* **1989**, *136*, C193–C203. [CrossRef]
50. Cole, I.S.; Paterson, D.A.; Ganther, W.D. Holistic model for atmospheric corrosion—Part 1—Theoretical framework for production, transportation and deposition of marine salts. *Corros. Eng. Sci. Technol.* **2003**, *38*, 129–134. [CrossRef]

51. Spence, J.; Haynie, F. Derivation of a damage function for galvanised steel structures: Corrosion kinetics and thermodynamic considerations. In *Corrosion Testing and Evaluation: SILVER Anniversary Volume*; ASTM: Philadelphia, PA, USA, 1990; Volume 1000, pp. 2008–2024.

52. Sherwood, D.; Emmanuel, B.; Cole, I. Moisture distribution in porous oxide and polymer over-layers and critical relative humidity and time of wetness for chloride and non-chloride-bearing atmospheres for atmospheric corrosion of metals. *J. Electrochem. Soc.* **2016**, *163*, C675–C685. [CrossRef]

53. Sherwood, D.; Reddy, M.V.; Cole, I.; Emmanuel, B. A model to estimate moisture distribution in porous oxides as a function of atmospheric conditions. *J. Electroanal. Chem.* **2014**, *725*, 1–6. [CrossRef]

54. Venkatraman, M.S.; Cole, I.S.; Emmanuel, B. Model for corrosion of metals covered with thin electrolyte layers: Pseudo-steady state diffusion of oxygen. *Electrochim. Acta* **2011**, *56*, 7171–7179. [CrossRef]

55. Simillion, H.; Van den Steen, N.; Terryn, H.; Deconinck, J. Geometry influence on corrosion in dynamic thin film electrolytes. *Electrochim. Acta* **2016**, *209*, 149–158. [CrossRef]

56. Todorova, M.; Neugebauer, J. Identification of bulk oxide defects in an electrochemical environment. *Faraday Discuss.* **2015**, *180*, 97–112. [CrossRef] [PubMed]

*materials* MDPI

*Article*

# ICP Materials Trends in Corrosion, Soiling and Air Pollution (1987–2014)

Johan Tidblad [1,*], Kateřina Kreislová [2], Markus Faller [3], Daniel de la Fuente [4], Tim Yates [5], Aurélie Verney-Carron [6], Terje Grøntoft [7], Andrew Gordon [1] and Ulrik Hans [3]

[1]   Swerea KIMAB, Dept Corrosion, 164 07 Kista, Sweden; andrew.gordon@swerea.se
[2]   Svuom Ltd., 17000 Prague, Czech Republic; kreislova@svuom.cz
[3]   Empa, Materials Science and Technology, 8600 Dübendorf, Switzerland; markus.faller@empa.ch (M.F.); ulrik.hans@empa.ch (U.H.)
[4]   CENIM—National Centre for Metallurgical Research, 28040 Madrid, Spain; delafuente@cenim.csic.es
[5]   BRE—Building Research Establishment Ltd., Watford WD25 9XX, UK; yatest@bre.co.uk
[6]   LISA (Laboratoire Interuniversitaire des Systèmes Atmosphériques), UMR 7583 CNRS/UPEC/UPD, 94010 Creteil, France; aurelie.verney@lisa.u-pec.fr
[7]   NILU—Norwegian Institute for Air Research, 2027 Kjeller, Norway; terje.grontoft@nilu.no
*   Correspondence: johan.tidblad@swerea.se; Tel.: +46-8-674-1733

Received: 14 July 2017; Accepted: 18 August 2017; Published: 19 August 2017

**Abstract:** Results from the international cooperative programme on effects on materials including historic and cultural monuments are presented from the period 1987–2014 and include pollution data ($SO_2$, $NO_2$, $O_3$, $HNO_3$ and $PM_{10}$), corrosion data (carbon steel, weathering steel, zinc, copper, aluminium and limestone) and data on the soiling of modern glass for nineteen industrial, urban and rural test sites in Europe. Both one-year and four-year corrosion data are presented. Corrosion and pollution have decreased significantly and a shift in the magnitude is generally observed around 1997: from a sharp decrease to a more modest decrease or to a constant level without any decrease. $SO_2$ levels, carbon steel and copper corrosion have decreased even after 1997, which is more pronounced in urban areas, while corrosion of the other materials shows no decrease after 1997, when looking at one-year values. When looking at four-year values, however, there is a significant decrease after 1997 for zinc, which is not evident when looking at the one-year values. This paper also presents results on corrosion kinetics by comparison of one- and four-year values. For carbon steel and copper, kinetics is relatively independent of sites while other materials, especially zinc, show substantial variation in kinetics for the first four years, which needs to be considered when producing new and possibly improved models for corrosion.

**Keywords:** atmospheric corrosion; soiling; pollution; carbon steel; weathering steel; zinc; copper; aluminium; limestone; glass

## 1. Introduction

"ICP Materials" or the "International co-operative programme on effects on materials including historic and cultural monuments" is an international project that has been run since the 1980's (www.corr-institute.se/icp-materials). The program started, together with other international cooperative programmes (ICP's) on effects on ecosystem and health, as a reaction to environmental problems faced in Europe and North America. The science produced within ICP Materials is in support of the Convention on Long-range Transboundary Air Pollution (LRTAP Convention), within the United Nations Economic Commission for Europe.

Over the years, almost eighty reports have been issued with results of the program, and many different scientific publications. Some scientific publications also reflect important environmental concerns. The first main publication of results was published in 2001 "Dose-response functions on dry and wet acid deposition effects after 8 years of exposure", and included results from the period 1987–1995 and empirical relations on how to calculate atmospheric corrosion attack based on environmental parameters [1]. The main environmental concern was acid rain and acidifying pollutants. The next important publication (2007) was named "Dose-response functions for the multi-pollutant situation" [2]. With decreasing levels of sulphur dioxide, it was realized that more complicated expressions were needed to successfully predict atmospheric corrosion, including other pollutants, such as nitric acid and particulate matter. During this period, climate change started to be high on the agenda and it was realized that dose–response functions from international exposure programs could also be used to assess the possible impact of long term changes in climate on corrosion. "Atmospheric corrosion of metals in 2010–2039 and 2070–2099" described this procedure, and was published in 2007 [3]. Finally, in 2012, the latest main publication from ICP Materials was released "ICP Materials Celebrates 25 Years of Research" [4]. This open source publication had the purpose of giving a complete metadata description of all the data available from the program, including citations to main data sources and publications.

ICP Materials is not the only international exposure program; other important programs include, for example, National Acid Precipitation Assessment Program (NAPAP) in the US [5], ISO CORRAG worldwide [6], Ibero-American Map of Atmospheric Corrosiveness (MICAT) in Ibero-American countries [7] and the corrosion network (CORNET) of the regional air pollution in developing countries (RAPIDC) in Asia/Africa [8]. However, ICP Materials is unique in its persistence, which has enabled it to perform exposures for thirty years (1987–2017) and there are currently plans for continuation of the program up to at least 2021.

The purpose of this publication is twofold. The first, and maybe most important, is to give an overview of the data, not only in normal publication format, but also to release a comprehensive database on corrosion, soiling and air pollution, available for download. This will enable independent researchers to quickly access the data, to check conclusions and to perform their own analysis. Naturally, this database does not include all data from the program, but this is the first time data from ICP Materials are released in this format, and releases will hopefully continue. The second is to provide an overview of the main trends in corrosion, soiling and pollution during the whole period (1987–2014) and during recent years. This gives the background for the decisions currently taken on how to develop the program in the coming years, which will include an increased focus on soiling of materials, mainly as a result of particulate matter deposition.

## 2. Results

ICP Materials exposure sites have changed over the years. In this publication, only data from the following sites are considered. The selection was made from sites that are currently active and have more than just a few years of data. Selected sites include three industrial sites, nine urban sites and seven rural sites, in total nineteen sites:

- Industrial sites: Kopisty, Bottrop and Katowice;
- Urban sites: Prague, Rome, Milan, Venice, Oslo, Stockholm, Madrid, Paris and Berlin; and
- Rural sites: Casaccia, Birkenes, Aspvreten, Toledo, Lahemaa, Svanvik and Chaumont.

Most of these sites were included from the beginning of the program (1987), except Katowice, Paris, Berlin, Svanvik and Chaumont, which were introduced later (1995–2000). It should be noted that these labels "Paris", "Berlin", etc. indicate that they are located in these cities. The values of corrosion, pollution and soiling at these sites should however not be considered representative of these cities, since the variation of corrosion and pollution within a city can be substantial.

Environmental parameters included in this publication are the pollutants $SO_2$, $NO_2$, $O_3$, $HNO_3$ and $PM_{10}$. Materials included in this publication are carbon steel, weathering steel, zinc, copper, aluminium, limestone and modern glass.

Before presenting the results, a special note on exposure periods is needed. All exposures performed in ICP Materials so far have started in the fall, usually in October, and then lasted for one year or several years, also ending in the fall. In this paper, all exposure periods are labelled with the start year so that, for example, "1987" in reality is a short hand notation for a one-year exposure between the fall of 1987 to the fall of 1988, and "2011–2014" is a short hand notation for a four-year exposure between the fall of 2011 and the fall of 2015.

## 2.1. Trends in Air Pollution

The concentrations of the pollutants $SO_2$, $NO_2$, $O_3$, $HNO_3$ and $PM_{10}$ in air ($\mu g\ m^{-3}$), as measured at the ICP Materials sites, are included in the description of trends in air pollution. Figure 1 shows that the average of the $SO_2$ concentrations at the industrial sites was considerably higher than for the urban and rural sites, in all the measurement years since 1987, except in 1995. In 1995, the value at the urban sites was much higher than in the measurement years just before and after, and nearly as high as for the industrial sites. The reason for this is technical and coincidental. It is due to a change of sites in the measurement programme, with only one urban site remaining in 1995, before a set of the former, and one new, urban sites were again included in 1996. This highlights a difficulty presenting averages and is the reason why this paper also presents trends for individual sites to illustrate characteristic trends. Until 1997, the average $SO_2$ concentration measured at the industrial sites was above 30 $\mu g/m^3$. The average $SO_2$ concentration measured at the urban sites was close to the average values measured for all the sites, in all years. By 1997, it had been reduced to below 10 $\mu g/m^3$, and, by 2005, to below 5 $\mu g/m^3$. The average $SO_2$ concentration measured at the rural sites was below 5 $\mu g/m^3$ in all the years.

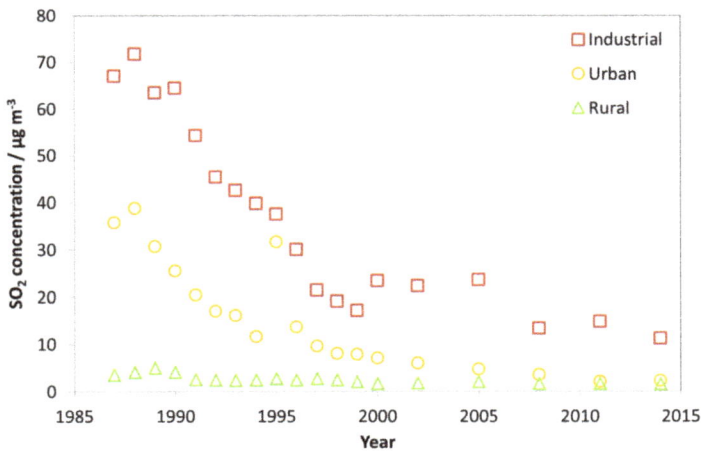

**Figure 1.** Average $SO_2$ concentration at industrial, urban and rural sites for individual years (1987–2014).

Figure 2 shows a quite different situation for $NO_2$ as compared to $SO_2$. The average $NO_2$ concentration has not changed much since 1987. Some decreases in concentrations were measured until approximately 2000, with little change thereafter, except some possible increase at the industrial sites. In contradiction to the situation for $SO_2$, the average concentration of $NO_2$ at the urban sites was higher than at the industrial sites in nearly all years, with the coincidental main exception being 1995, when only one urban site was included in the average. The higher average concentrations of $NO_2$ at the urban sites may have been due to more emissions from traffic and domestic heating than at the industrial sites. The average concentration of $NO_2$ at the rural sites was generally below 10 $\mu g/m^{-3}$, but with somewhat higher concentrations and more variation between years before 2000.

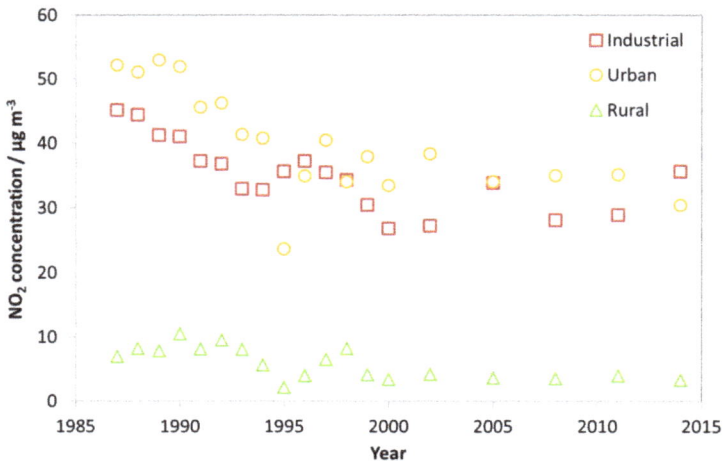

**Figure 2.** Average $NO_2$ concentration at industrial, urban and rural sites for individual years (1987–2014).

Figure 3 shows a quite different situation for $O_3$ as compared to $SO_2$ and $NO_2$. The trends for the average concentration of $O_3$ measured for the sites were slightly positive, and slightly more so from 1987 to about 2000 than from 2000 to 2014. The average concentration at the rural sites was always significantly higher than that at the urban and industrial sites. The difference between the average concentration at the urban and industrial sites was always small, with alternating ranking between them. In 1995, there was only one urban site in the calculated average, due to changing of sites in the programme, which explains the coincidental high "average" value in this year.

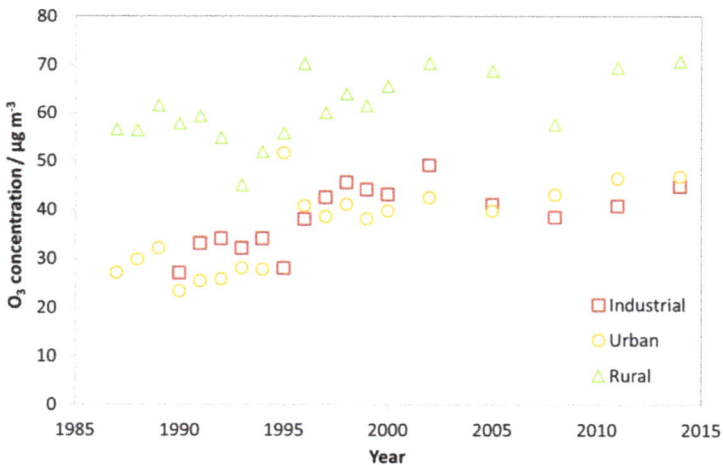

**Figure 3.** Average $O_3$ concentration at industrial, urban and rural sites for individual years (1987–2014).

Looking at Figures 1–3, there is a change in the trend of air pollution around 1997–1999 where a steep change is replaced by a more modest change (decrease for $SO_2$ and $NO_2$ and increase for $O_3$). Therefore, a compact way of summarising the data from all sites is to present diagrams with averages for the three periods 1987–1989, 1997–1999 and 2011–2014. This will more or less summarise the whole trend for all sites as well as give the opportunity to compare changes for different periods (1987–1989 vs. 1997–1999, 1987–1989 vs. 2011–2014, and 1997–1999 vs. 2011–2014). Figures 4–6 show this type of diagram for $SO_2$, $NO_2$ and $O_3$, respectively.

Figure 4 shows that the decrease of $SO_2$ at the industrial sites for the whole period of measurements, from 1987–1989 to 2011–2014, was dominated by a decrease in Kopisty, but from 1997–1999 by a decrease in Bottrop. The decrease at the urban sites from 1987–1988 to 2011–2014 was dominated by a decrease in Milan and Prague, with somewhat less decrease in Rome, Venice and Madrid. Since 1997–1999, the decrease was larger in Prague than in Milan and the ranking of the other sites also changed. The decreases at the new sites in Paris and Berlin were higher than in Madrid and Venice. The few measurement values for Rome showed a situation there after 1997–1999 was more similar to that of the low values in Stockholm and Oslo. It can however be noted that a relatively high $SO_2$ concentration as compared to previous years, of 4.4 µg/m$^3$, was measured in Oslo in 2014. This points to the need, still, for measurements of $SO_2$ in the present situation with generally low concentrations, and for attention to possible new or reappearing emission sources. Some notable decreases in $SO_2$ since 1987–1989 were also observed at the rural sites.

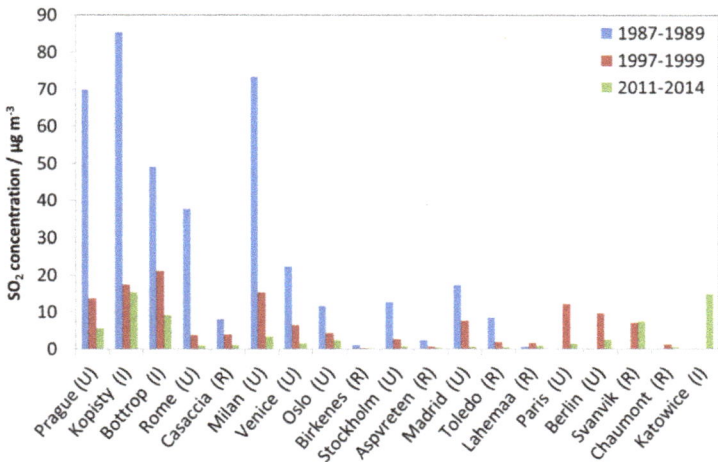

**Figure 4.** $SO_2$ concentration at individual sites based on averages for three selected periods, 1987–1989, 1997–1999 and 2011–2014.

Figure 5 shows a decrease in the $NO_2$ concentrations measured at most of the sites since 1987–1989, with an overall correspondence to the decrease for $SO_2$, with the highest decrease in Milan, but with some noteworthy exceptions. Foremost among them, in Prague and Madrid, there was a small increase in the concentration measured for $NO_2$ since 1987–1989 and a higher increase since 1997–1999. In both cases, $NO_2$ was observed to decrease until 2000 and increase thereafter. During 1997–1999, $NO_2$ was measured at four more sites, Paris, Berlin, Svanvik and Chaumont, than during 1987–1989. Measurements at the site Katowice started in 2000. In addition to the increase in $NO_2$ in Prague and Madrid, the concentration measured for $NO_2$ has also increased in Kopisty and Rome since 1997–1999. The reasons for these increases were high values for 2014 in Kopisty and 2011 in Rome.

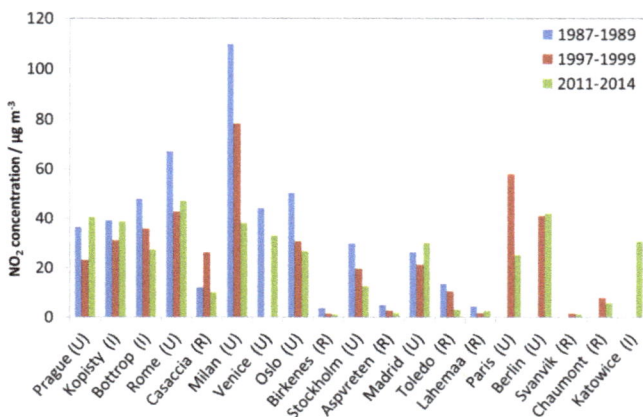

**Figure 5.** NO$_2$ concentration at individual sites based on averages for three selected periods, 1987–1989, 1997–1999 and 2011–2014.

The decreases in measured NO$_2$ in Stockholm and Oslo since 1987–1989, and in Casaccia and Toledo since 1997–1999, as compared to the other sites, were relatively larger for NO$_2$ than for SO$_2$. Among the new sites included from 1997–1999 to 2011–2014, a large decrease in NO$_2$ was measured in Paris.

Notable decrease in NO$_2$ was measured for the rural site of Toledo since 1992, and since 1997–1999. A general trend of slightly decreasing NO$_2$ was measured at the rural sites since 1987–1989 and 1997–1999, but with some high values for some years and variation in values for some sites, which could influence trend calculations. Most notably, at the site Casaccia, much higher values were measured in 1997 and 1998 than in other years, giving increase in concentration to 1997–1999 and decrease thereafter.

Figure 6 shows quite large increases in O$_3$ concentrations at the Italian sites Casaccia, Venice, Milan and Rome, and Spanish sites Madrid and Toledo from 1987–1989 to 2011–2014. The increases were mainly due to low start values in 1987–1989, but also partly due to high values during 2011–2014 in Casaccia, Toledo and Madrid.

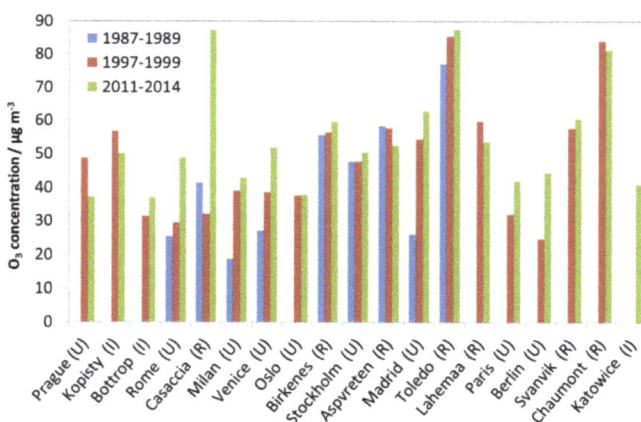

**Figure 6.** O$_3$ concentration at individual sites based on averages for three selected periods, 1987–1989, 1997–1999 and 2011–2014.

At the sites in Rome and Casaccia, lower values were measured in the mid-1990s, and then considerable increases were measured towards 2011–2014. Significant increases were also observed from 1997–1999 to 2011–2014 for some sites outside of Italy and Spain, for which $O_3$ measurement results then had become available: Berlin, Paris and Bottrop. The changes in $O_3$ showed no clear correlation with changes in $NO_2$ and $SO_2$. The largest variation and overall increase in $O_3$ concentration was measured for the rural site Casaccia. Except for this site, the sites with the largest increases were all urban sites. At the other rural sites, and the urban sites Stockholm and Oslo, small increases or decreases were measured. Exceptions from this were the two Czech sites, the industrial site Kopisty and the urban site Prague, where large decreases were measured from 1997–1999 to 2011–2014.

In contrast to $SO_2$, $NO_2$ and $O_3$, measurements of $HNO_3$ and $PM_{10}$ were not started in the programme until 2002. Furthermore, $PM_{10}$ is not a mandatory parameter to measure. Therefore, the data for $HNO_3$ and $PM_{10}$ do not permit evaluation of long-term trends in the same manner as for $SO_2$, $NO_2$ and $O_3$. Instead, data for individual sites are presented, in a similar way as for $SO_2$, $NO_2$ and $O_3$, but using only two periods: 2002–2005 and 2011–2014. With measurement results for $PM_{10}$ and $HNO_3$ for only four to five years, any interpretation of trends should be made with caution.

Figure 7 shows the concentration of gaseous nitric acid ($HNO_3$). Comparing the two periods, the average concentration measured for most sites has decreased. There is considerable variation between the values for the industrial, urban and rural sites, with higher values measured at some rural sites than at the urban and industrial sites. The highest values were measured at the urban sites Paris, Milan, Venice, Rome and Madrid, and the industrial site Katowice. Notably, low values were measured at the rural sites Svanvik, Aspvreten and Birkenes, and at the urban sites Oslo, Stockholm and Berlin. Details on the measurements of nitric acid have been presented elsewhere [9].

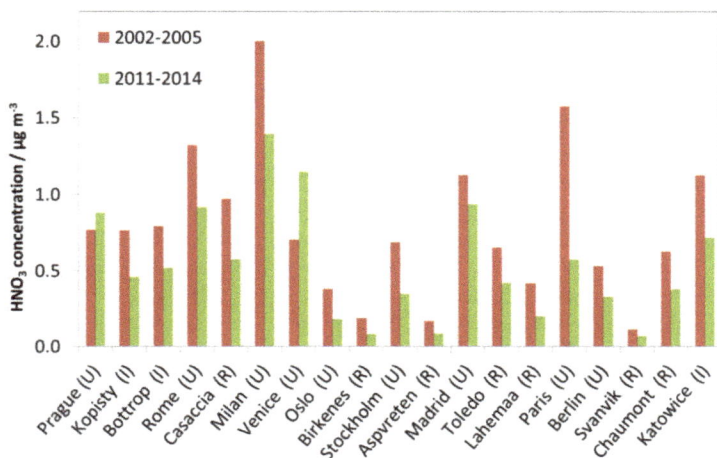

**Figure 7.** $HNO_3$ concentration at individual sites based on averages for two selected periods, 2002–2005 and 2011–2014.

Figure 8 shows the concentration in air of particulate matter with aerodynamic diameter smaller than 10 μm ($PM_{10}$). It is not possible to make any general conclusions regarding trends based on this limited dataset. Among the urban sites, notably low values were measured at the sites of Stockholm and Madrid. Among the rural sites, the highest concentrations measured overall were in Toledo, and the lowest concentrations were measured in Birkenes and Svanvik. Some sites showed notable trends of decreasing $PM_{10}$, including the urban site Berlin and the rural sites Chaumont and Lahemaa.

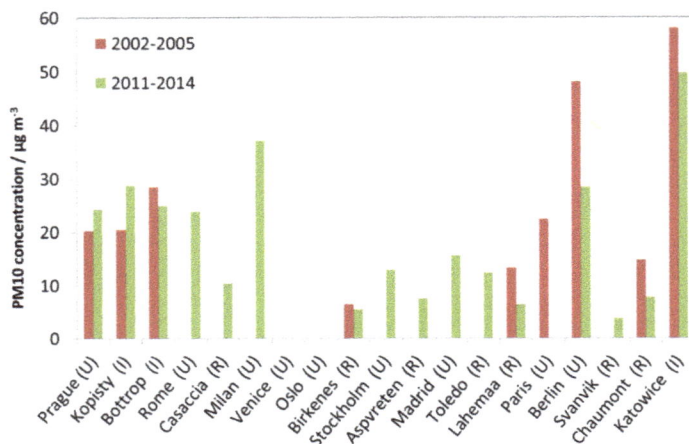

**Figure 8.** $PM_{10}$ concentration at individual sites based on averages for two selected periods, 2002–2005 and 2011–2014.

In summary, the trends in the changes of the average concentration of $SO_2$, $NO_2$ and $O_3$ for the ICP sites were all stronger in the first phase of the programme, from 1987 to about 2000, than thereafter. This was very apparent for $SO_2$, clearly seen for $NO_2$ for which there were no apparent trend after 2000, and slightly apparent for the positive trend for $O_3$.

From about 2000 to 2014, a trend of gradual decrease in the $SO_2$ concentration was measured at all sites, except Svanvik where a small increase was measured. The Svanvik site is located only 6 km away from the considerable $SO_2$ source of the nickel plant in the town of Nikel, Russia. There was no general trend in the measured concentration of $NO_2$ since 1997–1999, but for more of the sites decreases were measured than increases, with the largest changes in the measured concentrations being the decreases at the sites Milan ($-40$ $\mu g/m^3$) and Paris ($-33$ $\mu g/m^3$). Other sites where considerable decreases in $NO_2$ were measured were Casaccia, Bottrop, Toledo and Stockholm. Considerable increases in $NO_2$ were measured at the sites Prague, Madrid, Kopisty and Rome. A slight positive trend was measured for $O_3$ from 1997–1999 to 2011–2014 for nearly all the urban sites, except Prague, and for the rural site Casaccia. Since 2002, a clear and continuous decreasing trend in the concentration of $HNO_3$ was measured at all the industrial sites, at the urban sites Paris and Milan, and at the rural sites Casaccia and Toledo. Since 2002, a clear decreasing trend in $PM_{10}$ was measured at the Berlin and Chaumont sites.

Taken together, for air pollution, the largest recent (since about 2000) decreases in concentration at ICP sites were measured in Milan and Paris, then Bottrop and then the other urban and industrial sites. However, considerable increases in $NO_2$ were measured in Prague, Madrid and Kopisty, and considerable increases in $O_3$ in Rome and Berlin. The changes for the rural sites were minor compared to the industrial and urban sites, but with notable decrease of $NO_2$ in Toledo, and Casaccia, where the variation in $NO_2$ between years and the increase in $O_3$ was the largest.

*2.2. Trends in Carbon Steel Corrosion*

Figure 9 shows the average mass loss of unalloyed carbon steel (C < 0.2%, P < 0.07%, S < 0.05%, Cu < 0.07%) for one-year exposures for industrial test sites with one selected individual site, Kopisty. The corrosion decreased significantly between 1987 and 1997. The corrosion then remained on a level around 240 $g/m^2$ for the first exposure year, corresponding to a 50% decrease compared to the original value. In the last 25 years, the pollution at Kopisty reduced tremendously because of the decline of heavy industry. ISO 9223 corrosivity category C3 ranges from 200 to 400 $g/m^2$. With a recent value of

210 g/m$^2$ for the first year of exposure, this industrial test site has changed its corrosivity category during 1987–2014 from C4 to low C3.

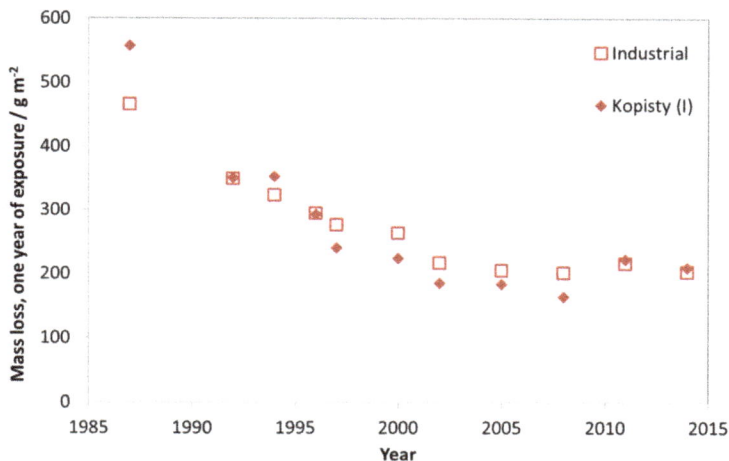

**Figure 9.** Carbon steel corrosion, average of industrial sites and the industrial site Kopisty.

In Figure 10, the trend of carbon steel corrosion is shown for the nine urban test sites with Prague as a typical example. As for the industrial sites, there is a strong decrease of mass loss between the exposures in 1987 and 1997, but also a small reduction of mass loss in the following periods.

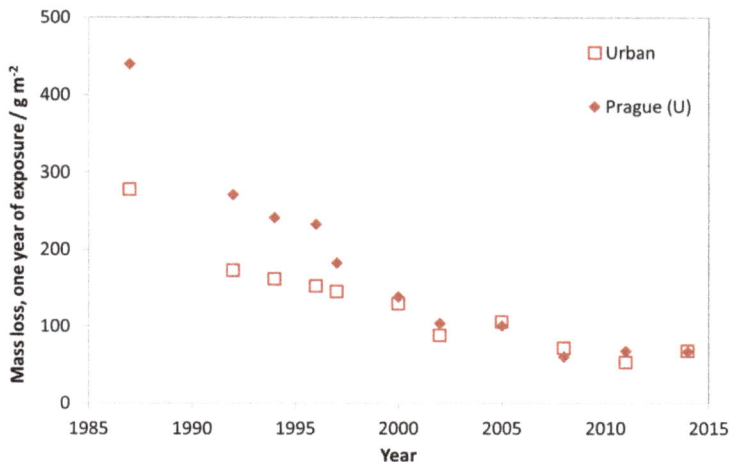

**Figure 10.** Carbon steel corrosion at urban sites and the urban site Prague.

The trend for rural test sites is similar to the trends of industrial and urban test sites. For rural test sites, a slight decrease of mass loss values was found (see Figure 11). Today, the urban test sites (C4 and C3, according to ISO 9223) are tending towards C2 with values typical for former rural atmospheres. This shows that labels such as "industrial", "urban", "rural" and "marine" can be useful for indicating the type of pollution (dominated by $SO_2$, $NO_2$, $O_3$ and chloride) but that they are not at all useful for classifying levels of corrosivity in a quantitative way.

Two repeated four-year exposures were performed, during 1997–2000 and 2011–2014. As for the one-year exposures, all test sites showed a decrease of corrosion values.

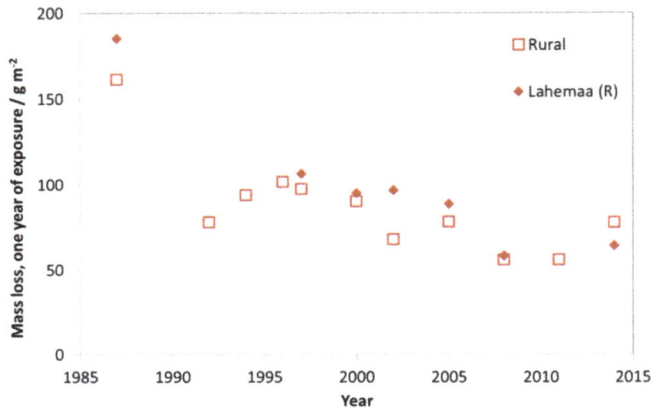

**Figure 11.** Carbon steel corrosion at rural sites and the rural site Lahemaa.

As an exposure was not performed during 1987–1990, a comparison of these two exposures does not show such a large decrease in corrosion rates, since the period of maximal reduction of air pollution (1987–1997) is not included. Figure 12 shows the mass loss after four years of exposure vs. mass loss after one year of exposure for two different exposure periods, 1997–2000 and 2011–2014. The clear relationship between one- and four-year periods shows that it is equivalent to show trends in corrosion based on four-year exposures instead of one-year exposures. The advantage of using four-year values for showing trends in corrosion is that it is less sensitive to year-to-year variation in climatic parameters (temperature, relative humidity, and precipitation). The higher corrosion values also make it easier to identify significant trends for individual sites. Note that in Figure 12, the relationship between one- and four-year values is practically 1:2, corresponding to a square-root kinetics for carbon steel during the first four years of exposure.

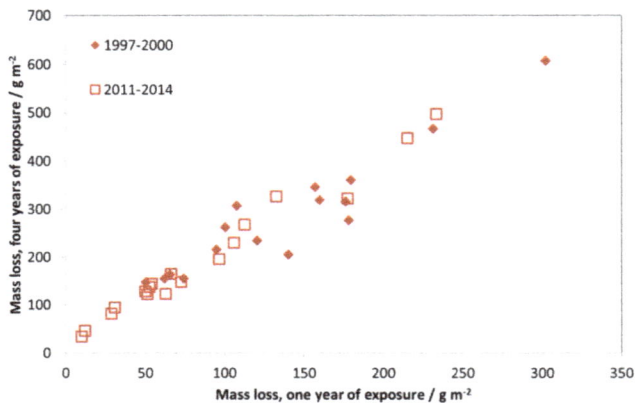

**Figure 12.** Carbon steel corrosion: one-year vs. four-year exposures for two different periods, 1997–2000 and 2011–2014. The one-year corrosion values are calculated as averages of two one-year exposures for the years 1997/2000 and 2011/2014, corresponding to the first and fourth years of the corresponding four-year period.

Figure 13 shows trends of four-year corrosion at the test sites and that corrosion has decreased at all sites corresponding to average levels for 2011–2014 of about 60% of those for 1997–2000.

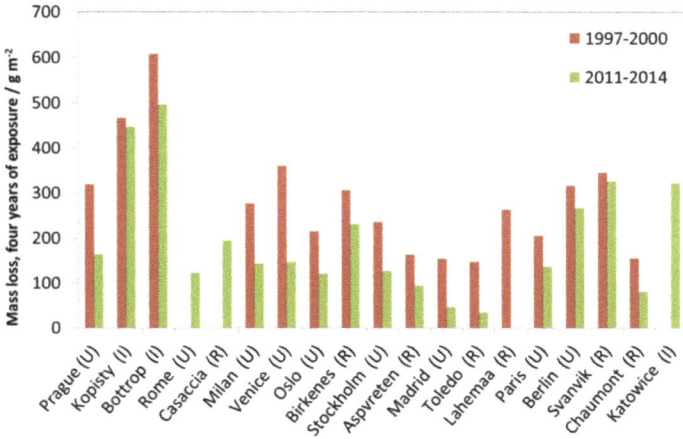

**Figure 13.** Carbon steel corrosion: four-year exposures at individual sites for two exposure periods, 1997–2000 and 2011–2014.

### 2.3. Trends in Weathering Steel Corrosion

The exposed weathering steel (C < 0.12%, Mn 0.3%–0.8%, Si 0.25%–0.7%, P 0.07%–0.15%, S < 0.04%, Cr 0.5%–1.2%, Ni 0.3%–0.6%, Cu 0.3%–0.55%, Al < 0.01%) is a low-alloyed steel with improved corrosion performance in polluted areas in unsheltered positions, especially after longer exposure times. Only two one-year exposures have been carried out, 1987 and 2011, and two four-year exposures, 1987–1990 and 2011–2014. Figure 14 shows the mass loss after four years of exposure vs. mass loss after one year of exposure for the two different pairs. The relationship between one- and four-year periods is not as clear as for carbon steel.

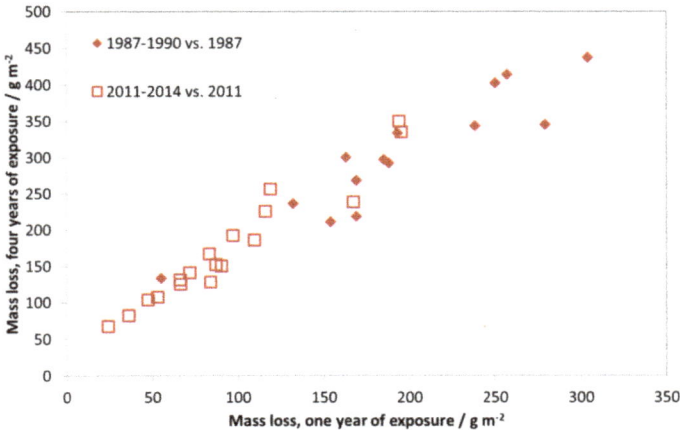

**Figure 14.** Weathering steel corrosion: one-year vs. four-year exposures for two different periods, 1987–1990 and 2011–2014. The one-year corrosion values are only available for the first years of the four-year periods, 1987 and 2011.

At lower corrosion values, the one- to four-year corrosion relationship is closer to 1:2, as for carbon steel, but, at higher corrosion values, the one- to four-year corrosion relationship approaches about 2:3, confirming the improved performance of weathering steel after longer exposure periods.

Figure 15 shows the four-year values for all sites. A general, significant decrease can be observed, and the corrosion was reduced by about 50%. As can be observed, the highest decreases correspond to urban sites, i.e., Madrid, Stockholm, Milan, Prague and Oslo, whereas the lowest correspond to both industrial sites, i.e., Kopisty and Bottrop. At the rural sites and some urban sites, an intermediate corrosion reduction was obtained. Similar to the case of carbon steel and other metals, the main reason for this corrosion diminishing is the general decrease in $SO_2$ levels, especially from 1987 to about 2000.

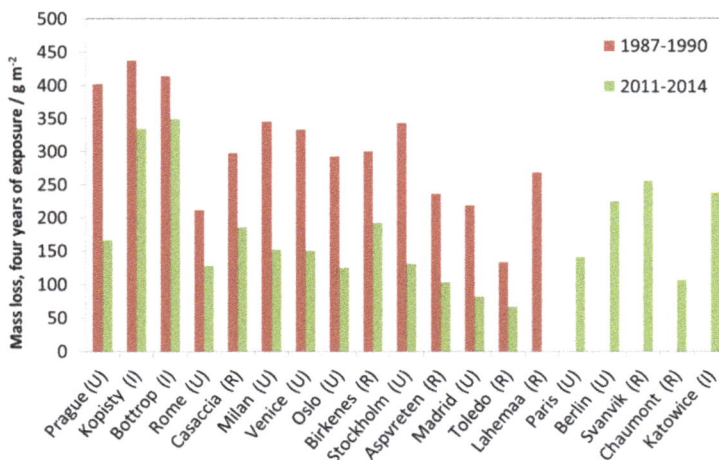

**Figure 15.** Weathering steel corrosion: four-year exposures at individual sites for two exposure, periods 1987–1990 and 2011–2014.

*2.4. Trends in Zinc Corrosion*

Zinc (>98.5%) with two different kinds of surface preparation, ground and glass blasted, has been exposed in the programme; the ground from the beginning of the programme, from 1987, which was then replaced with the blasted from 1997. Simultaneous exposures were performed in 2000 and 2008. The glass blasted zinc has a rougher surface leading to, at least initially, higher corrosion loss values.

As can be seen in Figure 16, the mass loss after one year for the investigated industrial test sites decreased significantly in the period 1987–1997. From 1997, the value then remained at a constant level (around 10 $g/m^2$) for the first exposure year (blasted zinc). The mass loss values for Kopisty are shown as an example for an industrial site, where the ISO 9223 corrosivity category changed from C4 to C3 (ISO 9223 corrosivity category C3 ranges from 5 to 15 $g/m^2$).

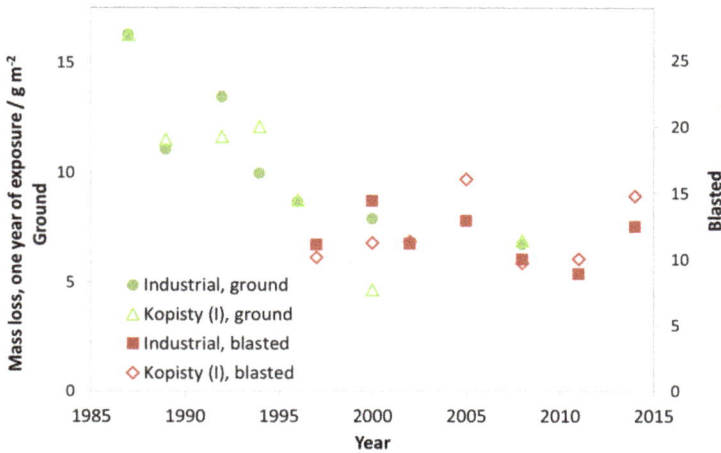

**Figure 16.** Zinc corrosion, ground and blasted at industrial sites and the industrial site Kopisty.

In Figure 17, the trend is shown for the nine investigated urban test sites with Prague as a typical example. As for the industrial sites, there is a strong decrease of mass loss between exposure periods 1987–1988 and 1997–1998, but also a small reduction of mass loss in the following periods. Today, the corrosivity category for the urban test sites (C3) is starting to approach C2 with values comparable to those in rural atmospheres.

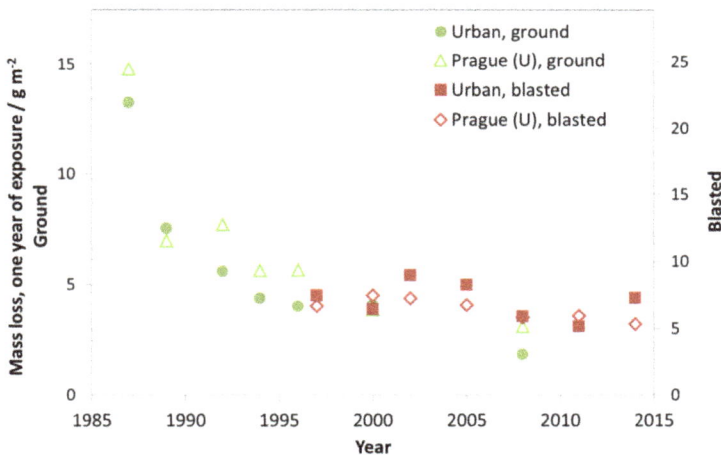

**Figure 17.** Zinc corrosion, ground and blasted at urban sites and the urban site Prague.

The trend for rural test sites is slightly less distinct compared to industrial and urban test sites. There is a higher fluctuation of mass loss values from year to year. For some rural test sites, a slight decrease of mass loss values can be found but there are other sites, such as Lahemaa (Estonia), with no clear trend (Figure 18). Higher mass loss values were sometimes measured at rural sites compared to urban sites. Overall, the mass loss values (blasted) at most sites now range 5–10 g/m$^2$. These values correspond to a thickness reduction of 0.7–1.4 μm, calculated based on the density of zinc 7.14 g cm$^{-3}$ (thickness reduction in μm = mass loss in g m$^{-2}$/7.14).

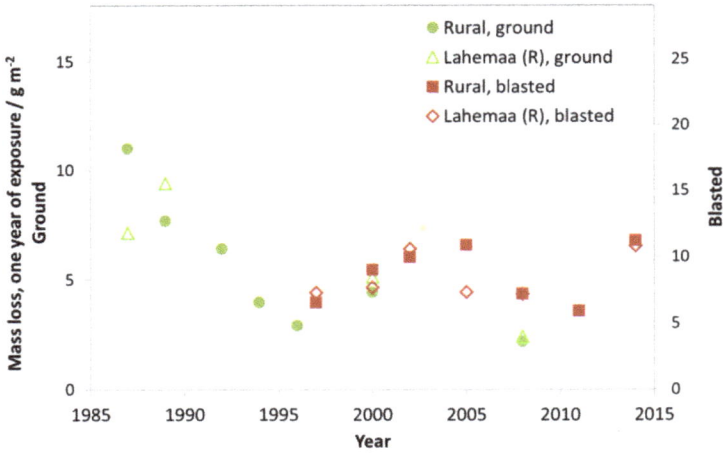

**Figure 18.** Zinc corrosion, ground and blasted at rural sites and the rural site Lahemaa.

Three repeated four-year exposures, starting from 1987, were undertaken with zinc samples in unsheltered exposure at different test sites. The first exposure period (1987–1988) was with ground surface condition and the other two with blasted surface condition. Figure 19 shows a comparison between four- and one-year corrosion, similar to for carbon steel (Figure 12) and weathering steel (Figure 14). In contrast to the other materials, there is a systematic difference when looking at the different periods. For ground zinc, the relationship between the mass loss of the one- and four-year samples is about 1:4, corresponding to a line going through the point 15 g m$^{-2}$ (one year of exposure) and 60 g m$^{-2}$ (four years of exposure), except for two sites. If the relationship between one- and four-year data were exactly 1:4, this would correspond to a linear development of corrosion with time, i.e., linear kinetics. For blasted zinc, the four-year values are lower than what would be expected from linear kinetics, indicating that the high corrosion values resulting from blasting as opposed to ground is an initial phenomenon most prominent after one year of exposure.

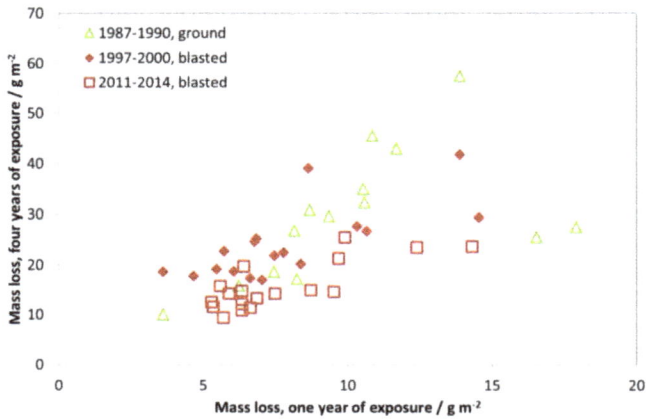

**Figure 19.** Zinc corrosion: one-year vs. four-year exposures for three different periods, 1987–1990 (ground zinc), 1997–2000 (blasted) and 2011–2014 (blasted). The one-year corrosion values are calculated as averages of two one-year exposures for the years 1987/1990, 1997/2000 and 2011/2014, respectively.

Figure 20 shows all four-year data for the individual sites. The first value (1987–1990) is not directly comparable to the two later values due to the different surface treatment. However, it is not possible to make a correction factor since, parallel four-year exposures have not been performed. Furthermore, results presented in Figure 19 show that it is not possible to use a comparison of one-year values to derive a correction value for four-year samples. Nevertheless, it has some merit to present ground zinc (uncorrected) in the same diagram as blasted zinc, even if the decrease in corrosion between 1987 and 1990 and the other periods will be underestimated in this way.

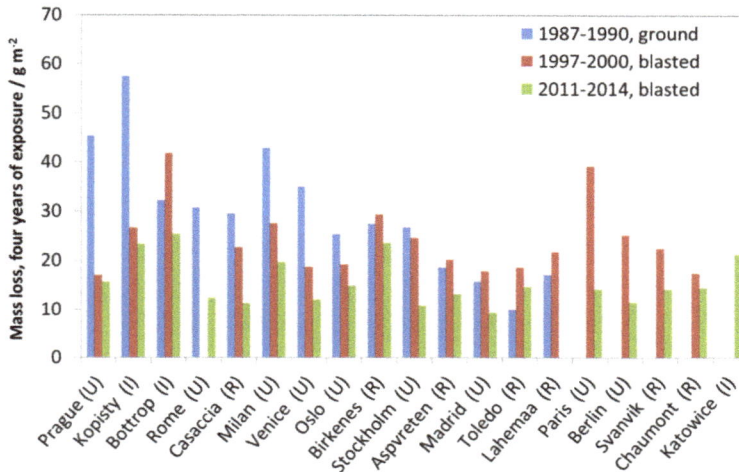

**Figure 20.** Zinc corrosion; four-year exposures at individual sites for three exposure periods, 1987–1990 (ground), 1997–2000 (blasted) and 2011–2014 (blasted).

An example of a strong decreasing trend is the test site Kopisty. The mass loss at the first exposure period (1987–1990) was around 60 g/m² during four years (which means an average yearly corrosion rate of 15 g/(m² year) or 2.1 µm/year). This value decreased to 23 g/m² for the third exposure period (2011–2014). The corrosion rate at this test site now lies at 5.8 g/(m² year) or 0.8 µm/a and can now be characterised as C3. This means a reduction of the corrosion rate of 66% from the value of the first exposure period. In practical terms, this means that a galvanized steel structure with a typical zinc layer thickness of 80 µm would previously have shown red rust (1987) after approximately 30 years, while it would now take 100 years. It should be noted that it is difficult to accurately estimate the long-term corrosion rate based on these values, especially considering the results presented in Figure 19, but, in any case the reduction is substantial. Furthermore, values at urban test sites are now in the same order of magnitude as rural test sites, which makes it more difficult to evaluate the effect of air pollution based on zinc corrosion.

When looking at Figure 20 and comparing the two last periods (1997–2000 and 2011–2014), there is a decrease in corrosion at all sites where a comparison is possible but it was not possible to capture this overall trend based on one-year data only (Figures 16–18). Thus, the advantage of using four-year values for showing trends in zinc corrosion is quite evident. Four-year values are less sensitive to year-to-year variation in climatic parameters. The higher corrosion values also make it easier to identify significant trends for individual sites.

## 2.5. Trends in Copper Corrosion

Copper (Cu 99%, P 0.015%–0.04%) has been exposed for one year in 1987, 1997, 2002, 2011 and 2014 and the trends for industrial, urban and rural sites (accompanied with examples for individual sites) are presented in Figures 21–23 in the same manner as for carbon steel (compare Figures 9–11). The trends are similar for carbon steel and copper. After 1997, the decreasing trend is less evident, and, at the industrial sites, there are no decreasing trends at all.

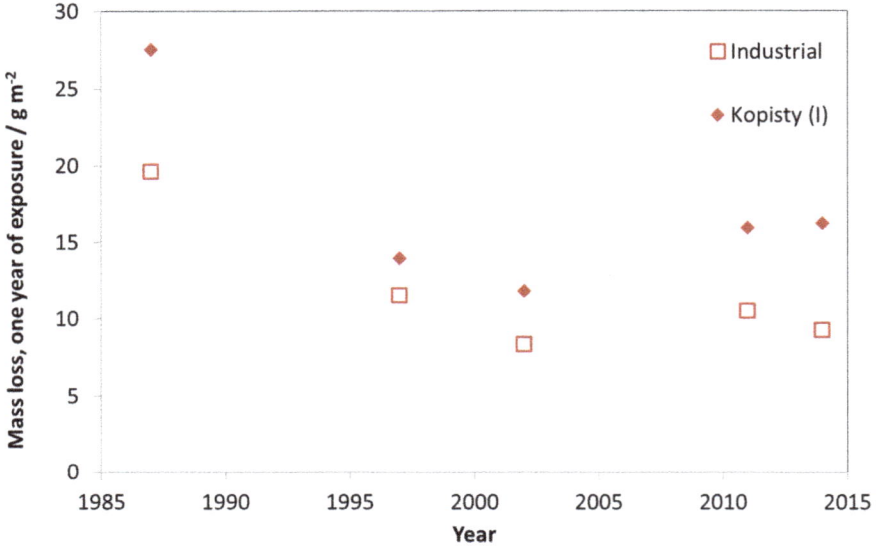

**Figure 21.** Copper corrosion at industrial sites and the industrial site Kopisty.

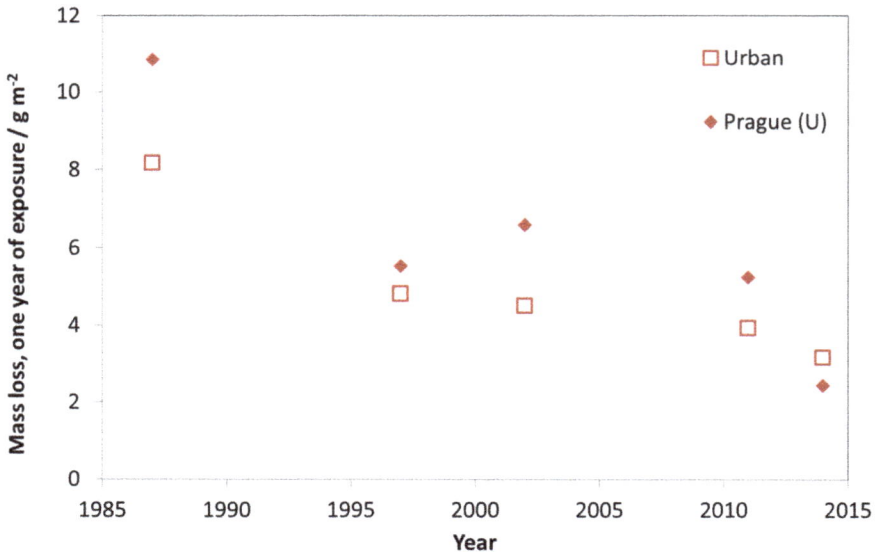

**Figure 22.** Copper corrosion at urban sites and the urban site Prague.

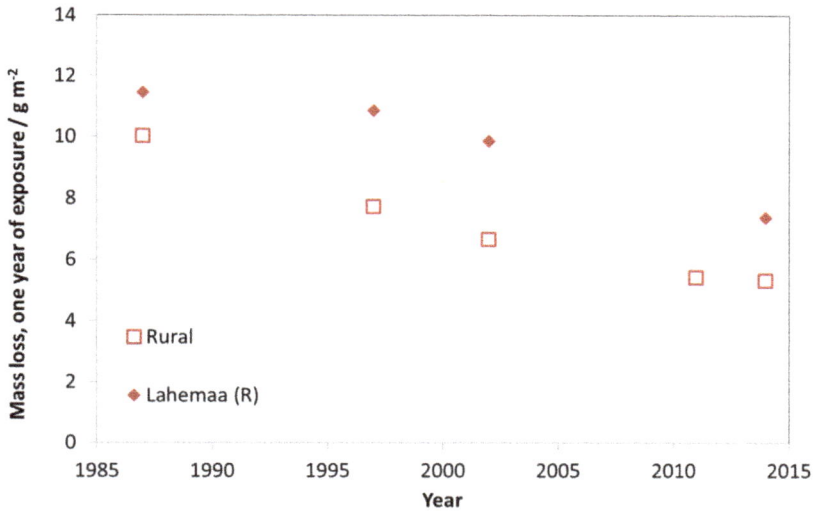

**Figure 23.** Copper corrosion at rural sites and the rural site Lahemaa.

Copper has been exposed for four years only for two exposure periods, 1987–1990 and 1997–2000. Figure 24 shows the mass loss after four years of exposure vs. mass loss after one year of exposure for the two periods. The relationship between one- and four-year data is about 1:3. Figure 25 summarises the trends in copper corrosion for all sites using three time periods, 1987, 1997–2002 average and 2011–2014 average, based on the available one-year data since four-year data are not available for the most recent period (2011–2014). The trends show decreasing corrosion for the most recent periods, except for Kopisty and Bottrop, both industrial sites.

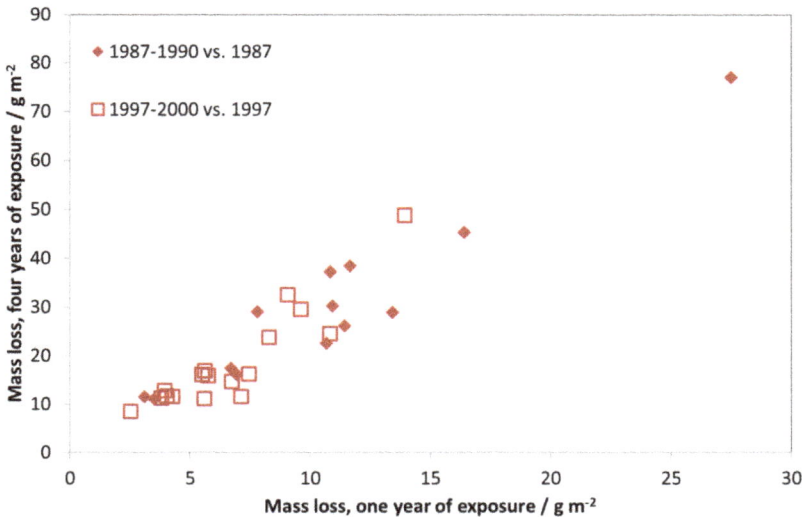

**Figure 24.** Copper corrosion one-year vs. four-year exposures for two different periods, 1987–1990 and 1997–2000.

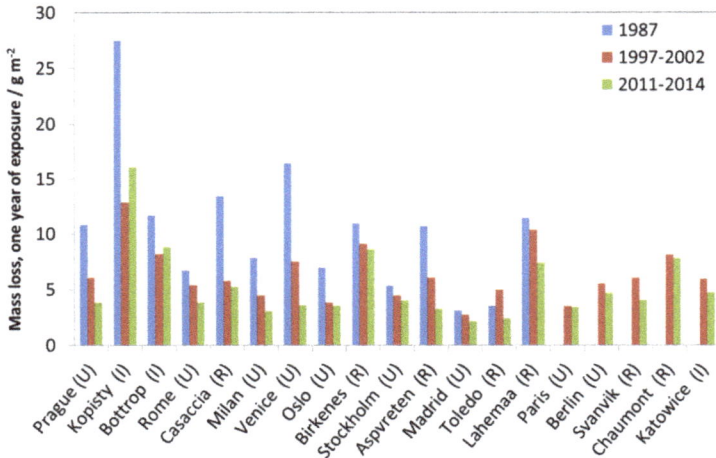

**Figure 25.** Copper corrosion at individual sites based on one-year data for 1987 and annual averages the periods 1997–2002 and 2011–2014.

### 2.6. Trends in Aluminium Corrosion

For aluminium (>99.5%), data from the period 1987–1994 exist for two-, four- and eight-year exposures. In 2011, one set of aluminium samples was exposed with the intention to make a withdrawal after two years of exposure. Inspections at sites after two years indicated very low corrosion rates and therefore withdrawal was made after four years of exposure. For evaluation of trends in aluminium corrosion, there is thus only two four-year periods available, 1987–1990 and 2011–2014. The data are shown in Figure 26. The decrease in corrosion is substantial, but it should be noted that some of the new sites, especially Berlin, show corrosion values comparable to those obtained in 1987–1990.

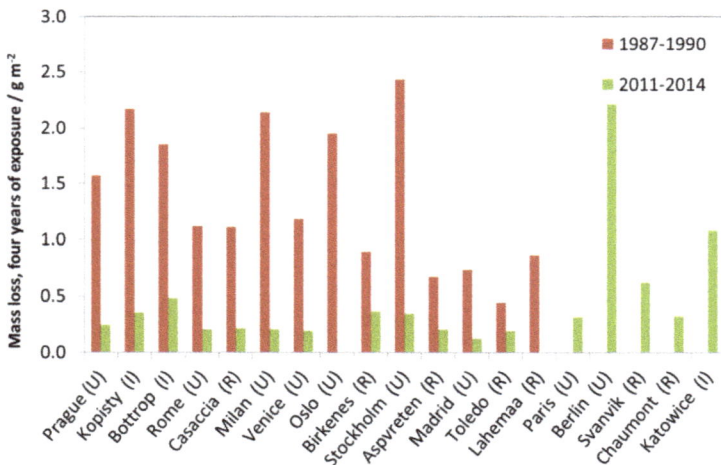

**Figure 26.** Aluminium corrosion: four-year exposures at individual sites for two exposure periods, 1987–1990 and 2011–2014.

## 2.7. Trends in Surface Recession of Limestone

Figures 27–29 show surface recession of limestone following the same model as for previous materials. Figure 27 shows industrial sites, Figure 28 urban sites and Figure 29 rural sites, each with one example site. The 1987 value is higher for industrial and urban sites but otherwise there is no evident decreasing trend after 1997, and the year-to-year fluctuations are substantial, indicating influence from varying climatic conditions.

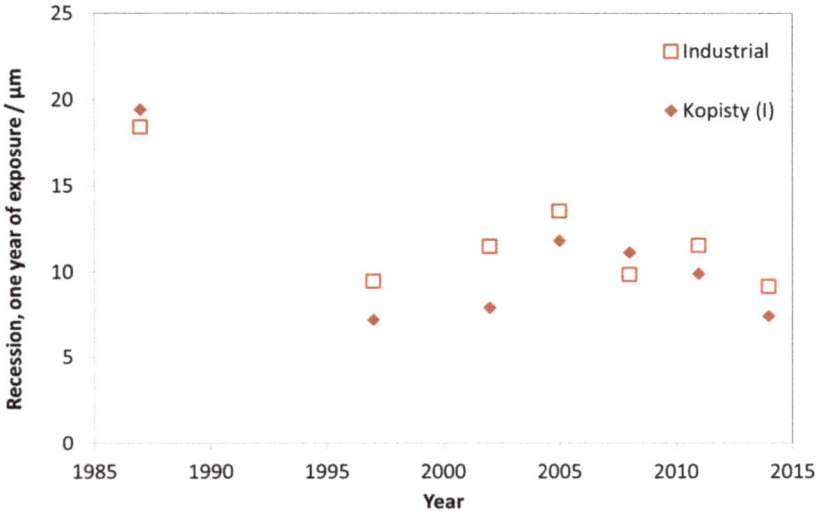

**Figure 27.** Limestone surface recession at industrial sites and the industrial site Kopisty.

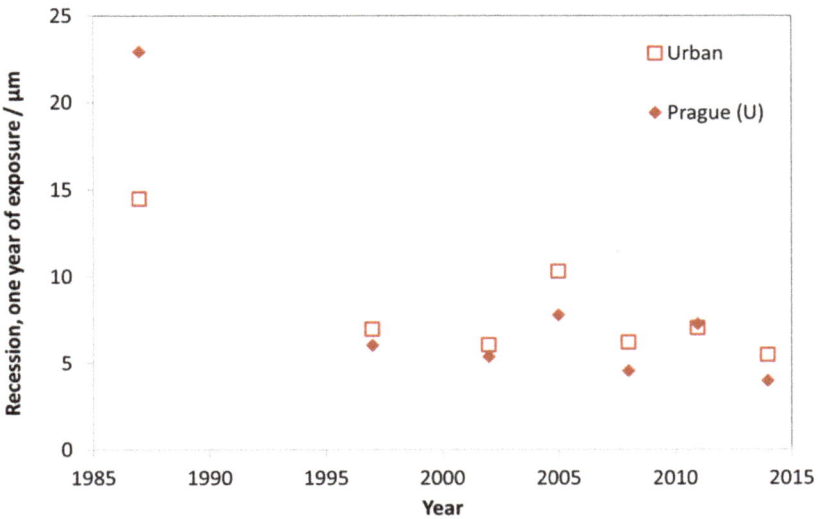

**Figure 28.** Limestone surface recession at urban sites and the urban site Prague.

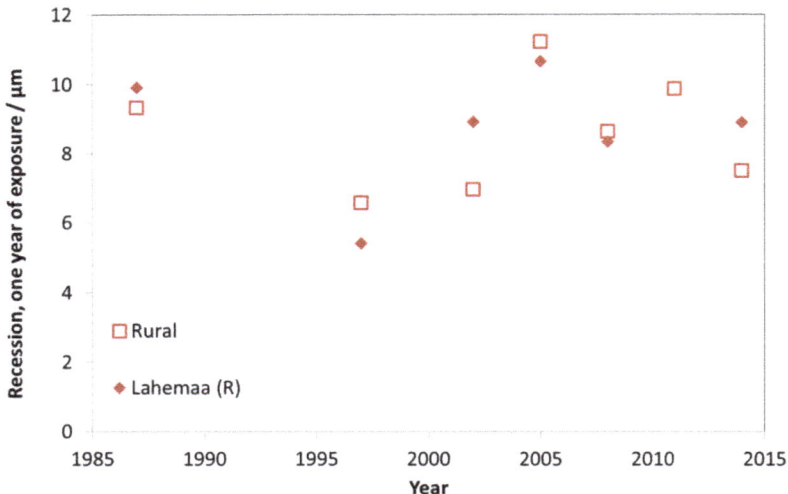

**Figure 29.** Limestone surface recession at rural sites and the rural site Lahemaa.

Figure 30 shows the four-year values vs. the one-year values, and shows almost a 1:4 correspondence, indicating linear kinetics, but with the one-year value being slightly higher, which is expected for limestone degradation.

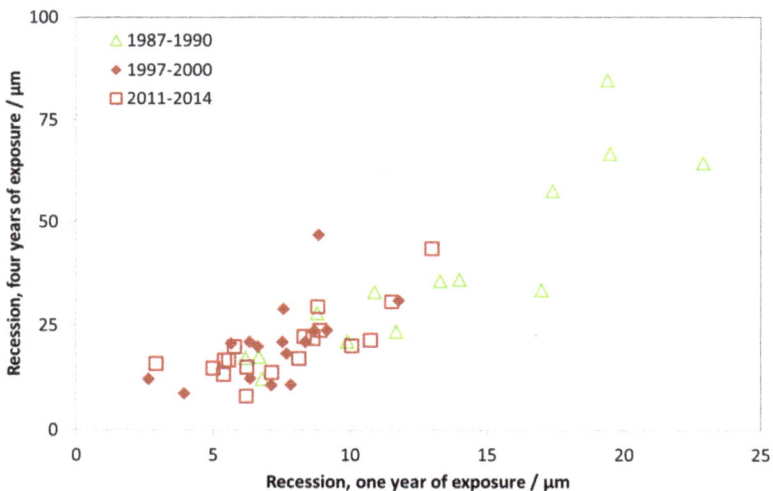

**Figure 30.** Limestone surface recession: one-year vs. four-year exposures for three different periods, 1987–1990, 1997–2000 and 2011–2014. The one-year corrosion values for comparison with 1987–1990 are taken from 1987, and for the other two periods calculated as averages of two one-year exposures for the years 1997/2002 and 2011/2014, respectively.

Figure 31 shows all four-year data for limestone. Compared to other materials, the 1997–2000 to 2011–2014 trend is less evident, except for Prague, Bottrop, Milan, Venice and Oslo.

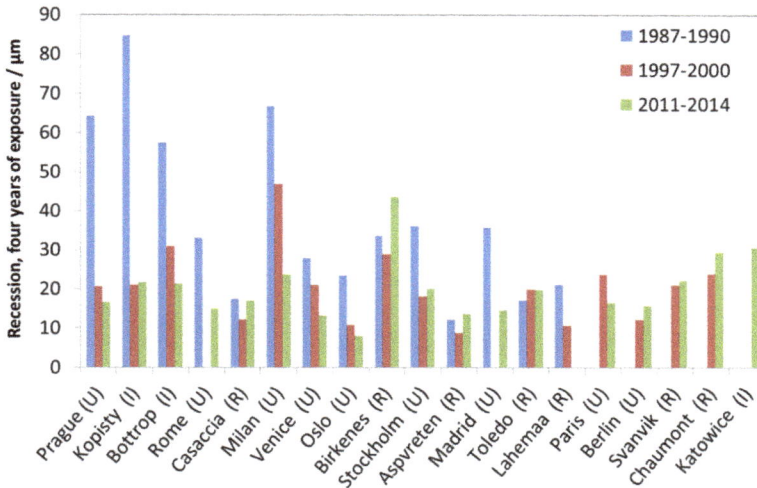

**Figure 31.** Limestone surface recession: four-year exposures at individual sites for three exposure periods, 1987–1990, 1997–2000 and 2011–2014.

## 2.8. Trends in Soiling of Modern Glass

Four one-year exposures of glass in sheltered conditions have been carried out between 2005 and 2014. Of the materials presented here, this is the only material exposed in sheltered position. The soiling is quantified using the haze parameter, which is the ratio between the diffuse and direct transmitted light [10]. Figure 32 shows that the haze increases, moderately for Birkenes and Aspvreten, and strongly for Casaccia, Venice and Paris. A sharp increase is difficult to explain for the site Casaccia, but is caused by the moving of the site for Paris in 2011. Haze is relatively constant or decreasing for the other sites.

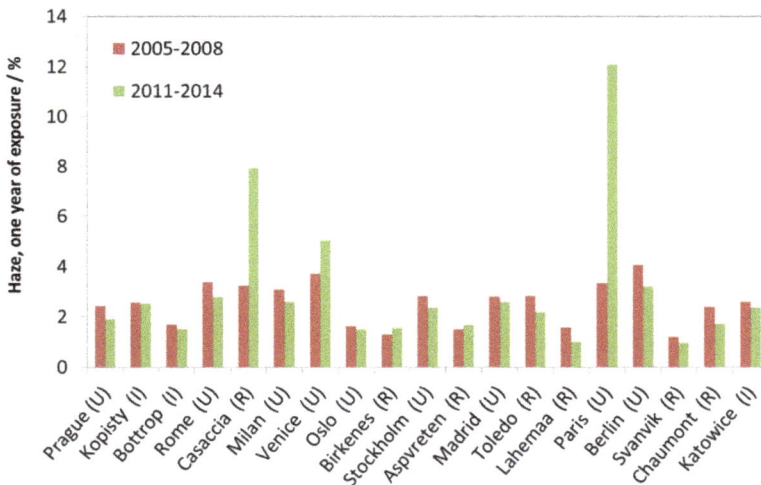

**Figure 32.** Soiling of modern glass after one-year of exposure, 2005–2008 and 2011–2014, in sheltered position expressed as haze (%).

## 3. Discussion

The results presented show that corrosion and pollution have decreased significantly during the period where data are available (1987–2014). When looking into the details of the decreasing trends, a shift in the magnitude generally occurs around 1997; a sharp decrease changed to a more modest decrease or to a constant level without any decrease. The levels of the pollutant $SO_2$ and the corrosion of the materials carbon steel and copper have decreased even after 1997, more pronounced in urban areas, while the other materials show no decreases in corrosion or soiling after 1997, when looking at one-year values. When looking at four-year values, however, there is a significant decrease after 1997 for zinc, which is not evident when looking at the one-year values. The advantage of using four-year values for showing trends in corrosion is that it is less sensitive to year-to-year variation in climatic parameters (temperature, relative humidity, and precipitation). The higher corrosion values make it also easier to identify significant trends for individual sites.

The reduction in corrosion has changed the way we look at "industrial", "urban" and "rural" sites from a corrosion point of view. In the past, these labels could in some way relate to the level of corrosion, but this is no longer the case, especially for some materials. The labels may be useful for indicating the type of pollution (dominated by $SO_2$, $NO_2$ or $O_3$, for example) but they are not useful at all for classifying levels of corrosivity in a quantitative way.

ISO 9223 provides a system for classification of corrosivity. When applying this to the data, the change in corrosion is often from C4 to C3 or even C2. This has practical implications, as it is expected that lifetimes of constructions affected by corrosion is significantly prolonged in some atmospheres. However, there are contemporary environments, such as the Berlin site (Figure 26), which still show corrosion values corresponding to past levels. This is a new type of site in the ICP Materials programme, close to the road with high levels of particulate deposition. This illustrated the need for continuous measurements of pollutants and the awareness to possible new or reappearing emission sources.

The paper also presents results on corrosion kinetics by comparison of one- and four-year corrosion values. Some materials, such as carbon steel and copper, show kinetics relatively independent of sites, while other materials, especially zinc, show substantial variation in kinetics for the first four years, which needs to be taken into account when producing new and possibly improved models for corrosion.

ICP Materials is now in the process of starting a new exposure (2017 and 2017–2020), which will provide new sets of one- and four-year data. Included in the programme will be, in addition to earlier exposures described in this paper, corrosion of stainless steel and soiling of two coil coated materials (white and brown) as well as soiling of two stone materials (limestone and marble). These additions show a direction of future development of the programme, i.e. more focused on soiling of materials and the effect of particulate matter.

## 4. Materials and Methods

Procedures used in ICP Materials are described in the technical manual, which includes information on materials, environment, how to run a test site (29 pages) and detailed descriptions of all test sites (86 pages). It is beyond the scope of this paper to repeat all this information, but, in general, exposures conform to procedures described in ISO 8565. The technical manual and all reports produced by ICP Materials are available for download at the ICP Materials home page.

In addition, all data discussed and presented in this paper can be found at the Supplementary File. They are also available for download, submitted as open access, at the ICP Materials home page.

**Supplementary Materials:** The following are available online at http://www.mdpi.com/1996-1944/10/8/969/s1.

**Acknowledgments:** The evaluation was based on data that were collected by members of the Task Force and organisations, including supporting organisations of the official UNECE ICP Materials network. These organisations are presented under the heading Acknowledgement at the ICP Materials web page.

**Author Contributions:** J.T. and A.G. were responsible for the overall compiling of the paper and for evaluation of copper and aluminium. K.K. was responsible for the evaluation of carbon steel, M.F. and U.H. for zinc, D.F. for weathering steel, T.Y. for limestone, A.V.-C. for modern glass and T.G. for environmental data.

**Conflicts of Interest:** The founding sponsors had no role in the design of the study; in the collection, analyses, or interpretation of data; in the writing of the manuscript, and in the decision to publish the results.

## References

1. Tidblad, J.; Kucera, V.; Mikhailov, A.A.; Henriksen, J.; Kreislova, K.; Yates, T.; Stöckle, B.; Schrener, M. UN ECE ICP Materials: Dose-response functions on dry and wet acid deposition effects after 8 years of exposure. *Water Air Soil Pollut.* **2001**, *130*, 1457–1462. [CrossRef]

2. Kucera, V.; Tidblad, J.; Kreislova, K.; Knotkova, D.; Faller, M.; Reiss, D.; Snethlage, R.; Yates, T.; Henriksen, J.; Schreiner, M.; et al. UN ECE ICP materials dose-response functions for the multi-pollutant situation. *Water Air Soil Pollut. Focus* **2007**, *7*, 249–258. [CrossRef]

3. Tidblad, J. Atmospheric corrosion of metals in 2010–2039 and 2070–2099. *Atmos. Environ.* **2012**, *55*, 1–6. [CrossRef]

4. Tidblad, J.; Kucera, V.; Ferm, M.; Kreislova, K.; Brüggerhoff, S.; Doytchinov, S.; Screpanti, A.; Grøntoft, T.; Yates, T.; De la Fuente, D.; et al. Effects of Air Pollution on Materials and Cultural Heritage: ICP Materials Celebrates 25 Years of Research. *Int. J. Corros.* **2012**, *2012*, 16. [CrossRef]

5. Baedecker, P.A. Effects of Acidic Disposition of Materials. In *Acidic Deposition: State of Science and Technology*; National Acid Preparation Assessment Program: Washington, DC, USA, 1990.

6. Knotkova, D. Atmospheric Corrosivity Classification. Results of the International Testing Program ISOCORRAG. In *Corrosion Control for Low-Cost Reliability: 12th International Corrosion Congress*; Progress Industries Plant Operations; NACE International: Houston, TX, USA, 1993; Volume 2, pp. 561–568.

7. Morcillo, M.; Almeida, E.M.; Rosales, B.M. (Eds.) Funciones de Dano (Dosis/Respuesta) de la Corrosion Atmospherica en Iberoamerica. In *Corrosiony Proteccion de Metales en las Atmosferas de Iberoamerica*; CYTED: Madrid, Spain, 1998; pp. 629–660.

8. Tidblad, J.; Kucera, V.; Samie, F.; Das, S.N.; Bhamornsut, C.; Peng, L.C.; So, K.L.; Dawei, Z.; Hong Lien, L.T.; Schollenberger, H.; et al. Exposure Programme on Atmospheric Corrosion Effects of Acidifying Pollutants in Tropical and Subtropical Climates. *Water Air Soil Pollut. Focus* **2007**, *7*, 241–247. [CrossRef]

9. Ferm, M.; De Santis, F.; Varotsos, C. Nitric acid measurements in connection with corrosion studies. *Atmos. Environ.* **2005**, *39*, 6664–6672. [CrossRef]

10. Lombardo, T.; Ionescu, A.; Lefèvre, R.A.; Chabas, A.; Ausset, P. Soiling of silica-soda-lime float glass in urban environment: measurements and modelling. *Atmos. Environ.* **2005**, *39*, 989–997. [CrossRef]

*materials*

MDPI

*Article*

# Annual Atmospheric Corrosion of Carbon Steel Worldwide. An Integration of ISOCORRAG, ICP/UNECE and MICAT Databases

Belén Chico, Daniel de la Fuente, Iván Díaz, Joaquín Simancas and Manuel Morcillo *

National Centre for Metallurgical Research (CENIM/CSIC), Av. de Gregorio del Amo, 8, 28040 Madrid, Spain; bchico@cenim.csic.es (B.C.); delafuente@cenim.csic.es (D.d.l.F.); ivan.diaz@cenim.csic.es (I.D.); jsimancas@cenim.csic.es (J.S.)
* Correspondence: morcillo@cenim.csic.es; Tel.: +34-915538900

Academic Editor: Yong-Cheng Lin
Received: 21 March 2017; Accepted: 24 May 2017; Published: 31 May 2017

**Abstract:** In the 1980s, three ambitious international programmes on atmospheric corrosion (ISOCORRAG, ICP/UNECE and MICAT), involving the participation of a total of 38 countries on four continents, Europe, America, Asia and Oceania, were launched. Though each programme has its own particular characteristics, the similarity of the basic methodologies used makes it possible to integrate the databases obtained in each case. This paper addresses such an integration with the aim of establishing simple universal damage functions (DF) between first year carbon steel corrosion in the different atmospheres and available environmental variables, both meteorological (temperature (T), relative humidity (RH), precipitation (P), and time of wetness (TOW)) and pollution ($SO_2$ and NaCl). In the statistical processing of the data, it has been chosen to differentiate between marine atmospheres and those in which the chloride deposition rate is insignificant ($<3$ mg/$m^2$.d). In the DF established for non-marine atmospheres a great influence of the $SO_2$ content in the atmosphere was seen, as well as lesser effects by the meteorological parameters of RH and T. Both NaCl and $SO_2$ pollutants, in that order, are seen to be the most influential variables in marine atmospheres, along with a smaller impact of TOW.

**Keywords:** atmospheric corrosion; carbon steel; damage function; ISOCORRAG; ICP/UNECE; MICAT

## 1. Introduction

The economic impact of corrosion of metallic structures is a matter of great relevance throughout the world. The World Corrosion Organisation (WCO) currently estimates the direct cost of corrosion worldwide at between €1.3 and 1.4 trillion, which is equivalent to 3.8% of the global Gross Domestic Product (GDP). More than half of the considerable damage due to corrosion is a result of atmospheric impacts on materials, which is logical considering that most metallic equipment and structures operate in the atmospheric environment. For this reason, the action of the atmosphere on metals is one of the major issues in corrosion science.

In a perfectly dry atmosphere, metallic corrosion progresses at an extremely low rate, and for practical purposes can be ignored. However, on wet surfaces, corrosion can be quite severe, as the atmospheric corrosion process is the sum of the individual corrosion processes that take place whenever an electrolyte layer forms on the metal surface. However, for the corrosion rate to be really significant, the atmosphere must also be polluted. Of all atmospheric pollutants, chlorides from marine aerosol and sulphur dioxide ($SO_2$) mainly from the combustion of fossil fuels, are the most common aggressive agents in the atmosphere.

It is a well-known fact, which has been proven by practical experience with real structure behaviour and the results of numerous tests, that the corrosion rate of metals in the atmosphere can be tens or even hundreds of times higher in some places than in others. Thus, it is of great interest to understand the basic variables that operate in atmospheric corrosion and in order to establish a classification of the aggressiveness of an atmosphere. The best possible knowledge of the factors that affect atmospheric corrosivity would obviously help to plan anticorrosive measures for metals in a given environment.

In the 1980s, three different cooperative studies involving the participation of a large number of countries were carried out:

- **ISOCORRAG cooperative programme.** This programme was designed by the Working Group/WG 4 of ISO 156 Technical Committee "Corrosion of metals and alloys", with the aim of standardising atmospheric corrosion tests) [1]. The Programme began in the year 1986 and, as a result of the efforts of WG 4, four international standards were developed: ISO 9223 [2,3], ISO 9224 [4], ISO 9225 [5] and ISO 9226 [6]. These standards were based on an extensive review of atmospheric exposure programmes carried out in Europe, North America, and Asia. The aim of drawing up these documents was to establish simple and practical guidelines for the technicians responsible for designing structures to be exposed to the atmosphere and for corrosion engineers responsible for adopting anticorrosive protection measures. ISO 9223 [2] provided a general classification system for atmospheres based either on 1-year coupon exposures or on measurements of environmental parameters to estimate time of wetness (TOW), sulphur dioxide concentration or deposition rate, and sodium chloride deposition rate. ISO 9224 provided an approach to calculating the extent of corrosion damage from extended exposures for five types of engineering metals based on application of guiding corrosion values (average and steady-state corrosion rates) for each corrosivity categories in ISO 9223. ISO 9225 provided the measurements techniques for the sulphur dioxide concentration or deposition rate, and sodium chloride deposition rate, needed as classification criteria in ISO 9223. ISO 9226 provided the procedure for obtaining one-year atmospheric corrosion measurements on standard coupons.

- **MICAT cooperative programme: "Ibero-American Atmospheric Corrosivity Map"** [7]. The MICAT programme was launched in 1988 as part of the Ibero-American CYTED "Science and Technology for Development" international programme and ended after six years of activities. Fourteen countries participated in the programme, whose goals were: (i) to obtain a greater knowledge of atmospheric corrosion mechanisms in the different environments of Ibero-America; (ii) to establish, by means of suitable statistical analysis of the results obtained, mathematical models that allow the calculation of atmospheric corrosion as a function of climate and pollution parameters; and (iii) to elaborate atmospheric corrosivity maps of the Ibero-American region.

- **ICP/UNECE cooperative programme** [8]. Airborne acidifying pollutants are known to be one of the major causes of corrosion of different materials, including the extensive damage that has been observed on historic and cultural monuments. In order to fill some important gaps in the knowledge of this field, the Executive Body for the Convention on Long-Range Transboundary Air Pollution (CLRTAP) decided to launch an International Cooperative Programme within the United Nations Economic Commission for Europe (ICP/UNECE). The programme started in September 1987 and initially involved exposure at 39 test sites in 11 European countries and in the United States and Canada. The aim of the programme was to perform a quantitative evaluation of the effect of sulphur pollutants in combination with $NOx$ and other pollutants as well as climatic parameters on the atmospheric corrosion of important materials.

Figure 1 shows the countries participating in each one of these programmes. The atmospheric corrosion stations are basically located in Europe, America and Asia, covering a broad range of meteorological and pollution conditions.

Though the three programmes, ISOCORRAG, ICP/UNECE and MICAT, each have their own particular characteristics, they nevertheless share a number of common objectives. The similarity of certain aspects of their methodologies allows a welcome meeting point between the three programmes, as was suggested by Morcillo in the 11th International Corrosion Congress held in Florence in April 1990, in the session on atmospheric corrosion where the three cooperative programmes were presented [9–11]. Such a meeting point would allow, for the first time, a worldwide perspective (38 countries) on the problem of atmospheric corrosion, covering a broad spectrum of climatological and atmospheric pollution conditions, never before considered in the abundant published literature on atmospheric corrosion. This idea was taken up at UNECE (Figure 2) by the "Working Group on Effects" of Executive Body for the Convention on Long-Range Transboundary Air Pollution [12].

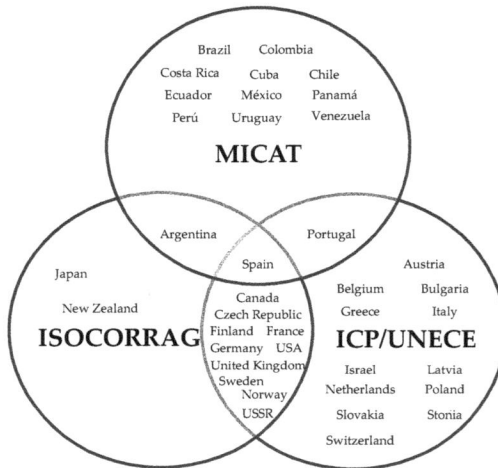

**Figure 1.** International Collaborative programmes on atmospheric corrosion and participant countries.

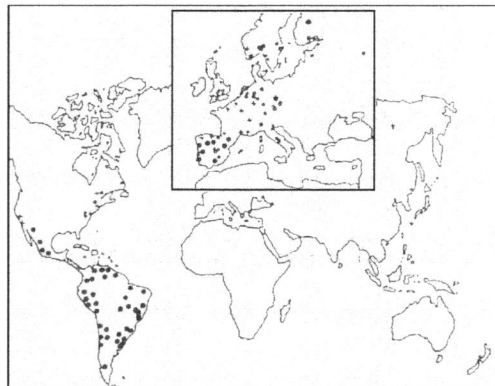

**Figure 2.** Atmospheric corrosion stations networks: ISOCORRAG (+), ICP (°) and MICAT (•) [12].

The statistical analysis of data obtained in atmospheric corrosion studies in order to obtain correlation equations that allow the estimation of annual corrosion rates from meteorological and pollution parameters is a matter of great interest. Such equations are known as damage or dose/response functions. They often incorporate the $SO_2$ concentration, the chloride concentration in areas close to the sea, and a parameter representing the wetness of the metallic surface (relative humidity, number of days of rain per year, time of wetness, etc.). Models for predicting the corrosion damage of metals in the atmosphere are useful when it comes to answering questions on the durability of metallic structures, determining the economic costs of damage associated with the degradation of materials, or acquiring knowledge about the effect of environmental variables on corrosion kinetics.

Abundant literature has been published on these models and damage functions. For instance, for long-term prediction of carbon steel atmospheric corrosion, mention may be made of the work of Benarie and Lipfert [13], Pourbaix et al. [14], Feliu et al. [15,16], Knotková and Barton [17], Kucera [18], Mc Cuen and Albrecht [19], Albrecht and Hall [20], Panchenko et al. [21,22], Melchers [23,24], etc. Recent reviews on corrosion models for long-term prediction of atmospheric corrosion has been made by Morcillo et al. [25] and Adikari and Munasinghe [26].

The purpose of this work is to bring together the three databases from the three international cooperative programmes (ICP/UNECE, MICAT and ISOCORRAG), carrying out a statistical analysis of the results they contain in order to establish mathematical expressions which allow an estimation of the extent of atmospheric corrosion of carbon steel during first-year exposure as a function of meteorological and pollution parameters.

## 2. Experimental

### 2.1. ICP/UNECE Programme

Twenty-four countries participated in the exposure programme with a total of 55 exposure sites. These sites included industrial, urban and rural atmospheres. Marine atmosphere exposures were not included. A list of the sites together with their code is given in Table 1.

Exposure always started in the autumn, typically from October of one year to September of the following year. The test site network originally consisted of 39 sites, which were all part of the original eight-year exposure between 1987 and 1995. Subsequently, in a four-year exposure programme carried out between 1997 and 2001, only part of the original sites were kept and eight new test sites were added. Since then, new sites have joined. Compared to the 2008–2009 exposure, the sites Lahemaa and Lincoln were withdrawn from the 2011 to 2012 exposure while a new site in St Petersburg (Russia) was added. In the 2014–2015 exposure, two new test sites, Hameenlina (Finland) and Zilina (Slovakia), were included.

Figure 3 shows a diagram of the exposure schedule. For each exposure and site, three identical flat samples were exposed. Average corrosion values for these three panels were obtained. A detailed description of the material and methods for measuring environmental parameters and the evaluation of corrosion attack is provided in Reference [8].

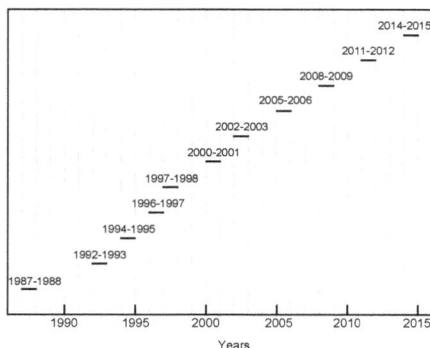

**Figure 3.** ICP/UNECE programme: Diagram showing the exposure sequences.

## 2.2. ISOCORRAG Programme

Fourteen countries participated in the exposure programme with a total of 53 exposure sites. These sites included industrial, urban, rural, marine and costal locations in temperate, tropical and arctic zones. A list of the sites together with their code is given in Table 2.

Flat carbon steel specimens were exposed in triplicate, fixing their size and thickness in accordance with the provisions of standard ISO 8565 [27]. A detailed description of the exposed material is provided in Reference [1].

A set of specimens was initially exposed for one-year exposure at each site. After six months, another set of specimens was exposed for one-year exposure. After one year, the first set of one-year exposed specimens was removed and another set of one-year specimens was exposed. Every six months, this process was repeated until six sets of specimens had been exposed for one year. Figure 4 shows a diagram of the exposure schedule. The original exposure was planned to begin in the autumn of 1986, but several delays occurred at various sites.

A detailed description of the material and methods for measuring environmental parameters and the evaluation of corrosion attack is provided in Reference [1].

**Table 1.** Atmospheric corrosion test sites included in the ICP/UNECE Programme.

| Code | Country | Test Site | Code | Country | Test Site |
|------|---------|-----------|------|---------|-----------|
| P01 | Czech Republic | Praha | P29 | United Kingdom | Clatteringshaws Loch |
| P02 | | Kasperske Hory | P30 | | Stoke Orchard |
| P03 | | Kopisty | P31 | Spain | Madrid |
| P04 | Finland | Espoo | P32 | | Bilbao |
| P05 | | Ahtari | P33 | | Toledo |
| P06 | | Helsinki | P34 | Russia | Moscow |
| P07 | Germany | Waldhof-Langenbrugge | P35 | Estonia | Lahemaa |
| P08 | | Aschaffenburg | P36 | Portugal | Lisbon-Jeronimo Mon. |
| P09 | | Langenfeld-Reusrath | P37 | Canada | Dorset |
| P10 | | Bottrop | P38 | USA | Steubenville |
| P11 | | Essen-Leithe | P39 | | Res. Triangle Park |
| P12 | | Garmisch-Partenkirchen | P40 | France | Paris |
| P13 | Italy | Rome | P41 | Germany | Berlin |
| P14 | | Casaccia | P43 | Israel | Tel Aviv |
| P15 | | Milan | P44 | Norway | Svanvik |
| P16 | | Venice | P45 | Switzerland | Chaumont |
| P17 | Netherlands | Vlaardingen | P46 | United Kingdom | London |
| P18 | | Eibergen | P47 | USA | Los Angeles |
| P19 | | Vredepeel | P49 | Belgium | Anvterps |
| P20 | | Wijnandsrade | P50 | Poland | Katowice |
| P21 | Norway | Oslo | P51 | Greece | Athens |
| P22 | | Borregaard | P52 | Latvia | Riga |
| P23 | | Birkenes | P53 | Austria | Vienna |
| P24 | Sweden | Stockholm S | P54 | Bulgaria | Sophia |
| P25 | | Stockholm C | P55 | Russia | St Petersburg |
| P26 | | Aspvreten | P57 | Finland | Hameelina |
| P27 | United Kingdom | Lincoln Catch. | P59 | Slovakia | Zilina |
| P28 | | Wells. Catch. | | | |

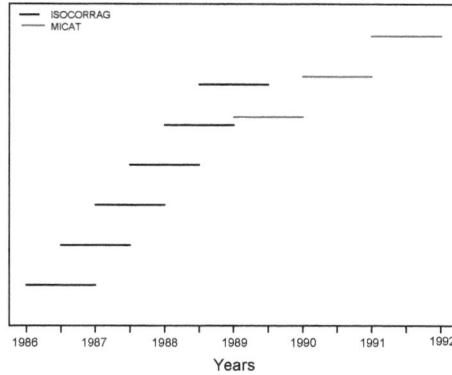

**Figure 4.** ISOCORRAG and MICAT programmes: diagrams showing the exposure sequences.

**Table 2.** Atmospheric corrosion test sites included in the ISOCORRAG Programme.

| Code | Country | Test Site | Code | Country | Test Site |
|------|---------|-----------|------|---------|-----------|
| I01 | | Iguazu | I29 | | Birkenes |
| I02 | | Camet | I30 | Norway | Tannanger |
| I03 | Argentina | Buenos Aires | I31 | | Bergen |
| I04 | | San Juan | I32 | | Svanvik |
| I05 | | Jubany Base | I33 | | Madrid |
| I06 | Canada | Bourcherville | I34 | Spain | El Pardo |
| I07 | | Kasperske Hory | I35 | | Lagoas-Vigo |
| I08 | Czech Republic | Praha-Bechovice | I36 | | Baracaldo, Vizcaya |
| I09 | | Kopisty | I37 | | Stockholm-Vanadis |
| I10 | Germany | Bergisch Gladbach | I38 | Sweden | Bohus Malmon, Kattesand |
| I11 | | Helsinki | I39 | | Bohus Malmon, Kvarnvik |
| I12 | Finland | Otaniemi | I40 | | Stratford, East London |
| I13 | | Ahtari | I41 | United Kingdom | Crowthorne, Berkshire |
| I14 | | Saint Denis | I42 | | Rye, East Sussex |
| I15 | | Ponteau Martigues | I43 | | Fleet Hall |
| I16 | | Picherande | I44 | | Kure Beach, N. Carolina |
| I17 | | Saint Remy | I45 | | Newark-Kerney, New Jersey |
| I18 | France | Salins de Giraud | I46 | USA | Panama Fort Sherman Costal Site |
| I19 | | Ostende, Belgium | I47 | | Research Triangle Park, N. Carolina |
| I20 | | Paris | I48 | | Point Reyes, California |
| I21 | | Auby | I49 | | Los Angeles, California |
| I22 | | Biarritz | I50 | | Mursmank |
| I23 | | Choshi | I51 | USSR | Batumi |
| I24 | Japan | Tokyo | I52 | | Vladivostok |
| I25 | | Okinawa | I53 | | Ojmjakon |
| I26 | New Zealand | Judgeford, Wellington | | | |
| I27 | Norway | Oslo | | | |
| I28 | | Borregaard | | | |

## 2.3. MICAT Programme

Fourteen countries participated in the exposure programme with a total of 75 exposure sites. These sites included industrial, urban, rural and marine atmospheres. A list of the sites together with their code is given in Table 3.

**Table 3.** Atmospheric corrosion test sites included in the MICAT Programme.

| Code | Country | Test Site | Code | Country | Test Site |
|------|---------|-----------|------|---------|-----------|
| M01 |  | Camet | M38 | Ecuador | Esmeraldas |
| M02 |  | Villa Martelli | M39 | | San Cristóbal |
| M03 | Argentina | Iguazú | M40 |  | León |
| M04 | | San Juan | M41 | | El Pardo |
| M05 | | Jubany | M42 | | Barcelona |
| M06 | | La Plata | M43 | Spain | Tortosa |
| M07 |  | Caratinga | M44 | | Granada |
| M08 | | Ipatinga | M45 | | Lagoas-Vigo |
| M09 | | Arraial do Cabo | M46 | | Labastida |
| M10 | | Cubatão | M47 | | Arties |
| M11 | Brazil | Ubatuba | M48 |  | Mexico |
| M12 | | São Paulo | M49 | México | Cuernavaca |
| M13 | | Río de Janeiro | M50 | | San Luis Potosí |
| M14 | | Belem | M51 | | Acapulco |
| M14 |  | Fortaleza | M52 |  | Panamá |
| M16 | | Brasilia | M53 | Panamá | Colon |
| M17 | | Paulo Afonso | M54 | | Veraguas |
| M18 | | Porto Velho | M55 | | Chiriquí |
| M19 |  | Isla Naval | M56 |  | Piura |
| M20 | Colombia | San Pedro | M57 | | Villa Salvador |
| M21 | | Cotové | M58 | Perú | San Borja |
| M22 |  | Puntarenas | M59 |  | Arequipa |
| M23 | Costa Rica | Limón | M60 | | Cuzco |
| M24 | | Arenal | M61 | | Pucallpa |
| M25 |  | Sabanilla | M62 |  | Leixões |
| M26 |  | Ciq | M63 | Portugal | Sines |
| M27 | Cuba | Cojímar | M64 | | Pego |
| M28 |  | Bauta | M65 |  | Trinidad |
| M29 |  | Cerrillos | M66 | Uruguay | Prado |
| M30 | | Valparaíso | M67 | | Melo |
| M31 | | Idiem | M68 | | Artigas |
| M32 | Chile | Petrox | M69 | | Punta del Este |
| M33 |  | Marsh | M70 |  | Tablazo |
| M34 | | Isla de Pascua | M71 | | Punto Fijo |
| M35 |  | Guayaquil | M72 | Venezuela | Coro |
| M36 | Ecuador | Riobamba | M73 | | Matanzas |
| M37 | | Salinas | M74 | | Barcelona, V |

Flat carbon steel panels were exposed in triplicate. A detailed description of the exposed material can be found in the book published with all the results of the project [7]. Figure 4 shows a diagram of the exposure schedule. The original exposure was planned to begin in 1989, but several delays occurred at various sites.

A detailed description of the material and methods for measuring environmental parameters and the evaluation of corrosion attack is provided in Reference [7].

### 2.4. Analysis of Data Properties

Before statistically analysing all the data collected from the various sources (data mining), data screening has been carried out. There follows a description of the criteria governing this screening:

1.  Extremely cold stations, with annual average temperatures below 0 °C, have been removed from the statistical analysis. Such is the case of the stations at Svanvik (Norway), Murmansk and Ojmjakon (USSR), Jubany (Argentina), Marsch (Chile) and Artigas (Uruguay), the latter three being Antarctic scientific bases. Low temperatures cause the metallic surface to be covered with

an ice layer for long time periods during the year, considerably impeding the development of corrosion processes. This ice layer reduces oxygen access to the metallic surface and its time of wetness, decreasing corrosion rates to extremely low values [28–31].

2.  In stations characterised as rural environments where $SO_2$ and $Cl^-$ deposition rates have not been determined due to being insignificant, values have been estimated for both pollutants. The figures indicated in Tables 4–6 correspond to the average value of the 0–3 mg $Cl^-/m^2$.d range (level $S_0$) and the 0–4 mg $SO_2/m^2$.d range (level $P_0$) according to standard ISO 9223 [3]. In those cases where both pollutants have been estimated, an average of the corrosion data from available annual series has been made.

3.  For test stations located in non-rural environments, all corresponding annual series data, or even the entirety of the available information, have been removed in those cases where, for some reason, meteorological or pollution data are not included.

4.  Chloride ion pollution data have not been determined for stations in the ICP/UNECE programme, which only considers non-marine test sites, unlike the other two exposure programmes (ISOCORRAG and MICAT). Therefore, the annual corrosion rate data and meteorological and $SO_2$ deposition rate obtained are only included in the statistical analysis for non-marine environments. In this respect, the criteria adopted has been to remove from the ICP/UNECE database all stations located at a distance of less than 2 km from the seashore, supposing in these cases a chloride ion deposition level of more than 3 mg/$m^2$.d (lower level $S_1$ according to standard ISO 9223 [3]). Bilbao station (Spain), despite being characterised by high $SO_2$ values, has been removed because of its location very close to the port.

On the other hand, a series of anomalous values have been observed at stations characterised as marine environments (ISOCORRAG and MICAT databases). Figure 5a shows the relationship between the variables of corrosion (μm/y) and salinity (mg $Cl^-/m^2$.d) in both databases. In this figure it is possible to see a cloud of points with very high salinity values (above 200 mg $Cl^-/m^2$.d) which does not seem to agree with the relatively low carbon steel corrosion values found (50–100 μm). It is also seen that a considerable rise in the marine chloride deposition rate (from 200 to 650 mg $Cl^-/m^2$.d) does not result in greater first-year corrosion of carbon steel, which is contradictory to the abundant literature on this matter recently reviewed by Alcántara et al. [32].These data have therefore been considered to be anomalous, and have been removed from the database. The corrosion stations removed for this reason are: Saint Remy (France), Tannanger (Norway) and Kvarnvik (Sweden). Figure 5b shows the linear relationship between these two variables after removing the aforementioned testing stations.

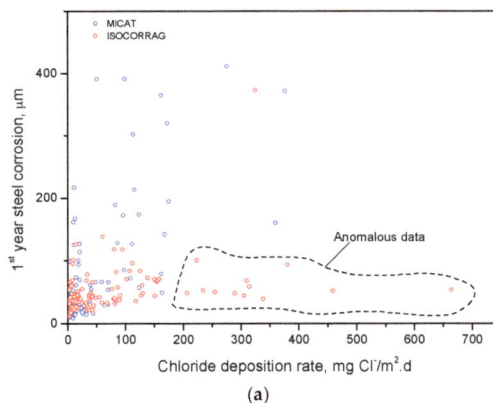

(a)

**Figure 5.** *Cont.*

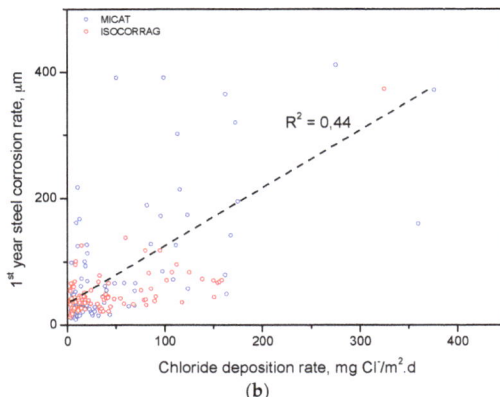

**Figure 5.** Relationship between annual steel corrosion and chloride deposition rate in marine test sites including in MICAT and ISOCORRAG databases (**a**); and the same relationship after anomalous data were eliminated (**b**).

## 2.5. Integration of ICP/UNECE, ISOCORRAG and MICAT Databases

Tables 4–6 present the databases finally considered for the ICP/UNECE, ISOCORRAG and MICAT programmes for statistical analysis after the screening mentioned in the preceding section.

ICP/UNECE: Table 4 presents the corrosion data obtained at the different testing stations along with the corresponding annual average values for the meteorological and pollution parameters measured in the programme: temperature ($^{\circ}$C), precipitation (mm/y), relative humidity (%) and $SO_2$ deposition rate (mg/m$^2$.d).

ISOCORRAG: Table 5 presents the corrosion data obtained at the different testing stations along with the corresponding annual average values for the meteorological and pollution parameters measured in the programme: temperature ($^{\circ}$C), relative humidity (%) (only for stations in the Czech Republic), time of wetness (annual fraction), $SO_2$ deposition rate (mg/m$^2$.d) and chloride ion deposition rate (mg/m$^2$.d).

MICAT: Table 6 presents the average corrosion value obtained at the different testing stations along with the annual average values for the meteorological and pollution parameters measured in the programme: temperature ($^{\circ}$C), relative humidity (%), time of wetness (annual fraction), precipitation (mm/y), $SO_2$ deposition rate (mg/m$^2$.d) and chloride ion deposition rate (mg/m$^2$.d).

There follows an indication of the similarities and differences between the experimental methods used in the three collaborative programmes and how this has affected the integration of the three databases for statistical analysis:

(a) Evaluation of the first-year corrosion (mass loss) of carbon steel according to ISO 9226 [6].

(b) Measurement of meteorological parameters (T, RH, and precipitation) according to standard conventional procedures. The ISOCORRAG programme does not consider precipitation or RH (except at the Czech Republic stations).

(c) Estimation of TOW according to ISO 9223 [2,3]. The ICP/UNECE programme does not consider this parameter.

(d) Measurement of $SO_2$ deposition rate according to ISO 9225 [5].

(e) Measurement of Cl deposition rate according to ISO 9225 [5]. The ICP/UNECE programme does not consider marine atmospheres.

Table 4. ICP/UNECE data considered in the study.

| Code | 1st Year Corrosion, μm | T, °C | RH, % | SO2 Deposition Rate mg/m².d | Precipitation, mm/y | Code | 1st Year Corrosion, μm | T, °C | RH, % | SO2 Deposition Rate, mg/m².d | Precipitation, mm/y |
|---|---|---|---|---|---|---|---|---|---|---|---|
| P01 | 55.6 | 9.5 | 79 | 62 | 639 | P23 | 12.09 | 6.8 | 77 | 0.16 | 1544 |
| P01 | 34.48 | 9.1 | 73 | 32.96 | 684 | P23 | 5.34 | 6.5 | 82 | 0.16 | 2195 |
| P01 | 30.66 | 9.8 | 77 | 25.68 | 581 | P24 | 33.97 | 7.6 | 78 | 13.44 | 531 |
| P01 | 29.52 | 8.6 | 78 | 18.88 | 475 | P24 | 15.27 | 7 | 70 | 4.56 | 577 |
| P01 | 23.16 | 9.9 | 76 | 12.24 | 522 | P24 | 13.1 | 7.5 | 73 | 3.36 | 581 |
| P01 | 17.56 | 9.5 | 79 | 7.04 | 601 | P24 | 13.61 | 7.4 | 68 | 2.64 | 556 |
| P01 | 13.1 | 9.3 | 72 | 5.12 | 513 | P24 | 15.9 | 6.7 | 76 | 2.08 | 463 |
| P01 | 12.98 | 9.3 | 74 | 8.88 | 491 | P24 | 14.76 | 8.1 | 81 | 1.52 | 635 |
| P01 | 7.51 | 10.1 | 74 | 5.2 | 525 | P24 | 10.31 | 7.1 | 80 | 1.28 | 384 |
| P01 | 8.52 | 10.2 | 70 | 5.12 | 534 | P24 | 11.7 | 8.9 | 74 | 1.44 | 273 |
| P01 | 8.4 | 11 | 73 | 3.68 | 414 | P24 | 7.76 | 7.8 | 76 | 0.64 | 270 |
| P02 | 28.5 | 7 | 77 | 15.76 | 850 | P24 | 5.47 | 7.8 | 77 | 0.72 | 428 |
| P02 | 19.47 | 6.6 | 73 | 14.32 | 921 | P24 | 7.38 | 8.3 | 81 | 0.4 | 330 |
| P02 | 18.83 | 7.2 | 74 | 9.76 | 941 | P25 | 33.46 | 7.6 | 78 | 15.68 | 531 |
| P03 | 70.87 | 9.6 | 73 | 66.64 | 426 | P25 | 13.1 | 7 | 70 | 3.76 | 577 |
| P03 | 44.53 | 8.9 | 71 | 39.2 | 432 | P25 | 12.09 | 7.5 | 73 | 2.72 | 581 |
| P03 | 44.78 | 9.7 | 75 | 39.36 | 513 | P26 | 18.7 | 6 | 83 | 2.64 | 543 |
| P03 | 37.28 | 8.5 | 73 | 24.48 | 431 | P26 | 9.54 | 6 | 81 | 1.04 | 468 |
| P03 | 30.41 | 9.9 | 76 | 14.64 | 420 | P26 | 10.31 | 6.8 | 82 | 0.88 | 525 |
| P03 | 28.5 | 9.2 | 80 | 14.32 | 510 | P26 | 8.78 | 6.5 | 83 | 0.64 | 409 |
| P03 | 23.41 | 8.7 | 73 | 8.96 | 463 | P26 | 7.89 | 5.9 | 86 | 0.48 | 479 |
| P03 | 23.28 | 8.3 | 76 | 14.48 | 442 | P26 | 8.78 | 7.2 | 86 | 0.48 | 772 |
| P03 | 20.74 | 9.3 | 80 | 10.8 | 521 | P26 | 5.09 | 5.6 | 82 | 0.48 | 562 |
| P03 | 28.24 | 9.6 | 79 | 15.2 | 417 | P26 | 5.22 | 6.3 | 84 | 0.48 | 435 |
| P03 | 26.59 | 10.8 | 71 | 9.12 | 433 | P26 | 3.56 | 7.1 | 82 | 0.32 | 452 |
| P04 | 34.48 | 5.9 | 76 | 14.88 | 626 | P26 | 2.54 | 6.7 | 86 | 0.32 | 511 |
| P04 | 16.67 | 5.6 | 79 | 1.84 | 755 | P26 | 14.89 | 7.2 | 82 | 0.24 | 784 |
| P04 | 15.39 | 6 | 80 | 2.08 | 698 | P27 | 40.08 | 9.2 | 84 | 14.16 | 365 |
| P05 | 16.79 | 3.1 | 78 | 5.04 | 801 | P27 | 39.31 | 9.6 | 82 | 14.24 | 530 |
| P05 | 6.11 | 3.4 | 81 | 0.72 | 610 | P27 | 30.15 | 10.5 | 78 | 5.44 | 515 |
| P05 | 7.51 | 3.9 | 83 | 0.64 | 675 | P27 | 34.35 | 10.2 | 81 | 7.5 | 708 |
| P05 | 6.87 | 3.2 | 76 | 0.48 | 618 | P27 | 24.81 | 9.7 | 81 | 6 | 831 |
| P05 | 6.74 | 3.5 | 80 | 0.72 | 742 | P27 | 22.9 | 10.4 | 78 | 3.92 | 548 |
| P05 | 6.49 | 4.8 | 82 | 0.64 | 845 | P28 | 32.19 | 10.8 | 86 | 5.76 | 447 |
| P05 | 4.83 | 4.5 | 80 | 0.64 | 713 | P28 | 25.95 | 10.5 | 82 | 2.56 | 614 |
| P06 | 34.35 | 6.3 | 78 | 16.56 | 673 | P28 | 25.19 | 11.2 | 79 | 2.64 | 696 |
| P06 | 20.74 | 6.2 | 78 | 3.84 | 702 | P30 | 39.06 | 10.2 | 78 | 12 | 610 |
| P06 | 24.94 | 6.6 | 76 | 4.4 | 649 | P30 | 29.26 | 10.3 | 76 | 7.44 | 549 |

Table 4. *Cont.*

| Code | 1st Year Corrosion, μm | T, °C | RH, % | SO₂ Deposition Rate mg/m².d | Precipitation, mm/y | Code | 1st Year Corrosion, μm | T, °C | RH, % | SO₂ Deposition Rate, mg/m².d | Precipitation, mm/y |
|---|---|---|---|---|---|---|---|---|---|---|---|
| P07 | 33.84 | 9.3 | 80 | 10.96 | 631 | P31 | 28.24 | 14.1 | 66 | 14.72 | 398 |
| P07 | 29.39 | 8.9 | 81 | 6.56 | 624 | P31 | 20.61 | 14.3 | 67 | 6.56 | 360 |
| P07 | 21.12 | 9.5 | 81 | 3.12 | 596 | P31 | 19.21 | 15.7 | 68 | 6.24 | 224 |
| P07 | 19.85 | 8.9 | 82 | 2.32 | 615 | P31 | 20.23 | 14.8 | 67 | 9.12 | 401 |
| P07 | 18.32 | 9.5 | 83 | 1.68 | 786 | P31 | 9.16 | 12.9 | 61 | 9.44 | 765 |
| P07 | 18.83 | 9.4 | 81 | 1.84 | 620 | P31 | 9.8 | 15 | 62 | 0.96 | 560 |
| P07 | 10.81 | 8.8 | 75 | 1.76 | 413 | P31 | 7.38 | 15.3 | 60 | 2.08 | 447 |
| P08 | 27.1 | 12.3 | 77 | 18.96 | 627 | P31 | 5.6 | 15.3 | 56 | 1.28 | 399 |
| P08 | 14.76 | 11.4 | 64 | 10.08 | 561 | P31 | 2.29 | 15.1 | 53 | 2.96 | 267 |
| P08 | 17.81 | 11.6 | 65 | 7.68 | 779 | P31 | 0.51 | 16.2 | 43 | 0.48 | 283 |
| P09 | 37.28 | 10.8 | 77 | 19.6 | 783 | P33 | 2.54 | 16 | 63 | 0.56 | 303 |
| P09 | 29.39 | 10.7 | 79 | 13.04 | 619 | P33 | 5.73 | 14 | 64 | 2.64 | 785 |
| P09 | 26.59 | 11.4 | 81 | 8.88 | 841 | P33 | 3.31 | 13.4 | 61 | 1.36 | 433 |
| P09 | 26.21 | 10 | 78 | 8.4 | 781 | P33 | 4.58 | 14.8 | 57 | 3.36 | 327 |
| P09 | 25.95 | 10.9 | 80 | 6.64 | 930 | P33 | 4.58 | 14 | 61 | 0.88 | 603 |
| P09 | 12.85 | 11.6 | 79 | 4 | 997 | P33 | 6.87 | 14 | 59 | 1.2 | 872 |
| P09 | 16.54 | 11.4 | 76 | 4.8 | 647 | P33 | 5.98 | 12.2 | 71 | 0.96 | 739 |
| P10 | 47.84 | 11.2 | 75 | 40.48 | 874 | P33 | 4.2 | 12.2 | 78 | 0.88 | 411 |
| P10 | 44.15 | 10.3 | 78 | 33.28 | 707 | P33 | 5.73 | 12.1 | 69 | 0.72 | 689 |
| P10 | 37.4 | 11.8 | 80 | 24.16 | 913 | P33 | 1.78 | 14.7 | 61 | 0.32 | 828 |
| P10 | 37.66 | 10.5 | 79 | 23.52 | 806 | P33 | 0.51 | 15.4 | 58 | 0.32 | 430 |
| P10 | 39.57 | 11.5 | 81 | 19.68 | 1044 | P33 | 2.04 | 12.8 | 60 | 0.4 | 516 |
| P10 | 37.28 | 11.7 | 81 | 14.32 | 791 | P34 | 23.03 | 5.5 | 73 | 15.36 | 575 |
| P10 | 28.24 | 11.3 | 77 | 13.52 | 780 | P34 | 17.94 | 5.7 | 74 | 22.96 | 881 |
| P10 | 27.48 | 10.8 | 81 | 8.88 | 663 | P34 | 15.39 | 5.6 | 71 | 13.12 | 667 |
| P10 | 28.63 | 11.1 | 75 | 7.44 | 849 | P34 | 17.18 | 6.5 | 74 | 13.7 | 838 |
| P10 | 30.92 | 11.4 | 78 | 7.04 | 880 | P34 | 17.3 | 7.4 | 69 | 13.7 | 812 |
| P11 | 43.51 | 10.5 | 79 | 24.24 | 713 | P34 | 11.7 | 5.9 | 71 | 3.28 | 750 |
| P11 | 37.28 | 10.1 | 79 | 18.32 | 684 | P35 | 23.54 | 5.5 | 83 | 0.72 | 448 |
| P11 | 30.66 | 10.9 | 78 | 12.96 | 889 | P35 | 13.49 | 5.4 | 82 | 1.1 | 859 |
| P12 | 17.56 | 8 | 82 | 7.52 | 1492 | P35 | 12.09 | 6.9 | 81 | 1.04 | 668 |
| P12 | 11.45 | 7.1 | 84 | 2.56 | 1552 | P35 | 12.21 | 5 | 81 | 1.36 | 655 |
| P12 | 10.81 | 7.4 | 83 | 1.92 | 1503 | P35 | 11.2 | 5.2 | 80 | 3.2 | 403 |
| P13 | 22.65 | 15.4 | 66 | 23.52 | 591 | P35 | 7.38 | 8.8 | 81 | 0.88 | 640 |
| P13 | 15.78 | 18.4 | 68 | 4.64 | 602 | P36 | 28.5 | 12.1 | 64 | 5.44 | 972 |
| P13 | 17.05 | 19.4 | 65 | 2.96 | 1125 | P36 | 39.19 | 18 | 62 | 12.88 | 545 |
| P13 | 8.02 | 17.8 | 53 | 0.88 | 625 | P36 | 25.95 | 19.1 | 67 | 3.76 | 443 |
| P13 | 8.02 | 18 | 66 | 0.64 | 1115 | P36 | 27.23 | 17.9 | 63 | 14.16 | 252 |

Table 4. *Cont.*

| Code | 1st Year Corrosion, μm | T, °C | RH, % | SO$_2$ Deposition Rate mg/m$^2$.d | Precipitation, mm/y | Code | 1st Year Corrosion, μm | T, °C | RH, % | SO$_2$ Deposition Rate, mg/m$^2$.d | Precipitation, mm/y |
|---|---|---|---|---|---|---|---|---|---|---|---|
| P14 | 29.9 | 14.6 | 71 | 6.64 | 650 | P37 | 18.96 | 5.5 | 75 | 2.64 | 961 |
| P14 | 18.83 | 14.9 | 76 | 4.16 | 717 | P37 | 13.99 | 4.3 | 80 | 1.68 | 1080 |
| P14 | 15.9 | 14.5 | 74 | 4.16 | 742 | P37 | 13.23 | 5.2 | 80 | 2.64 | 1023 |
| P14 | 8.65 | 16.3 | 63 | 0.56 | 600 | P37 | 14.76 | 7.4 | 75 | 1.92 | 788 |
| P14 | 10.81 | 14.5 | 67 | 0.16 | 742 | P37 | 11.96 | 7.2 | 76 | 0.48 | 964 |
| P14 | 6.49 | 15.9 | 69 | 2.96 | 857 | P38 | 27.23 | 14.6 | 69 | 7.68 | 847 |
| P14 | 10.31 | 15.7 | 71 | 0.88 | 585 | P38 | 23.54 | 15.5 | 64 | 8.08 | 982 |
| P14 | 14.38 | 15.5 | 73 | 0.96 | 1114 | P38 | 4.83 | 15.8 | 68 | 7.44 | 1038 |
| P15 | 46.56 | 15.3 | 72 | 57.76 | 1125 | P39 | 22.39 | 12.3 | 67 | 46.48 | 733 |
| P15 | 25.06 | 14.3 | 69 | 17.68 | 1092 | P39 | 36.9 | 11.8 | 65 | 34.48 | 729 |
| P15 | 22.01 | 14.5 | 69 | 12.32 | 1077 | P39 | 6.49 | 11.8 | 69 | 30.64 | 757 |
| P15 | 23.41 | 15.9 | 71 | 10.32 | 932 | P40 | 17.43 | 13.4 | 67 | 11.36 | 572 |
| P15 | 11.45 | 15.1 | 66 | 9.84 | 619 | P40 | 18.19 | 12.7 | 74 | 8.08 | 731 |
| P15 | 14.63 | 14 | 56 | 5.92 | 632 | P40 | 11.96 | 13.3 | 69 | 8.96 | 490 |
| P15 | 11.96 | 15 | 56 | 3.84 | 1179 | P40 | 11.58 | 12.6 | 73 | 5.28 | 571 |
| P15 | 5.85 | 13.9 | 63 | 1.76 | 583 | P40 | 7.51 | 12.7 | 70 | 2.48 | 427 |
| P15 | 8.02 | 15.8 | 63 | 3.52 | 1037 | P40 | 6.11 | 13.2 | 70 | 1.28 | 382 |
| P16 | 31.17 | 14.9 | 77 | 16.88 | 714 | P40 | 8.14 | 13.2 | 74 | 6.2 | 668 |
| P16 | 26.97 | 13.2 | 82 | 5.04 | 500 | P41 | 22.14 | 8.4 | 76 | 13.04 | 473 |
| P16 | 26.84 | 13.5 | 83 | 5.92 | 742 | P41 | 22.77 | 10.4 | 77 | 8.72 | 486 |
| P16 | 18.96 | 14.9 | 83 | 6.24 | 638 | P41 | 22.14 | 11.1 | 82 | 7.84 | 489 |
| P16 | 13.23 | 13.7 | 79 | 3.36 | 795 | P41 | 18.32 | 11.7 | 71 | 6.88 | 473 |
| P16 | 8.78 | 14.5 | 77 | 1.44 | 588 | P41 | 15.52 | 10.1 | 88 | 2.24 | 348 |
| P16 | 9.8 | 15 | 77 | 0.96 | 881 | P41 | 11.58 | 10 | 72 | 2.24 | 570 |
| P17 | 43.77 | 10.5 | 84 | 28.24 | 978 | P41 | 7.38 | 10.3 | 77 | 1.84 | 473 |
| P17 | 38.55 | 10.3 | 83 | 20.4 | 860 | P43 | 41.22 | 24.6 | 83 | 28 | 485 |
| P17 | 32.57 | 11 | 84 | 16.4 | 996 | P43 | 32.57 | 22 | 70 | 5.28 | 254 |
| P18 | 32.32 | 9.9 | 83 | 8.08 | 904 | P45 | 11.83 | 6.2 | 77 | 1.2 | 1135 |
| P18 | 25.95 | 9.5 | 82 | 5.92 | 873 | P45 | 8.52 | 6.9 | 77 | 1.04 | 1053 |
| P18 | 18.32 | 10.3 | 83 | 3.76 | 987 | P45 | 7.38 | 7.2 | 80 | 0.8 | 1281 |
| P19 | 36.01 | 10.3 | 81 | 10.4 | 845 | P45 | 4.58 | 7.3 | 75 | 1.04 | 1011 |
| P19 | 30.41 | 10 | 82 | 6.64 | 749 | P45 | 5.22 | 6.2 | 80 | 0.88 | 1404 |
| P19 | 22.9 | 10.9 | 83 | 3.6 | 829 | P45 | 3.31 | 6.3 | 80 | 0.56 | 950 |
| P20 | 33.08 | 10.3 | 81 | 10.96 | 801 | P45 | 2.8 | 7 | 79 | 0.32 | 1108 |
| P20 | 26.08 | 10.1 | 81 | 7.44 | 680 | P46 | 22.52 | 12.2 | 70 | 4.64 | 706 |
| P20 | 21.88 | 11.1 | 82 | 4.64 | 790 | P46 | 21.63 | 12.1 | 69 | 4.64 | 907 |
| P21 | 29.13 | 7.6 | 70 | 11.52 | 1024 | P46 | 19.08 | 12.7 | 66 | 4.64 | 494 |
| P21 | 17.18 | 7.7 | 68 | 4.8 | 440 | P47 | 17.3 | 17.4 | 61 | 0.48 | 33 |

**Table 4.** *Cont.*

| Code | 1st Year Corrosion, μm | T, °C | RH, % | SO₂ Deposition Rate mg/m².d | Precipitation, mm/y | Code | 1st Year Corrosion, μm | T, °C | RH, % | SO₂ Deposition Rate, mg/m².d | Precipitation, mm/y |
|---|---|---|---|---|---|---|---|---|---|---|---|
| P21 | 12.85 | 7.5 | 69 | 2.32 | 680 | P49 | 21.76 | 11.4 | 76 | 18.24 | 834 |
| P21 | 12.6 | 6.8 | 76 | 3.2 | 764 | P49 | 23.54 | 11.7 | 75 | 10.8 | 993 |
| P21 | 11.83 | 6.6 | 79 | 3.28 | 523 | P49 | 13.99 | 11.9 | 65 | 11.04 | 674 |
| P21 | 12.34 | 7.2 | 75 | 2.48 | 1050 | P50 | 34.48 | 9.4 | 81 | 27.52 | 870 |
| P21 | 7.12 | 6.4 | 74 | 1.36 | 794 | P50 | 30.79 | 8.2 | 76 | 30.88 | 702 |
| P21 | 9.54 | 7.2 | 74 | 1.04 | 869 | P50 | 28.75 | 7.5 | 76 | 28.88 | 674 |
| P21 | 7.51 | 6.9 | 76 | 1.6 | 737 | P50 | 28.37 | 7.7 | 84 | 12.24 | 651 |
| P21 | 5.22 | 7.4 | 74 | 0.48 | 715 | P50 | 25.32 | 8.8 | 74 | 12.96 | 676 |
| P21 | 7.89 | 7.5 | 76 | 3.36 | 805 | P50 | 4.58 | 10.7 | 71 | 10.72 | 484 |
| P22 | 54.71 | 6 | 78 | 28.64 | 1116 | P51 | 10.05 | 18.7 | 62 | 11.36 | 461 |
| P22 | 44.02 | 7 | 76 | 21.12 | 628 | P51 | 6.87 | 18.5 | 56 | 3.36 | 325 |
| P22 | 42.62 | 7.4 | 76 | 25.04 | 819 | P51 | 19.85 | 18.7 | 62 | 6.32 | 570 |
| P23 | 24.68 | 6.5 | 80 | 1.04 | 2144 | P52 | 10.56 | 8.2 | 77 | 2.8 | 633 |
| P23 | 16.79 | 5.9 | 75 | 0.56 | 1189 | P52 | 8.02 | 7.8 | 75 | 0.8 | 589 |
| P23 | 13.87 | 6.4 | 76 | 0.56 | 1420 | P53 | 10.43 | 11.2 | 73 | 2 | 855 |
| P23 | 14.38 | 5.6 | 75 | 0.32 | 1182 | P53 | 5.73 | 11.3 | 73 | 2.64 | 555 |
| P23 | 12.85 | 6.2 | 79 | 0.16 | 1744 | P53 | 10.56 | 12 | 71 | 3.36 | 527 |
| P23 | 14.5 | 6.6 | 83 | 0.24 | 2333 | P54 | 8.91 | 11.5 | 70 | 10.8 | 651 |
| P23 | 8.27 | 5.9 | 81 | 0.24 | 1390 | P55 | 12.6 | 6.1 | 76 | 2.48 | 636 |
| P23 | 13.61 | 6.2 | 79 | 0.4 | 1623 | P59 | 13.74 | 9.7 | 74 | 5.2 | 664 |
| P23 | 9.8 | 4.2 | 81 | 0.08 | 1392 | | | | | | |

**Table 5.** ISOCORRAG data considered in the study.

| Code | 1st Year Corrosion, μm | T, °C | RH, % | TOW, Annual Fraction | Deposition Rates, mg/m².d | | Code | 1st Year Corrosion, μm | T, °C | RH, % | TOW, Annual Fraction | Deposition Rates, mg/m².d | |
|---|---|---|---|---|---|---|---|---|---|---|---|---|---|
| | | | | | SO₂ | Cl | | | | | | SO₂ | Cl |
| I01 | 5.8 | 22.9 | 69 | 0.615 | 2 | 1.5 | I25 | 54.7 | 24 | | 0.478 | 8.48 | 75.85 |
| I02 | 24.9 | 14.1 | 76 | 0.682 | 2 | 18.21 | I25 | 57.2 | 23.4 | | 0.439 | 8.84 | 86.17 |
| I02 | 54.8 | 13.9 | 79 | 0.708 | 2 | 24.39 | I25 | 44.8 | 23.2 | | 0.338 | 9.76 | 91.03 |
| I02 | 78.2 | 14.3 | 75 | 0.725 | 2 | 33.38 | I25 | 39.2 | 23.5 | | 0.354 | 10.4 | 78.89 |
| I02 | 66 | 14.5 | 74 | 0.736 | 2 | 42.48 | I27 | 26.1 | 6.7 | | 0.299 | 13.84 | 1.21 |
| I02 | 68.3 | 14.2 | 81 | 0.711 | 2 | 32.16 | I27 | 26.6 | 6.2 | | 0.326 | 11.84 | 2.18 |

Table 5. *Cont.*

| Code | 1st Year Corrosion, μm | T, °C | RH, % | TOW, Annual Fraction | Deposition Rates, mg/m². d SO₂ | Cl | Code | 1st Year Corrosion, μm | T, °C | RH, % | TOW, Annual Fraction | Deposition Rates, mg/m². d SO₂ | Cl |
|---|---|---|---|---|---|---|---|---|---|---|---|---|---|
| I03 | 14.7 | 17.1 | | 0.529 | 9.7 | 1.5 | I27 | 30.2 | 7.4 | | 0.279 | 11.28 | 1.58 |
| I04 | 4.6 | 19.2 | | 0.104 | 2 | 1.5 | I27 | 21.5 | 8.5 | | 0.261 | 12.64 | 0.73 |
| I06 | 25.5 | 8 | | 0.287 | 11.28 | 33.38 | I27 | 26.5 | 7.9 | | 0.297 | 9.92 | 0.91 |
| I06 | 21.5 | 7.6 | | 0.267 | 12.8 | 42.48 | I27 | 20.1 | 8.5 | | 0.346 | 6.8 | 1.03 |
| I06 | 28.3 | 8 | | 0.238 | 12.4 | 33.38 | I28 | 68.4 | 4.9 | | 0.358 | 34.4 | 8.01 |
| I06 | 21.3 | 7.5 | | 0.283 | 12.24 | 37.02 | I28 | 60.8 | 5.4 | | 0.365 | 28.8 | 5.22 |
| I06 | 25.5 | 7 | | 0.287 | 12.72 | 35.2 | I28 | 66 | 5.7 | | 0.313 | 28.8 | 4.01 |
| I06 | 21.6 | 7 | | 0.317 | 14.8 | 33.98 | I28 | 60 | 7.5 | | 0.407 | 41.6 | 6.86 |
| I07 | 27.1 | 5.5 | 76 | 0.347 | 20.24 | 2.31 | I28 | 61.4 | 6.6 | | 0.423 | 42.4 | 6.07 |
| I07 | 23.1 | 7.1 | 77 | 0.414 | 13.68 | 1.64 | I28 | 53.6 | 6.7 | | 0.42 | 36.16 | 3.22 |
| I07 | 26 | 6.8 | 77 | 0.409 | 12.88 | 2 | I29 | 21.4 | 5.2 | | 0.42 | 1.44 | 0.61 |
| I07 | 23.3 | 7.1 | 76 | 0.353 | 10.48 | 2 | I29 | 18.6 | 5.9 | | 0.526 | 0.96 | 0.61 |
| I07 | 30.7 | 7 | 77 | 0.454 | 10.72 | 2 | I29 | 21.8 | 6.3 | | 0.478 | 0.8 | 0.61 |
| I07 | 25.7 | 7.3 | 77 | 0.503 | 13.92 | 2 | I29 | 17.1 | 7.5 | | 0.503 | 0.8 | 0.61 |
| I08 | 62.4 | 7.5 | 81 | 0.272 | 71.6 | 2.91 | I29 | 20.7 | 6.6 | | 0.453 | 0.8 | 0.61 |
| I08 | 44.3 | 8.8 | 79 | 0.285 | 52.32 | 1.21 | I29 | 18.6 | 6.2 | | 0.42 | 0.8 | 0.61 |
| I08 | 43.3 | 9.3 | 75 | 0.238 | 56.4 | 2.1 | I31 | 27.2 | 7.2 | | 0.372 | 7.84 | 2.61 |
| I08 | 42.1 | 9.7 | 74 | 0.226 | 52.64 | 2.1 | I31 | 22.3 | 7.7 | | 0.443 | 7.92 | 2.06 |
| I08 | 53.3 | 9.9 | 77 | 0.264 | 47.76 | 2.1 | I31 | 27.7 | 8.2 | | 0.495 | 7.92 | 2.12 |
| I08 | 38.9 | 10.1 | 77 | 0.299 | 43.04 | 2.1 | I31 | 25.7 | 8.9 | | 0.557 | 5.68 | 6.98 |
| I09 | 87.9 | 7.7 | 76 | 0.355 | 84 | 2.31 | I31 | 38 | 8.4 | | 0.584 | 5.44 | 6.92 |
| I09 | 66.1 | 9 | 73 | 0.279 | 66.64 | 1.09 | I31 | 26.3 | 8.4 | | 0.59 | 6.4 | 4.85 |
| I09 | 57.7 | 9.5 | 73 | 0.256 | 65.84 | 1.7 | I33 | 31.9 | 14.1 | | 0.15 | 22 | 1.5 |
| I09 | 59.1 | 9.8 | 72 | 0.266 | 71.6 | 1.7 | I33 | 29.8 | 14.3 | | 0.201 | 24.24 | 1.5 |
| I09 | 84.1 | 9.6 | 74 | 0.271 | 76.88 | 1.7 | I33 | 33.2 | 12.5 | | 0.301 | 36.88 | 1.5 |
| I09 | 69.2 | 9.9 | 74 | 0.235 | 66.48 | 1.7 | I33 | 22.4 | 14.1 | | 0.277 | 41.2 | 1.5 |
| I10 | 38.5 | 10.4 | | 0.545 | 18.72 | 1.52 | I33 | 26.1 | 14.9 | | 0.227 | 43.2 | 1.5 |
| I10 | 40.6 | 10.8 | | 0.535 | 17.84 | 1.09 | I33 | 22.7 | 14.9 | | 0.254 | 44.48 | 1.5 |
| I10 | 35.3 | 11.1 | | 0.506 | 12.08 | 1.09 | I34 | 16.3 | 25.3 | | 0.277 | 3.12 | 1.5 |
| I10 | 37.4 | 10.8 | | 0.486 | 10.56 | 1.4 | I34 | 17 | 25.3 | | 0.31 | 3.84 | 1.5 |
| I10 | 31.8 | 9.6 | | 0.428 | 14.64 | 1.03 | I34 | 17.4 | 25.3 | | 0.418 | 4.64 | 1.5 |
| I10 | 33.8 | 9.7 | | 0.424 | 12.48 | 0.61 | I34 | 12.9 | 25.3 | | 0.359 | 4.32 | 1.5 |
| I11 | 37.5 | 3.3 | | 0.339 | 17.12 | 2.18 | I34 | 15.6 | 25.2 | | 0.402 | 3.28 | 1.5 |
| I11 | 33 | 5.1 | | 0.395 | 17.12 | 2.49 | I34 | 13.7 | 25.5 | | 0.442 | 4.32 | 1.5 |
| I11 | 41.2 | 6.4 | | 0.394 | 16 | 2.55 | I35 | 34.4 | 15.2 | | 0.365 | 49.28 | 18.21 |
| I11 | 28.3 | 6.8 | | 0.42 | 14.72 | 2.43 | I35 | 24.7 | 16.2 | | 0.374 | 38.88 | 11.53 |
| I11 | 31.4 | 6.7 | | 0.439 | 13.28 | 2.41 | I35 | 25.2 | 16.2 | | 0.31 | 37.2 | 12.14 |
| I11 | 28.6 | 6.8 | | 0.464 | 12.24 | 2.41 | I35 | 27.6 | 15.8 | | 0.293 | 35.84 | 11.53 |

**Table 5.** *Cont.*

| Code | 1st Year Corrosion, μm | T, °C | RH, % | TOW, Annual Fraction | Deposition Rates, mg/m².d SO₂ | Cl | Code | 1st Year Corrosion, μm | T, °C | RH, % | TOW, Annual Fraction | Deposition Rates, mg/m².d SO₂ | Cl |
|---|---|---|---|---|---|---|---|---|---|---|---|---|---|
| I12 | 30.9 | 3 | | 0.297 | 16.24 | 2.55 | I35 | 22.7 | 16.6 | | 0.31 | 37.04 | 12.14 |
| I12 | 21.4 | 4.9 | | 0.325 | 13.04 | 1.52 | I35 | 26.8 | 17.2 | | 0.293 | 35.68 | 11.53 |
| I12 | 34.6 | 5.4 | | 0.388 | 15.2 | 1.09 | I36 | 45.9 | 14.5 | | 0.492 | 29.44 | 12.74 |
| I12 | 19.9 | 5.3 | | 0.348 | 11.2 | 1.72 | I36 | 51.1 | 15.8 | | 0.493 | 34.24 | 17.6 |
| I12 | 26.2 | 5.9 | | 0.434 | 8.48 | 1.72 | I36 | 45 | 16.7 | | 0.517 | 31.04 | 16.99 |
| I12 | 20.8 | 6.4 | | 0.491 | 9.04 | 1.72 | I36 | 44.3 | 16.2 | | 0.511 | 23.52 | 14.56 |
| I13 | 16.7 | 0.3 | | 0.378 | 4.72 | 1.94 | I36 | 33.3 | 16.1 | | 0.464 | 16.8 | 24.27 |
| I13 | 11 | 2.2 | | 0.345 | 4.24 | 1.5 | I37 | 28 | 5 | | 0.29 | 8.16 | 1.5 |
| I13 | 15.7 | 3.4 | | 0.313 | 4.08 | 1.5 | I37 | 26.9 | 6.8 | | 0.416 | 8.8 | 1.5 |
| I13 | 9.7 | 4 | | 0.347 | 2.8 | 1.5 | I37 | 28.1 | 7.1 | | 0.359 | 9.6 | 1.5 |
| I13 | 12.5 | 4 | | 0.357 | 2.48 | 1.5 | I37 | 21.6 | 8 | | 0.347 | 8 | 1.5 |
| I13 | 11.3 | 4.1 | | 0.386 | 1.52 | 1.5 | I37 | 23.5 | 8.4 | | 0.338 | 8.8 | 1.5 |
| I14 | 40.7 | 12.3 | | 0.473 | 42.24 | 15.17 | I37 | 18.1 | 8.4 | | 0.385 | 5.6 | 1.5 |
| I14 | 34.5 | 13.1 | | 0.546 | 37.04 | 15.17 | I38 | 43 | 6.1 | | 0.447 | 7.04 | 41.87 |
| I14 | 44.2 | 13.5 | | 0.52 | 31.04 | 18.21 | I38 | 28.8 | 8 | | 0.472 | 4 | 44.3 |
| I14 | 35 | 13 | | 0.511 | 32 | 18.81 | I38 | 33.1 | 8.7 | | 0.462 | 6.4 | 30.95 |
| I15 | 83.5 | 14.6 | | 0.423 | 120.8 | 125.01 | I38 | 33.3 | 9.5 | | 0.454 | 1.6 | 58.26 |
| I15 | 68.1 | 16.2 | | 0.488 | 77.04 | 155.96 | I38 | 41.8 | 9.5 | | 0.449 | 2.4 | 54.62 |
| I15 | 70.7 | 16.1 | | 0.427 | 61.04 | 158.38 | I38 | 31.2 | 9.7 | | 0.474 | 4 | 81.32 |
| I15 | 66.4 | 15.6 | | 0.349 | 64 | 154.14 | I40 | 42.3 | 11.4 | | 0.705 | 20.56 | 11.41 |
| I15 | 72.6 | 15.6 | | 0.503 | 35.84 | 138.36 | I40 | 35.1 | 11.4 | | 0.66 | 16.56 | 12.86 |
| I16 | 19.6 | 6.5 | | 0.493 | 14.4 | 4.85 | I40 | 36 | 11.4 | | 0.631 | 14.32 | 5.58 |
| I16 | 15.5 | 6.5 | | 0.474 | 9.04 | 3.64 | I40 | 37.6 | 11.4 | | 0.547 | 13.84 | 5.16 |
| I16 | 19.6 | 7.1 | | 0.542 | 8 | 4.25 | I40 | 42.9 | 11.4 | | 0.467 | 15.2 | 7.83 |
| I16 | 12.3 | 6.7 | | 0.47 | 6.48 | 3.03 | I40 | 38 | 11.4 | | 0.512 | 14.8 | 12.02 |
| I18 | 82.1 | 13.6 | | 0.352 | 32.5 | 83.74 | I41 | 36.4 | 10.5 | | 0.687 | 12.88 | 8.56 |
| I18 | 70.2 | 14.2 | | 0.37 | 32 | 149.89 | I43 | 39.6 | 9 | | 0.707 | 15.36 | 3.22 |
| I18 | 70.5 | 15.4 | | 0.45 | 31.44 | 101.95 | I43 | 35.4 | 9 | | 0.68 | 13.6 | 2.31 |
| I19 | 118 | 9.7 | | 0.691 | 8 | 95.27 | I43 | 38.1 | 9 | | 0.449 | 12.88 | 4.43 |
| I19 | 95.8 | 9.7 | | 0.664 | 24 | 112.26 | I43 | 41.7 | 9 | | 0.459 | 12.88 | 4.43 |
| I19 | 83.5 | 9.7 | | 0.728 | 25.6 | 107.41 | I43 | 41.8 | 9 | | 0.545 | 13.12 | 2.06 |
| I20 | 37.6 | 13 | | 0.494 | 42.88 | 1.5 | I44 | 37.7 | 9 | | 0.491 | 11.36 | 3.09 |
| I20 | 39.7 | 13 | | 0.38 | 42.88 | 1.5 | I44 | 40.2 | 13.3 | | 0.503 | 4.32 | 80.71 |
| I20 | 48 | 13 | | 0.218 | 42.4 | 1.5 | I44 | 32.5 | 18.1 | | 0.479 | 4.96 | 67.97 |

Table 5. *Cont.*

| Code | 1st Year Corrosion, μm | T, °C | RH, % | TOW, Annual Fraction | Deposition Rates, mg/m².d SO$_2$ | Cl | Code | 1st Year Corrosion, μm | T, °C | RH, % | TOW, Annual Fraction | Deposition Rates, mg/m².d SO$_2$ | Cl |
|---|---|---|---|---|---|---|---|---|---|---|---|---|---|
| I21 | 101 | 9.6 | | 0.471 | 171.68 | 8.92 | I44 | 37.6 | 17.8 | | 0.464 | 5.28 | 89.81 |
| I21 | 95.1 | 11.9 | | 0.527 | 147.68 | 8.5 | I44 | 35.6 | 17.4 | | 0.473 | 4.88 | 117.73 |
| I21 | 126 | 12.8 | | 0.567 | 133.6 | 14.93 | I44 | 43.8 | 18.2 | | 0.492 | 8.4 | 150.5 |
| I23 | 44 | 16 | | 0.654 | 5.84 | 36.41 | I45 | 26.4 | 11.8 | | 0.216 | 26.3 | 1.5 |
| I23 | 40.9 | 15.9 | | 0.639 | 6.48 | 47.94 | I46 | 373 | 27.3 | | 0.824 | 42.4 | 324.66 |
| I23 | 45.2 | 15.5 | | 0.644 | 6.64 | 41.87 | I51 | 32.2 | 13.2 | | 0.364 | 20 | 0.61 |
| I23 | 39.7 | 15.8 | | 0.643 | 6.48 | 37.62 | I51 | 33.6 | 13.4 | | 0.341 | 22.56 | 0.61 |
| I23 | 48.2 | 16.1 | | 0.644 | 6 | 39.44 | I51 | 29.4 | 13.1 | | 0.395 | 20.48 | 0.61 |
| I23 | 42.1 | 16.2 | | 0.683 | 5.68 | 40.05 | I51 | 30.2 | 13.3 | | 0.386 | 21.2 | 0.61 |
| I24 | 38 | 14.1 | | 0.18 | 11.28 | 2.61 | I51 | 22.5 | 13.7 | | 0.383 | 21.12 | 0.67 |
| I24 | 28.6 | 14.1 | | 0.221 | 11.6 | 2.49 | I51 | 24.2 | 13.3 | | 0.334 | 20.8 | 0.61 |
| I24 | 48.8 | 13.9 | | 0.275 | 11.28 | 2.85 | I52 | 39 | 3.9 | | 0.465 | 10.4 | 21.85 |
| I24 | 32.1 | 14 | | 0.262 | 11.36 | 3.34 | I52 | 26.4 | 4.2 | | 0.434 | 12.64 | 14.56 |
| I24 | 55.8 | 14.2 | | 0.258 | 12.24 | 3.16 | I52 | 22.4 | 5.8 | | 0.405 | 27.44 | 11.23 |
| I24 | 33.8 | 14.6 | | 0.291 | 12.4 | 3.09 | I52 | 23.9 | 5.9 | | 0.396 | 32.88 | 6.68 |
| I25 | 118 | 22.8 | | 0.538 | 8.88 | 80.1 | I52 | 17.4 | 6.8 | | 0.483 | 20.32 | 5.28 |
| I25 | 138 | 23.9 | | 0.525 | 6.64 | 60.08 | I52 | 26.3 | 6.2 | | 0.503 | 22.32 | 7.34 |

**Table 6.** MICAT data considered in the study.

| Code | 1st Year Corrosion, µm | T, °C | RH, % | TOW, Annual Fraction | Precipitation, mm/y | Deposition Rates, mg/m$^2$.d | |
|------|------|------|------|------|------|------|------|
| | | | | | | SO$_2$ | Cl |
| M01 | 54.8 | 13.9 | 79 | 0.708 | 805 | 2 | 40.2 |
| M01 | 66 | 14.5 | 80 | 0.736 | 1226 | 2 | 70 |
| M02 | 14.73 | 16.9 | 74 | 0.538 | 1377 | 9 | 1.5 |
| M03 | 5.7 | 21.2 | 75 | 0.643 | 2167 | 2 | 1.50 |
| M04 | 4.9 | 18.8 | 50 | 0.103 | 80 | 2 | 1.50 |
| M06 | 25.3 | 17 | 78 | 0.593 | 1178 | 6.2 | 1.5 |
| M06 | 28.8 | 16.7 | 77 | 0.565 | 1263 | 8.2 | 1.5 |
| M06 | 30.1 | 16.6 | 78 | 0.631 | 1361 | 6.2 | 1.5 |
| M07 | 8.6 | 21.5 | 74 | 0.482 | 847 | 0.8 | 8.9 |
| M07 | 11.5 | 20.9 | 75 | 0.482 | 1167 | 1.3 | 7.4 |
| M07 | 13.1 | 21.2 | 75 | 0.482 | 996 | 1.7 | 1.60 |
| M08 | 52.5 | 23.8 | 89 | 0.482 | 1122 | 23.8 | 8.6 |
| M08 | 47.3 | 22.9 | 91 | 0.482 | 1471 | 20.7 | 6.8 |
| M08 | 48.5 | 23 | 90 | 0.482 | 1444 | 24.5 | 5.2 |
| M09 | 159.8 | 24.8 | 77 | 0.582 | 605 | 9.5 | 359.8 |
| M09 | 194.7 | 24.5 | 79 | 0.582 | 985 | 5.3 | 174.8 |
| M09 | 141.7 | 24.2 | 77 | 0.582 | 716 | 4.4 | 167.70 |
| M10 | 98.7 | 22.7 | 73 | 0.579 | 960 | 40.4 | 4.50 |
| M10 | 161.2 | 22.9 | 71 | 0.579 | 870 | 57.4 | 9.20 |
| M10 | 216.9 | 22.6 | 79 | 0.579 | 1133 | 65.8 | 10.80 |
| M11 | 301.9 | 22.1 | 80 | 0.579 | 1689 | 2.6 | 113.20 |
| M13 | 127.1 | 20.1 | 80 | 0.598 | 1353 | 55.85 | 20.21 |
| M13 | 61.2 | 23.1 | 78 | 0.598 | 1369 | 44.09 | 14.22 |
| M13 | 73.1 | 21 | 82 | 0.598 | 1305 | 30.51 | 14.67 |
| M14 | 19.4 | 26.1 | 88 | 0.682 | 2395 | 2 | 1.50 |
| M16 | 12.9 | 20.4 | 69 | 0.442 | 1440 | 2 | 1.50 |
| M17 | 17.3 | 25.9 | 77 | 0.172 | 1392 | 2 | 1.50 |
| M18 | 4.9 | 26.6 | 90 | - | 2096 | 2 | 1.50 |
| M19 | 16 | 27.6 | 85 | 0.989 | 940 | 7.8 | 43.60 |
| M19 | 30.6 | 27.6 | 87 | 0.966 | 940 | 14.2 | 69.00 |
| M19 | 54 | 28.2 | 87 | 0.975 | 940 | 8.9 | 69.50 |
| M20 | 17 | 11.5 | 90 | 1 | 1800 | 0.6 | 1.50 |
| M21 | 19.6 | 27 | 76 | 0.33 | 900 | 0.3 | 1.50 |
| M22 | 61.6 | 27.6 | 80 | 0.562 | 1598 | 6.3 | 38.7 |
| M23 | 371.5 | 25.3 | 88 | 0.763 | 3531 | 3.5 | 376 |
| M24 | 69.3 | 22.9 | 88 | 0.838 | 3677 | 4 | 20.60 |
| M25 | 16.6 | 18.9 | 83 | 0.695 | 1780 | 2.4 | 12.10 |
| M26 | 36.1 | 25.2 | 80 | 0.571 | 1591 | 37.1 | 15.8 |
| M26 | 26.4 | 25.4 | 79 | 0.571 | 1303 | 36.5 | 10.9 |
| M26 | 29 | 24.7 | 79 | 0.571 | 1321 | 19.8 | 10.9 |
| M26 | 32.3 | 25.5 | 79 | 0.571 | 1129 | 25.6 | 14.3 |
| M26 | 31.3 | 25.4 | 79 | 0.571 | 1305 | 41 | 10.10 |
| M26 | 29.2 | 24.7 | 79 | 0.571 | 1540 | 24.7 | 8.20 |
| M26 | 27.3 | 25.2 | 79 | 0.571 | 1415 | 18.5 | 18.10 |
| M26 | 32.8 | 25.1 | 79 | 0.571 | 1064 | 49.2 | 7.40 |
| M27 | 391.1 | 25.2 | 80 | 0.571 | 1591 | 24.5 | 99.10 |
| M27 | 213.7 | 25.4 | 79 | 0.571 | 1303 | 13.5 | 115.60 |
| M27 | 173.6 | 24.7 | 79 | 0.571 | 1321 | 25.1 | 123.30 |
| M27 | 171.9 | 25.5 | 79 | 0.571 | 1129 | 20.4 | 96.00 |
| M27 | 391.2 | 25 | 79 | 0.571 | 1108 | 32.8 | 50.40 |
| M27 | 126.5 | 25.4 | 79 | 0.571 | 1305 | 18.9 | 111.40 |
| M27 | 71.6 | 25.7 | 79 | 0.571 | 1540 | 19.9 | 108.80 |
| M27 | 84.8 | 24.2 | 79 | 0.571 | 1415 | 17.7 | 97.70 |
| M27 | 188.9 | 25.1 | 80 | 0.571 | 1064 | 40.4 | 81.80 |

## 3. Discussion

It is a well known fact that the atmospheric corrosion of metals is influenced by many factors: (a) external conditions, meteorology and air pollution; (b) exposure conditions; (c) construction conditions; (d) internal conditions, such as nature of the metal and characteristics of corrosion products; among others.

Over the years, many models have been developed to assess the corrosion of carbon steel in the atmosphere. The specialised literature offers a large range of damage functions that relate atmospheric corrosion of carbon steel with environmental data. However, most of them are of limited applicability as they were obtained with minimal variations in meteorological parameters (small geographic areas). Special mention should be made of the efforts of Benarie and Lipfert [13] to develop universal corrosion functions in terms of atmospheric pollutants, meteorological parameters and the rain pH, as well as the work of Feliu et al. [15] compiling a comprehensive literature survey of worldwide atmospheric corrosion and environmental data that were statistically processed to establish general corrosion damage functions in terms of simple meteorological and pollution parameters. Reviews on this subject can be found in [7,26].

In recent decades, several international exposure programmes (ISOCORRAG [1], ICP/UNECE [8], and MICAT [7]) have been carried out with the aim of more systematically obtaining relationships (dose/response functions) between atmospheric corrosion rates and pollution levels in combination with climate parameters. Integration of the databases obtained in these three exposure programmes may make it possible to obtain universal damage functions based on a worldwide variety of meteorological and pollution conditions. This has been the chief aim of the work reported here.

The data have been fitted to the following linear equation:

$$C = a_1 + a_2\,RH + a_3\,P + a_4\,T + a_5\,TOW + a_6\,SO_2 + a_7\,Cl, \qquad (1)$$

This equation is quite simple. Other combinations between the different variables or other more sophisticated statistical treatments would likely yield better fits, but the aim of this work has been to use as simple as possible a relation.

According to this model, the dependent variable C (carbon steel annual corrosion in µm) is interpreted as a linear combination of a set of independent variables: RH, annual average relative humidity, in per cent; T, annual average temperature, in °C; P, annual precipitation, in mm; TOW, time of wetness, annual fraction of number of hours/year in which RH > 80% and T > 0 °C [3]; $SO_2$, $SO_2$ pollution, in mg/m$^2$.day; and Cl, chloride pollution, in mg/m$^2$.day. Each independent variable is accompanied by a coefficient ($a_2$–$a_7$) which indicates the relative weight of that variable in the equation. The equation also includes a constant $a_1$.

The minimum-quadratic regression equation is constructed by estimating the values of coefficients $a_1$–$a_7$ from the regression model. These estimates are obtained trying to keep the squared differences between the values observed and the forecast values to a minimum. In order to know the model's fitting quality in relation to the experimental data, the statistic $R^2$ is used, i.e., the square of the multiple correlation coefficient. $R^2$ expresses the proportion of variance of the dependent variable which is explained by the independent variables.

There are different methods to select the independent variables that a regression model must include. The most widely accepted is the stepwise regression model. With this method, the best variable is firstly selected (always with a statistical criterion); then the best of the rest is taken; and so on, until no variables that fulfil the selection criteria remain. A great change in $R^2$ when a new variable is inserted in the equation indicates that this variable provides unique information on the dependent variable that is not supplied by the other independent variables.

The study has been carried out with the assistance of a commercial computer programme (SSPS) [33]. Statistical processing has been carried out considering marine and non-marine atmospheres separately. The variability of the corrosion and environmental data is shown in Table 7.

**Table 7.** Characteristics of the corrosion and environmental data used in the statistical treatment.

| Type of Atmosphere | Variable | Smallest Value | Largest Value |
|---|---|---|---|
| Non marine | C (µm) | 0.51 | 87.9 |
| | T (°C) | 3.1 | 27 |
| | RH (%) | 35 | 90 |
| | SO$_2$ (mg/m$^2$.d) | 0.08 | 84 |
| Marine | C (µm) | 8.6 | 411.2 |
| | T (°C) | 3.9 | 28.2 |
| | TOW (annual fraction) | 0.16 | 0.99 |
| | SO$_2$ (mg/m$^2$.d) | 0.3 | 171.7 |
| | Cl (mg/m$^2$.d) | 3.03 | 376 |

*3.1. Non-Marine Atmospheres*

The three databases have been analysed together: ICP/UNECE database (Table 4) (all corrosion stations), and ISOCORRAG (Table 5) and MICAT (Table 6) databases (only those stations with a chloride ion deposition rate of Cl$^-$ < 3 mg/m$^2$.d).

The meteorological and pollution parameters common to all three databases and which have been included in the treatment are: temperature, relative humidity and SO$_2$ pollution, though relative humidity is only available in the ISOCORRAG database for stations in the Czech Republic [34]. The stepwise method has been used to select what independent variables are included in the treatment and which are significant. Statistically all the variables are significant, with SO$_2$ being the variable that contributes with an extremely high percentage ($R^2$ = 0.671) in the total recorded variance ($R^2$ = 0.725). RH and T also contribute, raising the $R^2$ by 0.037 and 0.017 units, respectively.

The resulting regression equation is:

$$C = -26.32 + 0.43\,T + 0.45\,RH + 0.82\,SO_2;\ (R^2 = 0.725)\ (N = 333), \tag{2}$$

where N is the number of data. The model explains 72.5% of the dependent variable. This is the regression equation with non-standard coefficients, partial regression coefficients which define the regression equation at direct scores. Figure 6 shows the relationship between predicted and observed carbon steel corrosion values by applying Equation (2).

The standardised partial regression coefficients are the coefficients that define the regression equation when it is obtained after standardising the original variables, i.e., after converting the direct scores into typical scores. These coefficients make it possible to evaluate the relative importance of each independent variable within the equation. The regression equation with standardised coefficients is shown in Equation (3):

$$C = 0.15\,T + 0.26\,RH + 0.82\,SO_2;\ (R^2 = 0.725)\ (N = 333), \tag{3}$$

Great caution must be used when making corrosion predictions with independent variable values that are much larger or smaller than those used to derive these equations (see Table 7).

The goodness of the fit is slightly higher than the damage function developed by ICP/UNECE for this type of atmospheres using a more sophisticated mathematical model (Table 8), with a notably lower number of data (N = 148) used in the statistical treatment.

If, instead of RH, time of wetness (TOW) were to be considered (No. of hours in which RH > 80% and T > 0 °C, and therefore less precise than RH), taking into consideration the ISOCORRAG and MICAT databases the resulting regression equation would be:

$$C = 6.58 + 0.75\,SO_2 + 20.85\,TOW;\ (R^2 = 0.684)\ (N = 138), \tag{4}$$

with a slightly lower regression coefficient than Equation (2). In this case, the temperature is a non-significant variable and the greatest specific weight again corresponds to $SO_2$, which contributes to the total recorded variance with an $R^2 = 0.646$ of a total $R^2 = 0.684$.

Considering another related parameter, such as precipitation (P) (in mm/y), instead of RH, the following regression equation would be obtained:

$$C = -29.26 + 0.87 \, SO_2 + 0.51 \, RH + 0.49 \, T - 0.003 \, P; \, (R^2 = 0.625) \, (N = 315), \qquad (5)$$

in which the resulting regression coefficient decreases even more.

**Table 8.** Published dose/response (D/R) functions for first-year corrosion of carbon steel.

| Type of Atmosphere | Programme | Ref. | D/R Function | N | $R^2$ |
|---|---|---|---|---|---|
| Non marine | ICP/UNECE (for weathering steels) | [18] | $C = 34[SO_2]^{0.13} \exp\{0.020 \, RH + f(T)\}$ where $f(T) = 0.059(T - 10)$ when $T \leq 10\,°C$, otherwise $f(T) = -0.036 \, (T - 10)$ | 148 | 0.68 |
| All atmospheres (marine and non-marine atmospheres) | ISOCORRAG | [35] | $C = 0.091 \, [SO_2]^{0.56} \, TOW^{0.52} \exp(f(T)) + 0.158[Cl]^{0.58} \, TOW^{0.25} \exp(0.050 \, T)$ where $f(T) = 0.103 \, (T - 10)$ when $T \leq 10\,°C$, otherwise $f(T) = -0.059 \, (T - 10)$ | 125 | 0.85 |
| | | | $C = 1.77 \, [SO_2]^{0.52} \exp(0.020 \, RH) \exp(f(T)) + 0.102[Cl]^{0.62} \exp(0.033 \, RH + 0.040 \, T)$ where $f(T) = 0.150 \, (T - 10)$ when $T \leq 10\,°C$, otherwise $f(T) = -0.054 \, (T - 10)$ | 128 | 0.85 |
| | MICAT | [7] | $C = -0.44 + 6.38 \, TOW + 1.58[SO_2] + 0.96[Cl]$ | 172 | 0.56 |
| | ISOCORRAG/MICAT Including data from Russian sites in frigid regions | [1] | $C = 0.085 \times SO_2^{0.56} \times TOW^{0.53} \times \exp(f) + 0.24 \times Cl^{0.47} \times TOW^{0.25} \times \exp(0.049 \, T)$ $f(T) = 0.098 \, (T - 10)$ when $T \leq 10\,°C$, otherwise $f(T) = -0.087 \, (T - 10)$ | 119 | 0.87 |

N = number of data.

## 3.2. Marine Atmospheres

As in the previous case, the stepwise method has been used to select what independent variables are included in the statistical treatment and are significant. In this case, given that the ICP/UNECE programme did not measure the Cl ion deposition rate as it only considered non-marine testing stations, only stations from the ISOCORRAG and MICAT databases have been considered, taking as independent variables the chloride deposition rate (Cl), $SO_2$ pollution, temperature and time of wetness. As has been noted above, data on relative humidity are not available for ISOCORRAG stations.

In this case Cl is the variable that contributes with the highest percentage ($R^2 = 0.411$) to the total recorded variance ($R^2 = 0.474$). $SO_2$ also contributes, raising the $R^2$ by 0.041 units. TOW is also a significant variable but raises the $R^2$ by only 0.022 units. Temperature is excluded as a significant variable.

The resulting regression equation is:

$$C = -24.50 + 0.75 \, Cl + 0.67 \, SO_2 + 77.32 \, TOW \, (R^2 = 0.474) \, (N = 206), \qquad (6)$$

while in standard coefficients the resulting equation would be:

$$C = 0.62 \, Cl + 0.22 \, SO_2 + 0.15 \, TOW \, (R^2 = 0.474) \, (N = 206), \qquad (7)$$

Cl and $SO_2$, in this order, are the variables with the greatest weight in the carbon steel corrosion rate. Figure 7 shows the relationship between predicted and observed carbon steel corrosion values by applying Equation (6).

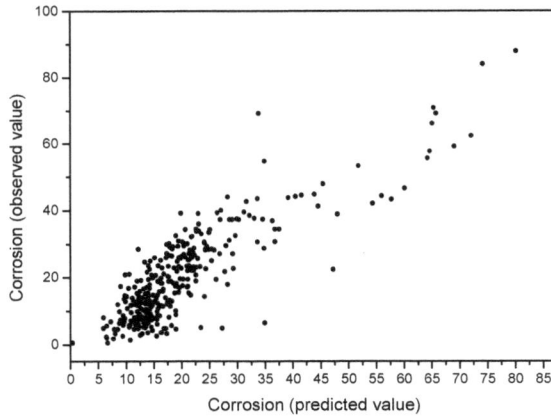

**Figure 6.** Relationship between predicted and observed carbon steel corrosion values by applying Equation (2).

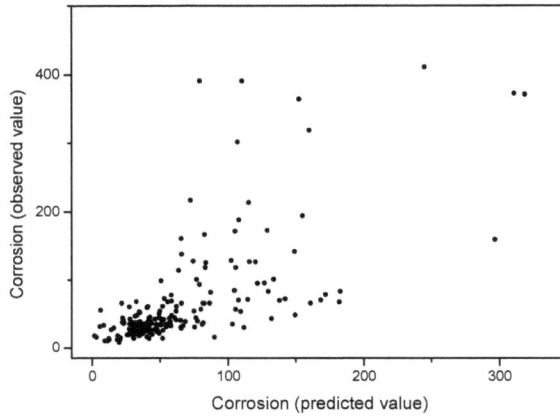

**Figure 7.** Relationship between predicted and observed carbon steel corrosion values by applying Equation (6).

The goodness of this fit is slightly lower than in the MICAT programme and notably lower than those obtained in the ISOCORRAG programme (Table 8). On the other hand, the number of data used in the statistical treatment to obtain the damage function (Equations (6) and (7)) is somewhat higher.

Replacing the TOW variable with RH, information that is available in MICAT and perhaps in ISOCORRAG records, could possibly have improved the fitting quality achieved.

It would be also helpful to make new fits using more sophisticated mathematical models, and we encourage experts in statistics to do this. For this purpose, a good starting point could be the perfected broad database that are presented in this work (Tables 4–6).

*3.3. Contribution of the Information Supplied for Each International Programme*

Having established the integrated database, it is of interest to compare the damage functions and correlation coefficients obtained using the combined information of the three programmes (Equations (3) and (7)) with those obtained separately using the information supplied each individual programme. In this way, it may be possible to determine the contribution of each programme to the general damage functions. The results obtained with this treatment of the information are shown in Table 9.

Considering non-marine atmospheres, it is seen that the greatest volume of information (81%) is supplied by the ICP/UNECE programme, which establishes $SO_2$, RH and T as the most significant variables, the former with the greatest weight. The contribution of the ISOCORRAG programme ($R^2$ = 0.867), which incorporates a relatively small amount of data as only the Czech Republic testing stations supplied information on these three variables, makes only a slight improvement to the general correlation coefficient (0.725) in relation to $R^2$ supplied by the ICP/UNECE programme (0.674). With regard to the information provided by the MICAT programme, the low correlation coefficient obtained (0.398) may be an indication of poorer data quality than in the other programmes, perhaps due to the participation in MICAT of countries with little or no prior experience in the field of atmospheric corrosion [7]. In this respect, it should not be overlooked that one of the MICAT research programme's main aims was precisely to promote the development of this line of research in some of the participating countries [11].

Considering marine atmospheres, information has been supplied only by the ISOCORRAG and MICAT programmes, with a similar amount of data in each case. These data show the important effect of Cl and $SO_2$ pollutants (in this order) on the magnitude of the atmospheric corrosion of carbon steel. The greatest contribution to the general damage function seems to correspond to the ISOCORRAG programme, which includes TOW as an also significant variable. The joining of data from the two programmes, whose individual damage functions have only low correlation coefficients, yields a combined damage function with an even lower correlation coefficient.

**Table 9.** D/R functions for first-year corrosion of carbon steel. Contribution of the information supplied by each of the international programmes to the overall D/R functions (Equations (3) and (7)) using the integrated database.

| Type of Atmosphere | Programme | N | Significant Variables | D/R Function | $R^2$ |
|---|---|---|---|---|---|
| Non-marine atmospheres | Overall D/R function (Equation (3)) (ICP/UNECE + ISOCORRAG + MICAT) | 333 | $SO_2$, RH, T | C= 0.82 $SO_2$ + 0.26 RH + 0.15 T | 0.725 |
| | ICP/UNECE | 269 | $SO_2$, RH, T | C= 0.74 $SO_2$ + 0.37 RH + 0.21 T | 0.674 |
| | ISOCORRAG | 18 | $SO_2$ | C= 0.93 $SO_2$ | 0.867 |
| | MICAT | 46 | $SO_2$, T | C= 0.49 $SO_2$ + 0.46 T | 0.398 |
| Marine atmospheres | Overall D/R function (Equation (7)) (ISOCORRAG + MICAT) | 206 | Cl, $SO_2$, TOW | C= 0.62 Cl + 0.22 $SO_2$ + 0.15 TOW | 0.474 |
| | ISOCORRAG | 97 | Cl, $SO_2$, TOW | C= 0.57 Cl + 0.31 $SO_2$ + 0.31 TOW | 0.582 |
| | MICAT | 109 | Cl, $SO_2$, | C= 0.69 Cl + 0.30 $SO_2$ | 0.538 |

N = number of data.

## 4. Goodness of the Fits

As we have seen, the environmental parameters considered in this work only partly explain the corrosion data. The goodness of fit of experimental data to a proposed model is often measured by the statistical $R^2$, i.e., the square of the correlation coefficient (R) between the observed values of the dependent variable and those predicted from the fitted line.

In previous work [15,16], the authors noted a series of causes that may affect the goodness of the fit obtained:

- Oversimplification of the mathematical model. In this sense, the best fits obtained in the ISOCORRAG programme, including or excluding the MICAT databases and data from Russian sites in frigid regions [1] (see Table 8), may have been at least partly due to considering interactions between the meteorological and pollution variables. One example of complex interactions involves the RH (or TOW), which in addition to its effect on the wetting of the metal surface, plays a major role in the mechanisms whereby air pollutants take part in corrosion.

- The lack of quality in corrosion and environmental data.
- Probable occurrence of other variables with marked effects on corrosion that were not considered in the statistical treatment. For instance, besides sulphur dioxide and chlorides, other pollutants not considered in the study may have played an important role in the corrosion data. In this sense, mention should be made of the effort made by ICP/UNECE to consider in the new damage functions (for the multi-pollutant situation) other important pollutants in terms of their effect on the corrosion of weathering steel [18].
- Many effects that have not been considered. To mention just a few: The magnitude of diurnal and seasonal changes in meteorological and pollution parameters, the frequency, duration and type of wetting and drying cycles, and the time of year when exposure is initiated.

Finally, it would be desirable, as noted by Leygraf et al. [36], to develop models on the basis of mechanistic considerations instead of statistical considerations, and recent work has attempted to take this into account [37–39]. Nevertheless, there is still a very long way to go to reach this desired goal. As Roberge et al. [40] note, the results obtained with mechanistic models reveal why statistical schemes have only a limited accuracy. There are many variables that can change from site to site which are not accounted for in the standard set of environmental variables. According to this researcher, a single transferable and comprehensive environmental corrosivity prediction model is yet to be published, and may ultimately not be possible due to the complexity of the issues involved.

## 5. Conclusions

The following may be considered the most relevant conclusions of this study:

- A highly complete and perfected database has been obtained from published data from the ISOCORRAG, ICP/UNECE and MICAT programmes.
- The number of data used in the statistical treatment has been much higher than that used in other damage functions previously published by the different programmes.
- The statistical treatment carried out has differentiated between two types of atmospheres: non-marine and marine, which may represent a significant simplification for persons with little knowledge of the atmospheric corrosion process who wish to estimate the corrosion of carbon steel exposed at a given location. Moreover, having considered a highly simple polynomial function (Equation (2)) in the work may also be an advantage in this sense.
- With regard to non-marine atmospheres, by joining the three databases (ISOCORRAG, ICP/UNECE and MICAT), the following damage function has been obtained:

$$C = -26.32 + 0.43\,T + 0.45\,RH + 0.82\,SO_2 \ (R^2 = 0.725) \ (N = 333),$$

where $SO_2$ is the variable of greatest significance. The inclusion of TOW (or precipitation) instead of RH leads to lower regression coefficients. The goodness of the fit obtained ($R^2$) is slightly higher than that obtained in the ICP/UNECE programme with a more sophisticated function.

- In relation with marine atmospheres, only the ISOCORRAG and MICAT databases have been considered (ICP/UNECE did not consider this type of atmospheres). The damage function obtained is:

$$C = -24.50 + 0.75\,Cl + 0.67\,SO_2 + 77.32\,TOW \ (R^2 = 0.474) \ (N = 206),$$

where Cl and SO$_2$, in this order, are the most significant variables. The goodness of the fit is slightly lower than that obtained in the MICAT programme, which uses a similar type of function, and notably lower than that obtained in the ISOCORRAG programme using a more sophisticated type of function.

**Acknowledgments:** The authors would like to express their gratitude to Johan Tidblad, from Swerea Kimab (Stockholm, Sweden) and Katerina Kreislova, from SVUOM (Prague, Czech Republic) for the information supplied about ICP/UNECE and ISOCORRAG programmes.

**Author Contributions:** Belén Chico and Manuel Morcillo conceived and designed the study; Iván Díaz and Joaquín Simancas collaborated to the data mining process; Daniel de la Fuente contributed to statistical analysis of data; Belén Chico and Manuel Morcillo analysed the data and wrote the paper.

**Conflicts of Interest:** The authors declare no conflict of interest.

1.  Knotkova, D.; Kreislova, K.; Dean, S.W.J. *ISOCORRAG. International Atmospheric Exposure Program: Summary of Results*; ASTM: West Conshohocken, PA, USA, 2010.
2.  *Corrosion of Metals and Alloys, Corrosivity of Atmospheres, Classification*; ISO 9223: 1992; International Organization for Standardization: Geneva, Switzerland, 1992.
3.  *Corrosion of Metals and Alloys, Corrosivity of Atmospheres, Classification, Determination and Estimation*; EN ISO 9223: 2012; European Committee for Standardization: Brussels, Belgium, 2012.
4.  *Corrosion of Metals and Alloys, Corrosivity of Atmospheres, Guiding Values for the Corrosivity Categories*; EN ISO 9224: 2012; European Committee for Standardization: Brussels, Belgium, 2012.
5.  *Corrosion of Metals and Alloys, Corrosivity of Atmospheres, Measurement of Environmental Parameters Affecting Corrosivity of Atmospheres*; EN ISO 9225: 2012; European Committee for Standardization: Brussels, Belgium, 2012.
6.  *Corrosion of Metals and Alloys, Corrosivity of Atmospheres, Determination of Corrosion Rate of Standard Specimens for the Evaluation of Corrosivity*; EN ISO 9226: 2012; European Committee for Standardization: Brussels, Belgium, 2012.
7.  Morcillo, M.; Almeida, E.; Rosales, B.; Uruchurtu, J.; Marrocos, M. *Corrosion y Protección de Metales en las Atmósferas de Iberoamérica. Parte I—Mapas de Iberoamérica de Corrosividad Atmosférica (Proyecto MICAT, XV.1/CYTED)*; CYTED: Madrid, Spain, 1998.
8.  *UN/ECE International Cooperative Programme on Effects on Materials Including Historic and Cultural Monuments, Report n. 01: Technical Manual*; Swedish Corrosion Institute: Stockholm, Sweden, 1988.
9.  Knotkova, D.; Vrobel, L. ISOCORRAG—The International testing program within ISO/TC 156/WG 4. In *Proceedings of the 11th International Corrosion Congress*; Associazione Italiana di Metallurgia: Florence, Italy, 1990; Volume 5, pp. 581–590.
10. Kucera, V.; Coote, A.T.; Henriksen, J.F.; Knotkova, D.; Leygraf, C.; Reinhardt, U. Effects of acidifying air pollutants on materials including historic and cultural monuments—An international cooperative programme within unece. In *Proceedings of 11th Internacional Corrosion Congress*; Associazione Italiana di Metallurgia: Florence, Italy, 1990; Volume 2, pp. 433–442.
11. Uller, L.; Morcillo, M. The setting-up of an iberoamerican map of atmospheric corrosion. MICAT study-CYTED-D. In *Proceedings of the 11th International Corrosion Congress*; Associazione Italiana di Metallurgia: Florence, Italy, 1990; pp. 35–45.
12. United Nations Economic Commission for Europe (UNECE). *Effects of Acid Deposition on Atmospheric Corrosion of Materials*; Executive Body for the Convention on Long-Range Transboundary Air Pollution, Working Group on Effects: Geneva, Switzerland, 1990.
13. Benarie, M.; Lipfert, F.L. A general corrosion function in terms of atmospheric pollutant concentrations and rain ph. *Atmos. Environ.* **1986**, *20*, 1947–1958. [CrossRef]
14. Pourbaix, M. The linear bilogaritmic law for atmospheric corrosion. In *Atmospheric Corrosion*; Ailor, W.H., Ed.; The Electrochemical Society, John Wiley and Sons: New York, NY, USA, 1982; pp. 107–121.
15. Feliu, S.; Morcillo, M.; Feliu, S., Jr. The prediction of atmospheric corrosion from meteorological and pollution parameters. 1. Annual corrosion. *Corros. Sci.* **1993**, *34*, 403–414. [CrossRef]
16. Feliu, S.; Morcillo, M.; Feliu, S., Jr. The prediction of atmospheric corrosion from meteorological and pollution parameters. 2. Long-term forecasts. *Corros. Sci.* **1993**, *34*, 415–422. [CrossRef]

17. Knotkova, D.; Barton, K. Corrosion aggressivity of atmospheres (derivation and classification). In *Atmospheric Corrosion of Metals*; Dean, S.W.J., Rhea, E.C., Eds.; American Society for Testing and Materials: Philadelphia, PA, USA, 1982; p. 225.

18. CLRTAP. Mapping of Effects on Materials, Chapter iv of Manual on Methodologies and Criteria for Modelling and Mapping Critical Loads and Levels and Air Pollution Effects, Risks and Trends. UNECE Convention on Long-Range Transboundary Air Pollution. Available online: www.icpmapping.org (accessed on 1 March 2017).

19. McCuen, R.H.; Albrecht, P.; Cheng, J.G. A new approach to power-model regression of corrosion penetration data. In *Corrosion Forms and Control for Infrastructure*; Chaker, V., Ed.; American Society for Testing and Materials: Philadelphia, PA, USA, 1992; Volume 1137, pp. 46–76.

20. Albrecht, P.; Hall, T.T. Atmospheric corrosion resistance of structural steels. *J. Mater. Civ. Eng.* **2003**, *15*, 2–24. [CrossRef]

21. Panchenko, Y.M.; Marshakov, A.I. Long-term prediction of metal corrosion losses in atmosphere using a power-linear function. *Corros. Sci.* **2016**, *109*, 217–229. [CrossRef]

22. Panchenko, Y.M.; Marshakov, A.I.; Igonin, T.N.; Kovtanyuk, V.V.; Nikolaeva, L.A. Long-term forecast of corrosion mass losses of technically important metals in various world regions using a power function. *Corros. Sci.* **2014**, *88*, 306–316. [CrossRef]

23. Melchers, R.E. A new interpretation of the corrosion loss processes for weathering steels in marine atmospheres. *Corros. Sci.* **2008**, *50*, 3446–3454. [CrossRef]

24. Melchers, R.E. Long-term corrosion of cast irons and steel in marine and atmospheric environments. *Corros. Sci.* **2013**, *68*, 186–194. [CrossRef]

25. Morcillo, M.; Chico, B.; Díaz, I.; Cano, H.; De la Fuente, D. Atmospheric corrosion data of weathering steels. A review. *Corros. Sci.* **2013**, *77*, 6–24. [CrossRef]

26. Adikari, M.; Munasinghe, N. Development of a corrosion model for prediction of atmospheric corrosion of mild steel. *Am. J. Constr. Build. Mater.* **2016**, *1*, 1–6.

27. Metals and Alloys. *Atmospheric Corrosion Testing, General Requirements*; ISO 8565; International Organization for Standardization: Geneva, Switzerland, 2011.

28. Chico, B.; de la Fuente, D.; Morcillo, M. Corrosión atmosférica de metales en condiciones climáticas extremas. *Bol. Soc. Esp. Ceram. Vidrio* **2000**, *39*, 329–332. [CrossRef]

29. Coburn, S.K.; Larrabee, C.P.; Lawson, H.H.; Ellis, O.B. Corrosivennes of various atmospheric test sites as measured by specimens of steel and zinc. In *Metal Corrosion in the Atmosphere*; American Society for Testing and Materials: Philadelphia, PA, USA, 1968; pp. 360–391.

30. Biefer, G.J. Atmospheric corrosion of steel in the canadian arctic. *Mater. Perform.* **1981**, *20*, 16–19.

31. Hughes, J.D.; King, G.A.; O'Brien, D.J. Corrosivity in Antarctica—Revelations on the nature of corrosion in the world coldest, driest, highest and purest continent. In *Proceedings of the 13th International Corrosion Congress*; Australasian Corrosion Association Inc: Melbourne, Australia, 1996.

32. Alcántara, J.; Chico, B.; Díaz, I.; de la Fuente, D.; Morcillo, M. Airborne chloride deposit and its effect on marine atmospheric corrosion of mild steel. *Corros. Sci.* **2015**, *97*, 74–88. [CrossRef]

33. International Business Machines Corporation. *SPSS Statistics*, version 23; IBM: North Castle, NY, USA, 2016.

34. Kreislova, K.; SVUOM, Prague, Czech Republic. Personal communication, 2017.

35. Knotkova, D.; Kreislova, K.; Kvapil, J.; Holubova, G.; Bubenickova. *CLRTAP, UNECE International Cooperative Programme on Effect on Materials, including Historic and Cultural Monuments, Report No. 42. Results from the Multi-pollutant Programme Corrosion Attack on Carbon Steel after 1, 2 and 4 Years of Exposure (1997–2001)*; Institute for Protection of Material (SVUOM): Prague, Czech Republic, 2003.

36. Leygraf, C.; Odnevall Wallinder, I.; Tidblad, J.; Graedel, T. *Atmospheric Corrosion*, 2nd ed.; The Electrochemical Society Series; John Wiley and Sons: Hoboken, NJ, USA, 2016.

37. Klassen, R.D.; Roberge, P.R. Aerosol transport modeling as an aid to understanding atmospheric corrosivity patterns. *Mater. Des.* **1999**, *20*, 159–168. [CrossRef]

38. Cole, I.S.; Furman, S.A.; Neufeld, A.K.; Ganther, W.D.; King, G.A. A holistic approach to modelling in atmospheric corrosion. In *Proceedings of the 14th International Corrosion Congress*; Corrosion Institute of Southern Africa: Cape Town, South Africa, 1999.

39. Klassen, R.D.; Roberge, P.R. The effects of wind on local atmospheric corrosivity. In *Corrosion 2001*; NACE International: Houston, TX, USA, 2001.

40. Roberge, P.R.; Klassen, R.D.; Haberecht, P.W. Atmospheric corrosivity modeling—A review. *Mater. Des.* **2002**, *23*, 321–330. [CrossRef]

*materials*

MDPI

*Article*

# Prediction of First-Year Corrosion Losses of Carbon Steel and Zinc in Continental Regions

Yulia M. Panchenko * and Andrey I. Marshakov

A.N. Frumkin Institute of Physical Chemistry and Electrochemistry, Russian Academy of Sciences, Moscow 119071, Russia; mar@ipc.rssi.ru
* Correspondence: panchenkoyum@mail.ru

Academic Editor: Douglas Ivey
Received: 18 February 2017; Accepted: 13 April 2017; Published: 18 April 2017

**Abstract:** Dose-response functions (DRFs) developed for the prediction of first-year corrosion losses of carbon steel and zinc ($K_1$) in continental regions are presented. The dependences of mass losses on $SO_2$ concentration, $K = f([SO_2])$, obtained from experimental data, as well as nonlinear dependences of mass losses on meteorological parameters, were taken into account in the development of the DRFs. The development of the DRFs was based on the experimental data from one year of testing under a number of international programs: ISO CORRAG, MICAT, two UN/ECE programs, the Russian program in the Far-Eastern region, and data published in papers. The paper describes predictions of $K_1$ values of these metals using four different models for continental test sites under UN/ECE, RF programs and within the MICAT project. The predictions of $K_1$ are compared with experimental $K_1$ values, and the models presented here are analyzed in terms of the coefficients used in the models.

**Keywords:** carbon steel; zinc; modeling studies; atmospheric corrosion

## 1. Introduction

Predictions of the corrosion mass losses ($K$) of structural metals, in general for a period not exceeding 20 years, are made using the power function:

$$K = K_1 \tau^n, \qquad (1)$$

where $K_1$ represents the corrosion losses for the first year, $g/m^2$ or $\mu m$; $\tau$ is the test time in years; and $n$ is a coefficient that characterizes the protective properties of corrosion products. The practical applications of Equation (1) for particular test locations in various regions of the world and the methods for $n$ calculation are summarized in [1–8].

The power linear function that is believed to provide the most reliable predictions for any period of time and in any region of the world was suggested in [9,10]. Corrosion obeys a power law (Equation (1)) during an initial period and a linear law after the stationary stage starts. The total corrosion losses of metals for any period of time during the stationary stage can be calculated using Equation (2):

$$K = K_{st} + \alpha(\tau - \tau_{st}), \qquad (2)$$

where $K_{st}$ stands for corrosion losses over the initial period calculated by Equation (1), $g/m^2$ or $\mu m$; $\tau_{st}$ is the year when stabilization begins; and $\alpha$ is the yearly gain in corrosion losses of metals during the stationary stage in $g/(m^2 year)$ or $\mu m/year$.

The differences in the predictions of corrosion losses by Equations (1) and (2) consist of different estimates of $\tau_{st}$, $\alpha$, and $n$ values for test locations with various corrosivity and atmosphere types. According to [10], $\tau_{st}$ equals 20 years. The $n$ values are given per atmosphere type, irrespective of the atmosphere corrosivity within a particular type. In [9], $\tau_{st} = 6$ years, and equations for $n$ calculations based on the corrosivity of various atmosphere types are suggested. In [9,10], the $\alpha$ values are equal to the instantaneous corrosion rate at $\tau_{st}$.

Furthermore, various types of dose-response functions (DRFs) have been developed for long-term predictions of $K$; these can be used for certain territories or for any region of the world [11–17]. It should be noted that DRFs are power functions and have an advantage in that they provide predictions of first-year corrosion losses ($K_1$) based on yearly-average meteorological and aerochemical atmosphere parameters. The power-linear function uses $K_1$ values that should match the yearly-average corrosivity parameters of the test site atmosphere. The $K_1$ values can be determined by repeated natural yearly tests in each location, which require significant expense and ISO 9223:2012(E) presents equations for the calculation of $K_1$ of structural metals for any atmosphere types [18].

Recently, one-year and long-term predictions have been performed using models based on an artificial neural network (ANN) [19–23]. Their use is undoubtedly a promising approach in the prediction of atmospheric corrosion. The ANN "training" stage is programmed so as to obtain the smallest prediction error. Linear and nonlinear functions are used for $K$ or $K_1$ prediction by means of an ANN. Using an ANN, the plots of $K$ ($K_1$) versus specific corrosivity parameters can be presented visually as 2D or 3D graphs [19]. Despite the prospects of $K$ prediction using ANNs, DRF development for certain countries (territories) is an ongoing task. The analytical form of DRFs is most convenient for application by a broad circle of experts who predict the corrosion resistance of materials in structures.

DRF development is based on statistical treatment, regression analysis of experimental data on $K_1$, and corrosivity parameters of atmospheres in numerous test locations. All DRFs involve a prediction error that is characterized, e.g., by the $R^2$ value or by graphical comparison in coordinates of predicted and experimental $K_1$. However, comparisons of the results on $K_1$ predictions based on different DRFs for large territories have not been available to date. Furthermore, the DRFs that have been developed assume various dependences of $K$ on $SO_2$ concentration; however, the shape of the $K = f(SO_2)$ function was not determined by analysis of data obtained in a broad range of atmosphere meteorological parameters.

The main purpose of this paper is to perform a mathematical estimate of the $K = f(SO_2)$ dependence for carbon steel and zinc, popular structural materials, and to develop new DRFs for $K_1$ prediction based on the $K = f(SO_2)$ dependences obtained and the meteorological corrosivity parameters of the atmosphere. Furthermore, we will compare the $K_1$ predictions obtained by the new and previously developed DRFs for any territories of the world, as well as analyze the DRFs based on the values of the coefficients in the equations.

## 2. Results

### 2.1. Development of DRFs for Continental Territories

To develop DRFs, we used the experimental data from all exposures for a one-year test period in continental locations under the ISO CORRAG international program [24], the MICAT project [11,25], the UN/ECE program [12,14], the Russian program [26], and the program used in [19]. The test locations for the UN/ECE program and the MICAT project are presented in Table 1. The corrosivity parameters of the test site atmospheres and the experimental $K_1$ values obtained in four one-year exposures under the UN/ECE program are provided in Table 2, those obtained in three one-year exposures under the MICAT project are given in Table 3, and those obtained in the RF program are provided in Table 4. Cai et al. [19] report a selection of data from various literature sources. Of this selection, we use only the experimental data for continental territories that are shown in Table 5. The test results under the ISO CORRAG program [24] are not included in this paper because they lack the atmosphere corrosivity parameters required for $K_1$ prediction. We used them simply to determine the $K = f(SO_2)$ dependences for steel and zinc.

**Table 1.** Countries, names, and codes of test locations.

| MICAT Project | | | UN/ECE Program | | |
|---|---|---|---|---|---|
| Country | Test Location | Designation | Country | Test Location | Designation |
| Argentina | Villa Martelli | A2 | Czech Republic | Prague | CS1 |
| Argentina | Iguazu | A3 | Czech Republic | Kasperske Hory | CS2 |
| Argentina | San Juan | A4 | Czech Republic | Kopisty | CS3 |
| Argentina | La Plata | A6 | Finland | Espoo | FIN4 |
| Brasil | Caratinga | B1 | Finland | Ähtäri | FIN5 |
| Brasil | Sao Paulo | B6 | Finland | Helsinki Vallila | FIN6 |
| Brasil | Belem | B8 | Germany | Waldhof Langenbrügge | GER7 |
| Brasil | Brasilia | B10 | Germany | Aschaffenburg | GER8 |
| Brasil | Paulo Afonso | B11 | Germany | Langenfeld Reusrath | GER9 |
| Brasil | Porto | B12 | Germany | Bottrop | GER10 |
| Colombia | San Pedro | CO2 | Germany | Essen Leithe | GER11 |
| Colombia | Cotove | CO3 | Germany | Garmisch Partenkirchen | GER12 |
| Ecuador | Guayaquil | EC1 | Netherlands | Eibergen | NL18 |
| Ecuador | Riobamba | EC2 | Netherlands | Vredepeel | NL19 |
| Spain | Leon | E1 | Netherlands | Wijnandsrade | NL20 |
| Spain | Tortosa | E4 | Norway | Oslo | NOR21 |
| Spain | Granada | E5 | Norway | Birkenes | NOR23 |
| Spain | Arties | E8 | Sweden | Stockholm South | SWE24 |
| Mexico | Mexico (a) | M1 | Sweden | Stockholm Centre | SWE25 |
| Mexico | Mexico (b) | M2 | Sweden | Aspvreten | SWE26 |
| Mexico | Cuernavaca | M3 | Spain | Madrid | SPA31 |
| Mexico | San Luis Potosi | PE4 | Spain | Toledo | SPA33 |
| Peru | Arequipa | PE5 | Russian Federation | Moscow | RUS34 |
| Peru | Arequipa | PE6 | Estonia | Lahemaa | EST35 |
| Peru | Pucallpa | U1 | Canada | Dorset | CAN37 |
| Uruguay | Trinidad | U3 | USA | Research Triangle Park | US38 |
| - | - | - | USA | Steubenville | US39 |

**Table 2.** Atmosphere corrosivity parameters of test locations, first-year corrosion losses of carbon steel and zinc ($K_1$, g/m$^2$) under the UN/ECE program, and numbers of test locations in the order of increasing $K_1$.

| Designation | *T*, °C | *RH*, % | *TOW*, Hours/a | *Prec*, mm/a | [SO$_2$], µg/m$^3$ | [H$^+$], mg/L | Steel | | Zinc | |
|---|---|---|---|---|---|---|---|---|---|---|
| | | | | | | | g/m$^2$ | No. | g/m$^2$ | No. |
| CS1 | 9.5 | 79 | 2830 | 639.3 | 77.5 | - | 438.0 | 76 | 14.89 | 92 |
| CS1 | 10.3 | 74 | 2555 | 380.8 | 58.1 | 0.0221 | - | - | 6.98 | 45 |
| CS1 | 9.1 | 73 | 2627 | 684.3 | 41.2 | 0.0714 | 270.7 | 64 | 7.78 | 53 |
| CS1 | 9.8 | 77 | 3529 | 581.1 | 32.1 | 0.0342 | 241.0 | 58 | 5.69 | 31 |
| CS2 | 7.0 | 77 | 3011 | 850.2 | 19.7 | - | 224.0 | 51 | 8.95 | 65 |
| CS2 | 7.4 | 76 | 3405 | 703.4 | 25.6 | 0.045 | - | - | 7.99 | 58 |
| CS2 | 6.6 | 73 | 2981 | 921 | 17.9 | 0.1921 | 152.9 | 33 | 6.77 | 44 |
| CS2 | 7.2 | 74 | 3063 | 941.2 | 12.2 | 0.0366 | 148.2 | 30 | 3.46 | 4 |
| CS3 | 9.6 | 73 | 2480 | 426.4 | 83.3 | - | 557.0 | 77 | 16.41 | 94 |
| CS3 | 9.9 | 72 | 2056 | 416.6 | 78.4 | 0.0242 | - | - | 11.59 | 87 |
| CS3 | 8.9 | 71 | 2866 | 431.6 | 49 | 0.058 | 350.2 | 73 | 11.74 | 88 |
| CS3 | 9.7 | 75 | 2759 | 512.7 | 49.2 | 0.0567 | 351.8 | 74 | 12.17 | 89 |

**Table 2.** *Cont.*

| Designation | T, °C | RH, % | TOW, Hours/a | Prec, mm/a | [SO$_2$], µg/m$^3$ | [H$^+$], mg/L | Steel g/m$^2$ | Steel No. | Zinc g/m$^2$ | Zinc No. |
|---|---|---|---|---|---|---|---|---|---|---|
| FIN4 | 5.9 | 76 | 3322 | 625.9 | 18.6 | - | 271.0 | 63 | - | - |
| FIN4 | 6.4 | 80 | 4127 | 657 | 13.9 | 0.0392 | - | - | 8.42 | 62 |
| FIN4 | 5.6 | 79 | 3446 | 754.6 | 2.3 | 0.0231 | 130.3 | 21 | 5.18 | 25 |
| FIN4 | 6.0 | 80 | 3607 | 698.1 | 2.6 | 0.0334 | 120.9 | 20 | 4.68 | 19 |
| FIN5 | 3.1 | 78 | 2810 | 801.3 | 6.3 | - | 132.0 | 23 | 8.92 | 66 |
| FIN5 | 3.9 | 80 | 3342 | 670.7 | 1.8 | 0.0271 | - | - | 7.70 | 52 |
| FIN5 | 3.4 | 81 | 2994 | 609.7 | 0.9 | 0.0201 | 48.4 | 4 | 6.62 | 41 |
| FIN5 | 3.9 | 83 | 3324 | 675.4 | 0.8 | 0.0247 | 59.3 | 5 | 4.61 | 16 |
| FIN6 | 6.3 | 78 | 3453 | 673.1 | 20.7 | - | 273.0 | 65 | - | - |
| FIN6 | 6.8 | 80 | 4017 | 665.6 | 15.3 | 0.0554 | - | - | 9.29 | 70 |
| FIN6 | 6.2 | 78 | 3360 | 702.4 | 4.8 | 0.0221 | 162.2 | 34 | 5.69 | 33 |
| FIN6 | 6.6 | 76 | 3288 | 649.2 | 5.5 | 0.0139 | 195.8 | 44 | 5.62 | 30 |
| GER7 | 9.3 | 80 | 4561 | 630.6 | 13.7 | - | 264.0 | 62 | - | - |
| GER7 | 10.2 | 80 | 4390 | 499.7 | 11 | 0.0358 | - | - | 7.85 | 56 |
| GER7 | 8.9 | 81 | 4382 | 624.4 | 8.2 | 0.0342 | 230.9 | 53 | 9.07 | 68 |
| GER7 | 9.5 | 81 | 4676 | 595.6 | 3.9 | 0.0265 | 166.1 | 36 | 4.25 | 13 |
| GER8 | 12.3 | 77 | 4282 | 626.9 | 23.7 | - | 213.0 | 48 | - | - |
| GER8 | 12.2 | 67 | 2541 | 655.4 | 14.2 | 0.0411 | - | - | 4.68 | 18 |
| GER8 | 11.4 | 64 | 3563 | 561.2 | 12.6 | 0.0183 | 116.2 | 17 | 5.18 | 26 |
| GER8 | 11.6 | 65 | 2359 | 779 | 9.6 | - | 141.2 | 27 | 4.10 | 12 |
| GER9 | 10.8 | 77 | 4220 | 782.9 | 24.5 | - | 293.0 | 69 | - | - |
| GER9 | 11.7 | 80 | 4940 | 697.6 | 20.3 | 0.0366 | - | - | 6.62 | 40 |
| GER9 | 10.7 | 79 | 4437 | 619.1 | 16.3 | 0.0291 | 230.9 | 54 | 9.07 | 69 |
| GER9 | 11.4 | 81 | 5210 | 841 | 11.1 | 0.0278 | 209.8 | 47 | 7.63 | - |
| GER10 | 11.2 | 75 | 4077 | 873.8 | 50.6 | - | 373.0 | 75 | - | - |
| GER10 | 12 | 76 | 4107 | 696.6 | 48.5 | 0.0253 | - | - | 10.66 | 81 |
| GER10 | 10.3 | 78 | 4201 | 707.3 | 41.6 | 0.0211 | 347.1 | 72 | 15.34 | 93 |
| GER10 | 11.8 | 80 | 4930 | 912.9 | 30.2 | 0.0334 | 294.1 | 70 | 7.85 | 55 |
| GER11 | 10.5 | 79 | 4537 | 713.1 | 30.3 | - | 342.0 | 71 | - | - |
| GER11 | 11.5 | 77 | 4040 | 644.5 | 25.6 | 0.042 | - | - | 9.72 | 73 |
| GER11 | 10.1 | 79 | 4120 | 683.6 | 22.9 | 0.0253 | 293.3 | 68 | 11.45 | 86 |
| GER11 | 10.9 | 78 | 4632 | 889.3 | 16.2 | 0.0247 | 241.0 | 57 | 7.06 | 46 |
| GER12 | 8.0 | 82 | 4989 | 1491.5 | 9.4 | - | 133.0 | 24 | 8.35 | 61 |
| GER12 | 7.3 | 82 | 4201 | 1183.1 | 6.1 | 0.0171 | - | - | 7.27 | 49 |
| GER12 | 7.1 | 84 | 4545 | 1552.4 | 3.2 | 0.0018 | 89.7 | 9 | 7.20 | 48 |
| GER12 | 7.4 | 83 | 4375 | 1503 | 2.4 | - | 85.0 | 8 | 3.74 | 9 |
| NL18 | 9.9 | 83 | 5459 | 904.2 | 10.1 | - | 232.0 | 55 | 9.93 | 76 |
| NL18 | 10.9 | 79 | 4482 | 705.9 | 8.5 | 0.0046 | - | - | 8.14 | 59 |
| NL18 | 9.5 | 82 | 4808 | 872.8 | 7.4 | 0.004 | 204.4 | 45 | 7.92 | 57 |
| NL18 | 10.3 | 83 | 5358 | 987.1 | 4.7 | 0.0366 | 144.3 | 28 | 4.75 | 20 |
| NL19 | 10.3 | 81 | 5354 | 845 | 13 | - | 283.0 | 66 | - | - |
| NL19 | 11 | 81 | 4969 | 569.1 | 9.9 | 0.0049 | - | - | 9.07 | 67 |
| NL19 | 10 | 82 | 5084 | 749.2 | 8.3 | 0.0021 | 238.7 | 56 | 11.09 | 84 |
| NL19 | 10.9 | 83 | 5454 | 828.9 | 4.5 | - | 180.2 | 39 | - | - |
| NL20 | 10.3 | 81 | 5125 | 801.3 | 13.7 | - | 259.0 | 59 | - | - |
| NL20 | 11.1 | 77 | 4424 | 608.8 | 10.3 | 0.0106 | - | - | 10.22 | 77 |
| NL20 | 10.1 | 81 | 4688 | 679.6 | 9.3 | 0.0113 | 205.1 | 46 | 11.38 | 85 |
| NL20 | 11.1 | 82 | 5141 | 789.9 | 5.8 | 0.0038 | 172.4 | 37 | 6.34 | 37 |
| NOR21 | 7.6 | 70 | 2673 | 1023.8 | 14.4 | - | 229.0 | 52 | - | - |
| NOR21 | 8.8 | 70 | 2864 | 526.6 | 7.9 | 0.0326 | - | - | 5.69 | 32 |
| NOR21 | 7.7 | 68 | 2471 | 440.1 | 6 | 0.0156 | 134.9 | 25 | 6.70 | 43 |
| NOR21 | 7.5 | 69 | 2827 | 680 | 2.9 | 0.0136 | 100.6 | 11 | 3.53 | 7 |

**Table 2.** *Cont.*

| Designation | $T$, °C | $RH$, % | $TOW$, Hours/a | $Prec$, mm/a | $[SO_2]$, $\mu g/m^3$ | $[H^+]$, mg/L | Steel $g/m^2$ | No. | Zinc $g/m^2$ | No. |
|---|---|---|---|---|---|---|---|---|---|---|
| NOR23 | 6.5 | 80 | 4831 | 2144.3 | 1.3 | - | 194.0 | 43 | - | - |
| NOR23 | 7.4 | 77 | 4193 | 1762.2 | 0.9 | 0.042 | - | - | 8.50 | 63 |
| NOR23 | 5.9 | 75 | 3341 | 1188.6 | 0.7 | 0.0374 | 131.8 | 22 | 10.58 | 80 |
| NOR23 | 6.4 | 76 | 3779 | 1419.7 | 0.7 | 0.0326 | 109.2 | 15 | 5.04 | 24 |
| SWE24 | 7.6 | 78 | 3959 | 531 | 16.8 | - | 264.0 | 61 | 10.36 | 79 |
| SWE24 | 8.7 | 70 | 3074 | 473.2 | 8.4 | 0.0366 | - | - | 6.12 | 35 |
| SWE24 | 7 | 70 | 2580 | 577 | 5.7 | 0.043 | 120.1 | 18 | 4.54 | 15 |
| SWE24 | 7.5 | 73 | 3160 | 580.6 | 4.2 | 0.0231 | 103.0 | 13 | 4.25 | 14 |
| SWE25 | 7.6 | 78 | 3959 | 531 | 19.6 | - | 263.0 | 60 | 9.76 | 74 |
| SWE25 | 8.7 | 70 | 3074 | 473.2 | 10.3 | 0.0366 | - | - | 5.62 | 29 |
| SWE25 | 7 | 70 | 2580 | 577 | 4.7 | 0.043 | 103.0 | 12 | 3.53 | 5 |
| SWE25 | 7.5 | 73 | 3160 | 580.6 | 3.4 | 0.0231 | 95.2 | 10 | 3.53 | 8 |
| SWE26 | 6.0 | 83 | 4534 | 542.7 | 3.3 | - | 147.0 | 29 | 8.31 | 60 |
| SWE26 | 7.6 | 77 | 3469 | 342.3 | 2 | 0.043 | - | - | 6.70 | 42 |
| SWE26 | 6 | 81 | 3592 | 467.8 | 1.3 | 0.043 | 74.9 | 6 | 4.90 | 23 |
| SWE26 | 6.8 | 82 | 4118 | 525.2 | 1.1 | 0.0278 | 81.1 | 7 | 6.05 | 34 |
| SPA31 | 14.1 | 66 | 2762 | 398 | 18.4 | - | 222.0 | 50 | 7.74 | 54 |
| SPA31 | 15.2 | 56 | 1160 | 331.5 | 15.3 | 0.0073 | - | - | 4.82 | 22 |
| SPA31 | 14.3 | 67 | 2319 | 360.1 | 8.2 | 0.0003 | 162.2 | 35 | 3.53 | 6 |
| SPA31 | 15.7 | 68 | 2766 | 223.9 | 7.8 | 0.0002 | 151.3 | 32 | 2.30 | 2 |
| SPA33 | 14.0 | 64 | 2275 | 785 | 3.3 | - | 45.0 | 3 | 3.37 | 3 |
| SPA33 | 15.5 | 61 | 2147 | 610.4 | 13.5 | 0.0006 | - | - | 3.89 | 11 |
| SPA33 | 13.4 | 61 | 1888 | 432.5 | 1.7 | 0.0012 | 25.7 | 1 | 3.89 | 10 |
| SPA33 | 14.8 | 57 | 1465 | 327.4 | 4.2 | 0.0006 | 35.9 | 2 | 1.66 | 1 |
| RUS34 | 5.5 | 73 | 2084 | 575.4 | 19.2 | - | 181.0 | 40 | 10.32 | 78 |
| RUS34 | 5.7 | 76 | 2894 | 860.2 | 30.8 | 0.0006 | - | - | 8.64 | 64 |
| RUS34 | 5.7 | 74 | 2444 | 880.6 | 28.7 | 0.0009 | 141.2 | 26 | 6.48 | 39 |
| RUS34 | 5.6 | 71 | 1514 | 666.7 | 16.4 | 0.0008 | 120.9 | 19 | 4.61 | 17 |
| EST35 | 5.5 | 83 | 4092 | 447.8 | 0.9 | - | 185.0 | 41 | 7.18 | 47 |
| EST35 | 6.7 | 81 | 4332 | 532.7 | 0.6 | 0.0226 | - | - | 9.43 | 71 |
| CAN37 | 5.5 | 75 | 3252 | 961.1 | 3.3 | - | 149.0 | 31 | 9.88 | 75 |
| CAN37 | 5 | 79 | 3431 | 1103 | 3 | 0.042 | - | - | 6.26 | 38 |
| CAN37 | 4.3 | 80 | 3302 | 1080 | 2.1 | 0.0482 | 110.0 | 16 | 5.26 | 27 |
| CAN37 | 5.2 | 80 | 3386 | 1022.8 | 3.3 | 0.0461 | 103.7 | 14 | 6.19 | 36 |
| US38 | 14.6 | 69 | 3178 | 846.7 | 9.6 | - | 176.0 | 38 | 10.72 | 82 |
| US38 | 16.3 | 66 | 3026 | 1106.7 | 9.2 | 0.0358 | - | - | 12.46 | 90 |
| US38 | 15.5 | 64 | 2644 | 982.3 | 10.1 | 0.0349 | 184.9 | 42 | 9.72 | 72 |
| US38 | 15.8 | 68 | - | 1037.6 | 9.3 | 0.0482 | - | - | 4.75 | 21 |
| US39 | 12.3 | 67 | 2111 | 733.1 | 58.1 | - | 214.0 | 49 | 13.61 | 91 |
| US39 | 11.2 | 61 | 1391 | 967.4 | 55.2 | 0.0838 | - | - | 11.02 | 83 |
| US39 | 11.8 | 65 | 1532 | 729.4 | 43.1 | 0.0941 | 290.2 | 67 | 7.34 | 50 |
| US39 | 11.8 | 69 | - | 756.8 | 38.3 | 0.0765 | - | - | 5.26 | 28 |

**Table 3.** Atmosphere corrosivity parameters of test locations, first-year corrosion losses of carbon steel and zinc ($K_1$, g/m$^2$) under the MICAT program and those reported in [20], and numbers of test locations in the order of increasing $K_1$. Adapted from [20], with permission from © 2000 Elsevier.

| Designation | T, °C | RH, % | Rain, mm/a | [$SO_2$], μg/m$^3$ | $Cl^-$, mg/(m$^2$·Day) | TOW, h/a | Steel | | Zinc | |
|---|---|---|---|---|---|---|---|---|---|---|
| | | | | | | | g/m$^2$ | No. | g/m$^2$ | No. |
| A2 * | 16.7 | 75 | 1729 | 10 | Ins | 5063 | 122.5 | 36 (34) | 8.06 | 41 |
| A2 | 17.1 | 72 | 983 | 10 | Ins | 4222 | 125.6 | 38 | 7.56 | 39 |
| A2 | 17.0 | 74 | 1420 | 9 | Ins | 4862 | 96.7 | 25 | 10.15 | 47 |
| A3 | 20.6 | 76 | 2158 | Ins (5) ** | Ins (1.5) | 5825 | 44.5 | 12 (11) | 14.76 | 53 |
| A3 | 20.9 | 74 | 2624 | Ins (5) | Ins (1.5) | 5528 | 45.2 | 13 (12) | 8.42 | 43 |
| A3 | 22.1 | 75 | 1720 | Ins (5) | Ins (1.5) | 5545 | 43.7 | 10 (9) | 8.50 | 44 |
| A4 | 18.0 | 51 | 35 | Ins (5) | Ins (1.5) | 999 | 35.9 | 6 (6) | 2.02 | 15 |
| A4 | 20.0 | 49 | 111 | Ins (5) | Ins (1.5) | 850 | 35.1 | 5 (5) | 0.94 | 3 |
| A4 | 18.3 | 51 | 93 | Ins (5) | Ins (1.5) | 867 | 43.7 | 11 (10) | 1.58 | 10 |
| A6 | 17.0 | 78 | 1178 | 6.22 | Ins | 5195 | 197.3 | 55 (51) | 5.54 | 28 |
| A6 * | 16.7 | 77 | 1263 | 8.21 | Ins | 4949 | 224.6 | 59 (55) | 6.70 | 32 |
| A6 * | 16.6 | 78 | 1361 | 6.2 | Ins | 5528 | 234.8 | 61 (57) | 7.49 | 37 |
| B1 | 21.2 | 75 | 996 | 1.67 | 1.57 | 4222 | 102.2 | 28 (26) | 4.32 | 26 |
| B6 | 19.7 | 75 | 1409 | 67.2 (28) | Ins (1.5) | 5676 | 113.9 | 31 (29) | 8.57 | 45 |
| B6 | 19.5 | 76 | 1810 (1910) | 66.8 (28) | Ins (1.5) | 5676 | 182.5 | 53 (49) | 10.66 | 48 |
| B6 | 19.6 | 75 | 1034 | 48.8 (28) | Ins (1.5) | 5676 | 188.8 | 54 (50) | 6.98 | 34 |
| B8 | 26.1 | 88 | 2395 | Ins (5) | Ins (1.5) | 5974 | 151.3 | 44 (40) | 7.92 | 40 |
| B10 | 20.4 | 69 (72) | 1440 | Ins (5) | Ins (1.5) | 3872 | 100.6 | 26 (24) | 12.82 | 50 |
| B11 | 25.9 | 77 | 1392 | Ins | Ins | 1507 | 134.9 | 41 | 11.52 | 49 |
| B12 | 26.6 | 90 | 2096 | Ins | Ins | 4222 | 38.2 | 8 | 23.83 | 57 |
| CO2 | 9.6 (14.1) | 98 (81) | 1800 | 0.56 (5) | Ins (1.5) | 8760 (7008) | 106.9 | 30 (28) | 24.48 | 58 |
| CO2 | 11.4 | 90 | 1800 | 0.56 (5) | Ins (1.5) | 8760 (7808) | 138.1 | 42 (38) | 25.78 | 60 |
| CO2 | 13.5 (14.2) | 81 (73) | 1800 | 0.56 (5) | Ins (1.5) | 8760 (7808) | 152.9 | 46 (42) | 20.88 | 55 |
| CO3 * | 27.0 | 76 | 900 | 0.33 | Ins | 2891 | 120.9 | 35 (33) | 18.65 | 54 |
| CO3 * | 27.0 | 76 | 900 | 0.33 | Ins | 2891 | 204.4 | 57 (53) | 27.00 | 61 |
| CO3 * | 27.0 | 76 | 900 | 0.33 | Ins | 2891 | 132.6 | 40 (37) | 25.56 | 59 |
| EC1 | 26.1 | 71 | 936 | 4.20 | 1.5 | 4853 | 152.1 | 45 (41) | 1.08 | 5 |
| EC1 | 26.9 | 82 | 635 | 2.72 | 1.31 | 5790 | 176.3 | 52 (48) | 1.15 | 6 |
| EC1 * | 24.8 | 75 | 564 | 2.1 | 1.66 | 3101 | 201.2 | 56 (52) | 2.38 | 17 |
| EC2 | 12.9 | 66 | 554 | 1.0 | 0.4 | 3583 | 60.8 | 17 (16) | - | - |
| EC2 * | 13.2 | 71 | 598 | 1.35 | 1.14 | 4932 | 70.2 | 21 (20) | - | - |
| E1 | 12.0 | 69 | 652 | 1.18 (16.2) | 1.5 | 3364 | 158.3 (150.5) | 48 (44) | 3.02 | 20 |

Table 3. Cont.

| Designation | T, °C | RH, % | Rain, mm/a | [SO₂], µg/m³ | Cl⁻, mg/(m²·Day) | TOW, h/a | Steel | | Zinc | |
| | | | | | | | g/m² | No. | g/m² | No. |
|---|---|---|---|---|---|---|---|---|---|---|
| E1 * | 10.6 | 65 | 495 | 1.18 | 1.5 | 2374 | 175.5 | 51 (47) | 2.88 | 18 |
| E1 | 11.1 | 63 | 334 | 1.18 (16.2) | 1.5 | 2111 | 153.7 | 47 (43) | 2.09 | 16 |
| E4 | 18.1 | 65 | 554 | 8.3 | 1.5 | 3416 | 158.3 | 49 (45) | 1.94 | 14 |
| E4 | 17.0 | 63 | 521 | 5.7 | 1.5 | 2646 | 151.3 | 43 (39) | 1.51 | 8 |
| E4 | 17.2 | 62 | 374 | 1.9 | 1.5 | 2768 | 163.8 | 50 (46) | 1.94 | 13 |
| E5 | 16.3 | 59 | 416 | 10.3 | 1.5 | 1323 | 95.9 | 24 (23) | 1.01 | 4 |
| E5 | 15.0 (15.8) | 59 (58) | 258 (239) | 5.4 | 1.5 | 1104 | 53.0 | 16 (15) | 0.65 | 2 |
| E5 | 15.6 | 58 | 266 | 2.8 | 1.5 | 2400 | 49.9 | 15 (14) | 0.65 | 1 |
| E8 | 8.8 | 52 (72) | 738 | 9.1 | 1.8 | 876 | 25.7 | 3 (3) | 1.66 | 11 |
| E8 | 6.9 | 52 (72) | 624 | 8.9 | 1.6 | 876 | 28.1 | 4 (4) | 1.22 | 7 |
| E8 | 7.8 | 52 (72) | 681 | 9.0 | 1.7 | 876 | 37.4 | 7 (7) | 3.10 | 21 |
| M1 | 16.0 | 62 | 743 | 15.6 | 1.5 | 2523 (2321) | 120.1 | 34 (32) | 5.83 | 29 |
| M1 | 14.8 (15.2) | 66 (65) | 747 | 7.7 (5.6) | 1.5 | 2523 | 67.1 | 20 (19) | 5.98 | 31 |
| M1 | 15.4 | 64 (63) | 747 | 17.5 | 1.5 | 2523 (2427) | 39.8 | 9 (8) | 5.83 | 30 |
| M2 | 21.0 | 56 | 1352 | 6.7 | 1.5 | 1664 | 118.6 | 33 (31) | 8.35 | 42 |
| M2 | 21.0 | 56 | 1724 | 9.9 | Ins (1.5) | 1857 | 88.9 | 22 (21) | 14.33 | 52 |
| M2 | 21.0 | 56 | 1372 | 7.1 | Ins (1.5) | 1752 | 106.9 | 29 (27) | 6.84 | 33 |
| M3 | 18.0 | 51 | 374 | 31.1 | Ins | 1410 | 292.5 | 62 (58) | 10.01 | 46 |
| M3 * | 18.0 | 62 | 374 | 10.9 | Ins | 1410 | 205.9 | 58 (54) | 21.24 | 56 |
| M3 * | 18.0 | 60 | 374 | 14.6 | Ins | 2646 | 229.3 | 60 (56) | 7.06 | 35 |
| PE4 | 16.4 | 37 | 17 | Ins (5) | Ins (1.5) | 26 | 117.0 | 32 (30) | 1.66 | 12 |
| PE4 | 17.2 | 33 | 34 (89) | Ins (5) | Ins (1.5) | 175 (26) | 128.7 | 39 (36) | 1.58 | 9 |
| PE5 | 12.2 | 67 | 632 | Ins (0) | Ins (0) | 2847 | 7.8 | 1 (1) | 3.89 | 23 |
| PE5 | 12.2 | 67 | 672 (792) | Ins (0) | Ins (0) | 2689 (2847) | 13.3 | 2 (2) | 2.88 | 19 |
| PE6 | 25.4 | 84 | 1523 | Ins (5) | Ins (1.5) | 5037 (4580) | 122.5 | 37 (35) | 7.06 | 36 |
| PE6 | 25.8 | 83 | 1158 (1656) | Ins (5) | Ins (1.5) | 5790 (4380) | 100.6 | 27 (25) | 7.49 | 38 |
| U1 | 16.8 | 74 | 1182 | 0.6 (1) | 1.8 (2.2) | 5133 | 64.0 | 19 (18) | 4.03 | 24 |
| U1 * | 16.6 | 73 | 1324 | 0.8 | 1.2 | 4976 | 62.4 | 18 (17) | 3.74 | 22 |
| U1 * | 16.7 | 76 | 1306 | Ins | Ins | 4792 | 47.6 | 14 (13) | 4.10 | 25 |
| U3 * | 17.7 | 79 | 1490 | Ins | Ins | 5764 | 94.4 | 23 (22) | 4.39 | 27 |
| CH1 | 14.2 | 71 | 355 | 20 | 2.18 | 3469 | 221.5 | 63 | 12.89 | 51 |

* the test locations not used in [20]; ** the values reported in [20] are shown in parentheses.

**Table 4.** Atmosphere corrosivity parameters of test locations and first-year corrosion losses of carbon steel and zinc ($K_1$, g/m$^2$) in Russian Federation test locations and their numbers in the order of increasing $K_1$.

| Test Location | T, °C | RH, % | Prec, mm/a | [SO$_2$], μg/m$^3$ | Steel | | Zinc | |
|---|---|---|---|---|---|---|---|---|
| | | | | | g/m$^2$ | No. | g/m$^2$ | No. |
| Bilibino | −12.2 | 80 | 218 | 3 | 5.4 | 1 | 1.64 | 1 |
| Oimyakon | −16.6 | 71 | 175 | 3 | 8.1 | 2 | 1.81 | 3 |
| Ust-Omchug | −11 | 70 | 317 | 5 | 12.4 | 3 | 2.91 | 5 |
| Atka | −12 | 72 | 376 | 3 | 15.2 | 4 | 1.69 | 2 |
| Susuman | −13.2 | 71 | 283 | 10 | 17.0 | 5 | 3.07 | 6 |
| Tynda | −6.5 | 72 | 525 | 5 | 21.2 | 6 | 5.30 | 10 |
| Klyuchi | 1.4 | 69 | 253 | 3 | 23.4 | 7 | 2.03 | 4 |
| Aldan | −6.2 | 72 | 546 | 5 | 24.6 | 8 | 5.47 | 11 |
| Pobedino | −0.9 | 77 | 604 | 3 | 36.5 | 9 | 4.30 | 7 |
| Yakovlevka | 2.5 | 70 | 626 | 3 | 40.6 | 10 | 4.64 | 9 |
| Pogranichnyi | 3.6 | 67 | 595 | 3 | 49.0 | 11 | 4.32 | 8 |
| Komsomolsk-on-Amur | −0.7 | 76 | 499 | 10 | 63.2 | 12 | 6.35 | 12 |

**Table 5.** Atmosphere corrosivity parameters and first-year corrosion losses of carbon steel in test locations. Adapted from [19], with permission from © 1999 Elsevier.

| [SO$_2$], μg/m$^3$ | Cl$^-$, mg/(m$^2$·Day) | $K_1$, g/m$^2$ |
|---|---|---|
| 3 | 2 | 137.7 |
| 5 | 0,3 | 46.1 |
| 5 | 0,7 | 130.7 |
| 8 | 1 | 137.7 |
| 8 | 0 | 140.0 |
| 14 | 2 | 193.8 |
| 15 | 2 | 228.4 |
| 15 | 1 | 236.1 |
| 17 | 0,16 | 136.1 |
| 26 | 1 | 236.1 |
| 32 | 2 | 276.1 |
| 116 | 0,62 | 232.2 |

## 2.2. Predictions of First-Year Corrosion Losses

To predict $K_1$ for steel and zinc, we used the new DRFs presented in this paper (hereinafter referred to as "New DRFs"), in the standard [18] (hereinafter referred to as "Standard DRFs"), in [13] (hereinafter referred to as "Unified DRFs"), and the linear model [20] (hereinafter referred to as "Linear DRF").

The Standard DRFs are intended for the prediction of $K_1$ ($r_{corr}$ in the original) in SO$_2$- and Cl$^-$-containing atmospheres in all climatic regions of the world. The $K_1$ values are calculated in μm.

For carbon steel, Equation (3):

$$K_1 = 1.77 \times P_d^{0.52} \times \exp(0.020 \times RH + f_{St}) + 0.102 \times S_d^{0.62} \times \exp(0.033 \times RH + 0.040 \times T), \quad (3)$$

where $f_{St} = 0.150 \cdot (T - 10)$ at $T \leq 10\,°C$; $f_{St} = -0.054 \cdot (T - 10)$ at $T > 10\,°C$.

For zinc, Equation (4):

$$K_1 = 0.0129 \times P_d^{0.44} \times \exp(0.046 \times RH + f_{Zn}) + 0.0175 \times S_d^{0.57} \times \exp(0.008 \times RH + 0.085 \times T), \quad (4)$$

where $f_{Zn} = 0.038 \times (T - 10)$ at $T \leq 10\,°C$; $f_{Zn} = -0.071 \times (T - 10)$ at $T > 10\,°C$, where $T$ is the temperature (°C) and $RH$ (%) is the relative humidity of air; $P_d$ and $S_d$ are SO$_2$ and Cl$^-$ deposition rates expressed in mg/(m$^2$day), respectively.

In Equations (3) and (4), the contributions to corrosion due to $SO_2$ and $Cl^-$ are presented as separate components; therefore, only their first components were used for continental territories.

Unified DRFs are intended for long-term prediction of mass losses $K$ (designated as *ML* in the original) in $SO_2$-containing atmospheres in all climatic regions of the Earth. It is stated that the calculation is given in $g/m^2$.

For carbon steel, Equation (5):

$$K = 3.54 \times [SO_2]^{0.13} \times \exp\{0.020 \times RH + 0.059 \times (T\text{-}10)\} \times \tau^{0.33} \ T \leq 10\ °C;$$
$$K = 3.54 \times [SO_2]^{0.13} \times \exp\{0.020 \times RH - 0.036 \times (T\text{-}10)\} \times \tau^{0.33} \ T > 10\ °C. \tag{5}$$

For zinc, Equation (6):

$$K = 1.35 \times [SO_2]^{0.22} \times \exp\{0.018 \times RH + 0.062 \times (T\text{-}10)\} \times \tau^{0.85} + 0.029 \times Rain[H^+] \times \tau\ T \leq 10\ °C;$$
$$K = 1.35 \times [SO_2]^{0.22} \times \exp\{0.018 \times RH - 0.021 \times (T\text{-}10)\} \times \tau^{0.85} + 0.029 \times Rain[H^+] \times \tau\ T > 10\ °C. \tag{6}$$

where $T$ is the temperature (°C) and $RH$ (%) is the relative humidity of air; $[SO_2]$ is the concentration of $SO_2$ in $\mu g/m^3$; "Rain" is the rainfall amount in mm/year; $[H^+]$ is the acidity of the precipitation; and $\tau$ is the exposure time in years.

To predict the first-year corrosion losses, $\tau = 1$ was assumed.

The standard DRFs and Unified DRFs were developed on the basis of the results obtained in the UN/ECE program and MICAT project using the same atmosphere corrosivity parameters (except from Rain[H$^+$]). If $\tau = 1$, the models have the same mathematical form and only differ in the coefficients. Both are intended for $K_1$ predictions in any regions of the world, hence it is particularly interesting to compare the results of $K_1$ predictions with actual data.

The linear model was developed for $SO_2$- and $Cl^-$-containing atmospheres. It is based on the experimental data from the MICAT project only and relies on an artificial neural network. It is of special interest since it has quite a different mathematical form and uses different parameters. In the MICAT project, the air temperature at the test sites is mainly above 10 °C (Table 3). Nevertheless, we used this model, like the other DRFs, also for test locations with any temperatures.

The first-year corrosion losses of carbon steel (designated as "Fe" in the original) are expressed as Equation (7):

$$K_1 = b_0 + Cl^- \times (b_1 + b_2 \times P + b_3 \times RH) + b_4 \times TOW \times [SO_2], \tag{7}$$

where $b_0 = 6.8124$, $b_1 = -1.6907$, $b_2 = 0.0004$, $b_3 = 0.0242$, and $b_4 = 2.2817$; $K_1$ is the first-year corrosion loss in $\mu m$; $Cl^-$ is the chloride deposition rate in $mg/(m^2 \cdot day)$; P is the amount of precipitation in mm/year; $RH$ is the air relative humidity in %; $TOW$ is the wetting duration expressed as the fraction of a year; and $[SO_2]$ is the $SO_2$ concentration in $\mu g/m^3$. The prediction results for the first year are expressed in $\mu m$.

To predict $K_1$ in continental regions, only the component responsible for the contribution to corrosion due to $SO_2$ was used.

The $K_1$ values in $\mu m$ were converted to $g/m^2$ using the specific densities of steel and zinc, 7.8 and 7.2 $g/cm^3$, respectively. Furthermore, the relationship $P_{d,p}$ $mg/(m^2 \cdot day) = 0.67\ P_{d,c}$ $\mu g/m^3$ was used, where $P_{d,p}$ is the $SO_2$ deposition rate and $P_{d,c}$ is the $SO_2$ concentration [18].

The calculation of $K_1$ is given for continental test locations at background $Cl^-$ deposition rates $\leq 2$ $mg/(m^2 \cdot day)$ under UN/ECE and RF programs and MICAT project. The $R^2$ values characterizing the prediction results as a whole for numerous test locations are not reported here. The $K_1$ predictions obtained were compared to the experimental values of $K_1$ for each test location, which provides a clear idea about the specific features of the DRFs.

## 3. Results

### 3.1. DRF Development

Corrosion of metals in continental regions depends considerably on the content of sulfur dioxide in the air. Therefore, development of a DRF primarily requires that this dependence, i.e., the mathematical relationship $K = f(SO_2)$, be found. The dependences reported in graphical form in [20,27] differ from each other. The relationship is non-linear, therefore the decision should be made on which background $SO_2$ concentration should be selected, since the calculated $K_1$ values would be smaller than the experimental ones at $[SO_2] < 1$ if non-linear functions are used. $[SO_2]$ values $< 1$ can only be used in linear functions. The background values in Tables 2–4 are presented as "Ins." (Insignificant), $\leq 1$, 3, 5 $\mu g/m^3$, which indicates that there is no common technique in the determination of background concentrations. For $SO_2$ concentrations of "Ins." or $\leq 1\ \mu g/m^3$, we used the value of 1 $\mu g/m^3$, whereas the remaining $SO_2$ concentrations were taken from the tables.

In finding the $K = f(SO_2)$ relationship, we used the actual test results of all first-year exposures under each program rather than the mean values, because non-linear functions are also used.

The $K = f(SO_2)$ relationships obtained for each program are shown in Figure 1 for steel and in Figure 2 for zinc. In a first approximation, this relationship can be described by the following function for experimental $K_1$ values obtained in a broad range of meteorological atmosphere parameters:

$$K_1 = K_1^\circ \times [SO_2]^\alpha, \tag{8}$$

where $K_1^\circ$ are the average corrosion losses over the first year $(g/m^2)$ in a clean atmosphere for the entire range of $T$ and $RH$ values; and $\alpha$ is the exponent that depends on the metal.

The $K_1^\circ$ values corresponding to the mean values of the parameter range of climatic conditions in clean atmospheres were found to be the same for the experimental data of all programs, namely, 63 and 4 $g/m^2$, while $\alpha = 0.47$ and 0.28 for carbon steel and zinc, respectively. A similar $K_1^\circ$ value for carbon steel was also obtained from the Linear DRF, Equation (6). In fact, at background $SO_2$ concentrations = 1 $\mu g/m^3$ in PE4 test location (Table 3) at TOW = 26 h/year (0.002 of the year), the calculated $K_1^\circ$ is to 53 $g/m^2$, while for CO2 test location at TOW = 8760 h/year (entire year) it is 71 $g/m^2$; the mean value is 62 $g/m^2$.

Based on Equation (8), it may be accepted in a first approximation that the effect of $[SO_2]$ on corrosion is the same under any climatic conditions and this can be expressed in a DRF by an $[SO_2]^\alpha$ multiplier, where $\alpha = 0.47$ or $\alpha = 0.28$ for steel or zinc, respectively. The $K_1^\circ$ values in Equation (8) depend on the climatic conditions and are determined for each test location based on the atmosphere meteorological parameters.

In the development of New DRF, the $K_1$ values were determined using the DRF mathematical formula presented in the Standard DRF and in the Unified DRF, as well as meteorological parameters $T$, $RH$, and $Prec$ (*Rain* for warm climate locations or *Prec* for cold climate locations). The complex effect of $T$ was taken into account: corrosion losses increase with an increase in $T$ to a certain limit, $T_{lim}$; its further increase slows down the corrosion due to radiation heating of the surface of the material and accelerated evaporation of the adsorbed moisture film [12,28]. It has been shown [29] that $T_{lim}$ is within the range of 9–11 °C. Similarly to Equations (3)–(6), it is accepted that $T_{lim}$ equals 10 °C. The need to introduce $Prec$ is due to the fact that in northern RF regions, the $K_1$ values are low at high $RH$, apparently owing not only to low $T$ values but also to the small amount of precipitation, including solid precipitations. The values of the coefficients reflecting the effect of $T$, $RH$ and $Prec$ on corrosion were determined by regression analysis.

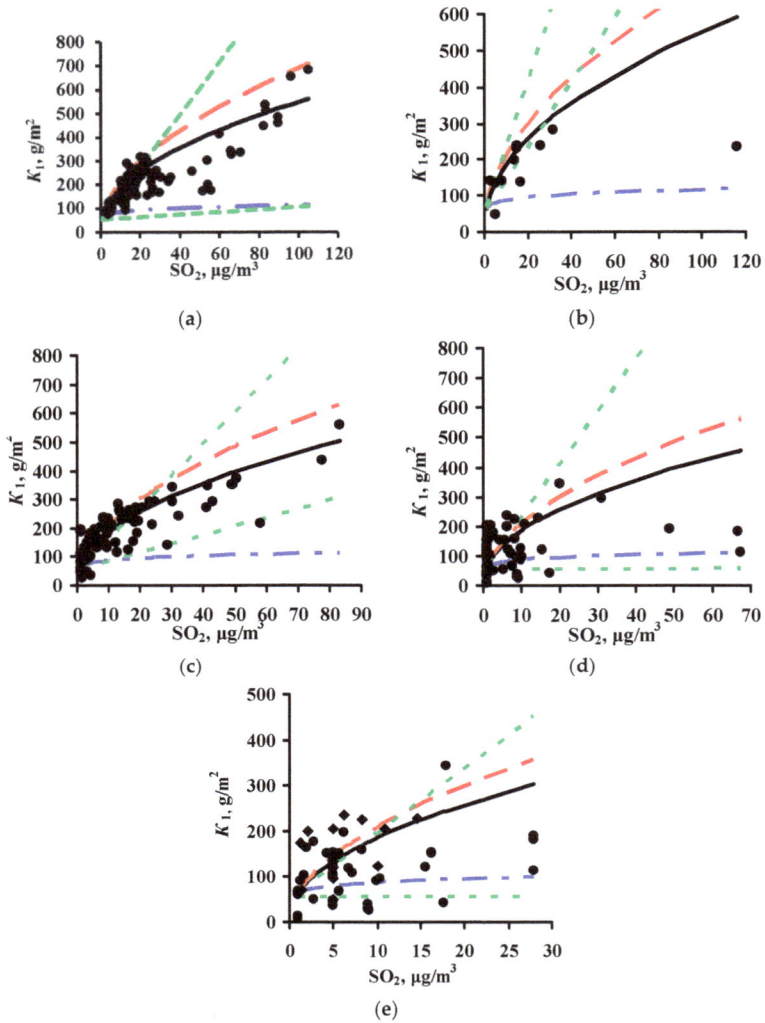

**Figure 1.** Dependence of first-year corrosion losses of steel ($K_1$) on $SO_2$ concentration based on data from ISO CORRAG program (**a**), Ref. [19] (**b**), UN/ECE program (**c**), MICAT project (**d**), and data from MICAT project cited in [20] (**e**). ▬▬$\alpha$ = 0.47 (New DRF), ▬▬$\alpha$ = 0.52 (Standard DRF), ▬•▬$\alpha$ = 0.13 (Unified DRF), ▬▬model [20] for TOW ranges in accordance with the data in Tables 2–5.

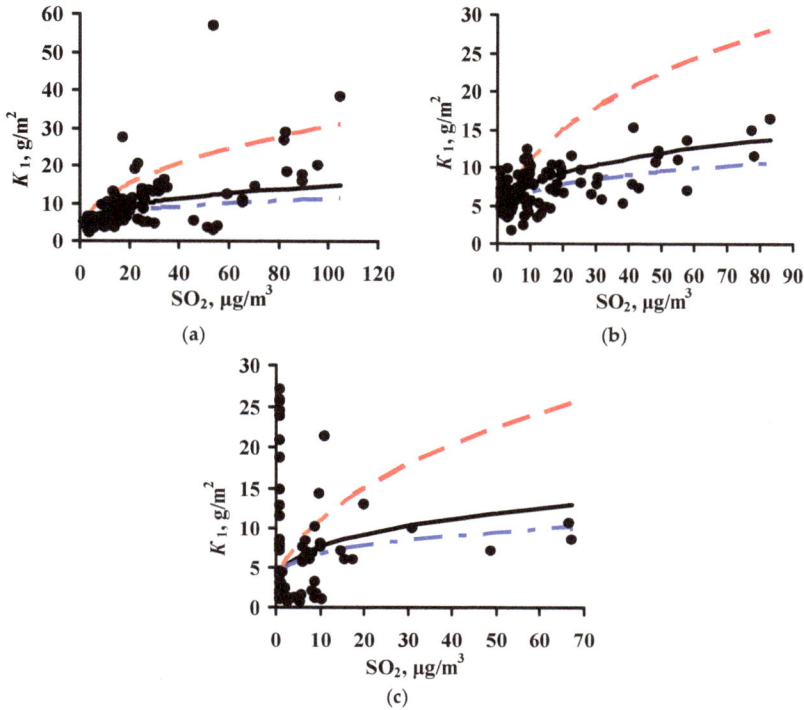

**Figure 2.** Dependence of first-year corrosion losses of zinc ($K_1$) on $SO_2$ concentration based on the data from ISO CORRAG program (**a**), UN/ECE program (**b**), and MICAT project (**c**). ------$\alpha$ = 0.28 (New DRF), -----$\alpha$ = 0.44 (Standard DRF), -•-•-$\alpha$ = 0.22 (Unified DRF).

The New DRFs developed for the prediction of $K_1$ (g/m$^2$) for the two temperature ranges have the following forms:
for carbon steel:

$$K_1 = 7.7 \times [SO_2]^{0.47} \times \exp\{0.024 \times RH + 0.095 \times (T\text{-}10) + 0.00056 \times Prec\}\ T \leq 10\,^\circ C;$$
$$K_1 = 7.7 \times [SO_2]^{0.47} \times \exp\{0.024 \times RH - 0.095 \times (T\text{-}10) + 0.00056 \times Prec\}\ T > 10\,^\circ C, \quad (9)$$

and for zinc:

$$K_1 = 0.71 \times [SO_2]^{0.28} \times \exp\{0.022 \times RH + 0.045 \times (T\text{-}10) + 0.0001 \times Prec\}\ T \leq 10\,^\circ C;$$
$$K_1 = 0.71 \times [SO_2]^{0.28} \times \exp\{0.022 \times RH - 0.085 \times (T\text{-}10) + 0.0001 \times Prec\}\ T > 10\,^\circ C. \quad (10)$$

### 3.2. Predictions of $K_1$ Using Various DRFs for Carbon Steel

Predictions of $K_1$ were performed for all continental test locations with chloride deposition rates $\leq 2$ mg/(m$^2$·day). The results of $K_1$ prediction ($K_1^{pr}$) from Equations (3)–(7), (9), and (10) are presented separately for each test program. To build the plots, the test locations were arranged by increasing experimental $K_1$ values ($K_1^{exp}$). Their sequence numbers are given in Tables 2–4. The increase in $K_1$ is caused by an increase in atmosphere corrosivity due to meteorological parameters and $SO_2$ concentration. All the plots are drawn on the same scale. All plots show the lines of prediction errors $\delta = \pm 30\%$ (the 1.3 $K_1^{exp}$–0.7 $K_1^{exp}$ range). This provides a visual idea of the comparability of $K_1^{pr}$ with $K_1^{exp}$ for each DRF. The scope of this paper does not include an estimation of the discrepancy between the $K_1{}^{pr}$ values obtained using various DRFs with the $K_1{}^{exp}$ values obtained for each test location under the UN/ECE and RF programs. The scatter of points is inevitable. It results from the imperfection of each DRF and the inaccuracy of experimental data on meteorological parameters, $SO_2$ content, and $K_1{}^{exp}$ values. Let us just note the general regularities of the results on $K_1{}^{pr}$ for each DRF.

The results on $K_1{}^{pr}$ for carbon steel for the UN/ECE program, MICAT project, and RF program are presented in Figures 3–5, respectively. It should be noted that according to the Unified DRF (Equation (5)), the $K_1{}^{pr}$ of carbon steel in RF territory [30] had low values. It was also found that the $K_1{}^{pr}$ values are very low for the programs mentioned above. Apparently, the $K_1{}^{pr}$ values (Equation (5)) were calculated in μm rather than in $g/m^2$, as the authors assumed. To convert $K_1{}^{pr}$ in μm to $K_1{}^{pr}$ in $g/m^2$, the 3.54 coefficient in Equation (6) was increased 7.8-fold.

In the UN/ECE program, the $K_1{}^{pr}$ values match $K_1{}^{exp}$ to various degrees; some $K_1{}^{pr}$ values exceed the error $\delta$ (Figure 3). Let us describe in general the locations in which $K_1{}^{pr}$ values exceed $\delta$. For the New DRFs (Figure 3a) there are a number of locations with overestimated $K_1{}^{pr}$ and with underestimated $K_1{}^{pr}$ values at different atmosphere corrosivities. For the Standard DRF (Figure 3b) and Linear DRF (Figure 3d), locations with underestimated $K_1{}^{pr}$ values prevail, also at different $K_1{}^{exp}$. For the Unified DRF (Figure 3c), $K_1{}^{pr}$ are overestimated for locations with small $K_1{}^{exp}$ and underestimated for locations with high $K_1{}^{exp}$. The possible reasons for such regular differences for $K_1{}^{pr}$ from $K_1{}^{exp}$ will be given based on an analysis of the coefficients in the DRFs.

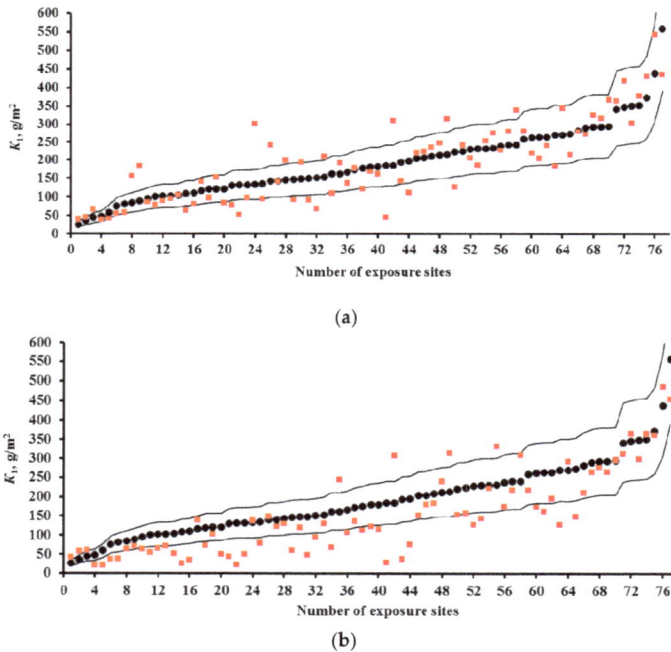

(a)

(b)

**Figure 3.** *Cont.*

(c)

(d)

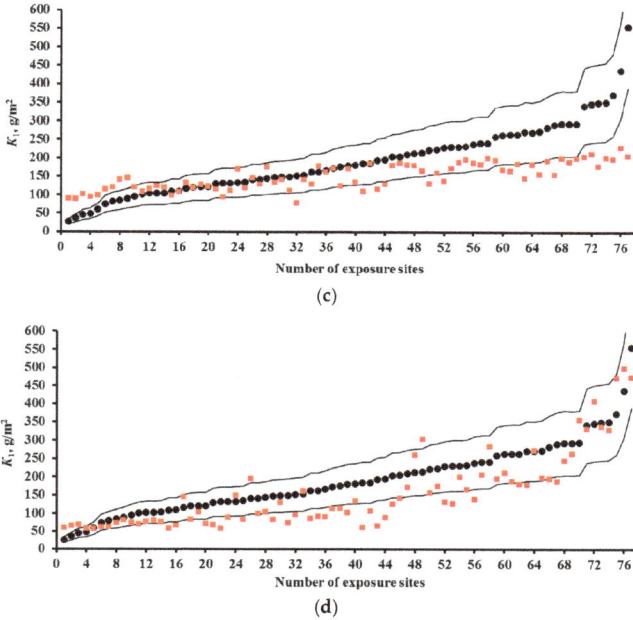

**Figure 3.** Carbon steel. UN/ECE program. $K_1$ predictions by the New DRF (**a**); Standard DRF (**b**); Unified DRF (**c**); and Linear DRF [20] (**d**). ●—experimental $K_1$ data; ■—$K_1$ predictions. Thin lines show the calculation error (± 30%). The numbers of the exposure sites are given in accordance with Table 2.

For the MICAT project, $K_1^{pr}$ considerably exceeds $\delta$ for all DRFs in many locations (Figure 4). Overestimated and considerably overestimated $K_1^{pr}$ values are mainly observed in locations with small $K_1^{exp}$, while underestimated $K_1^{pr}$ values are mainly observed for locations with high $K_1^{exp}$. Furthermore, for the Linear DRF (Figure 4d), particularly overestimated values are observed in location B6 (No. 31, No. 53, and No. 54) at all exposures. This test location should be noted. The corrosivity parameters under this program reported in [20] are different for some test locations (Table 3). In fact, for B6, the [$SO_2$] value for all exposures is reported to be 28 µg/m$^3$ instead of 67.2; 66.8 and 48.8 µg/m$^3$. Figure 4e presents $K_1^{pr}$ for the Linear DRF with consideration for the parameter values reported in [20]. Naturally, $K_1^{pr}$ for B6 decreased considerably in comparison with the values in Figure 4d but remained rather overestimated with respect to $K_1^{exp}$.

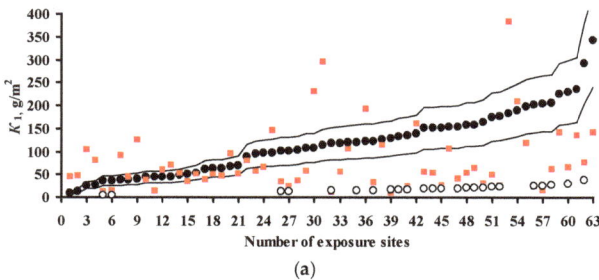

(a)

**Figure 4.** *Cont.*

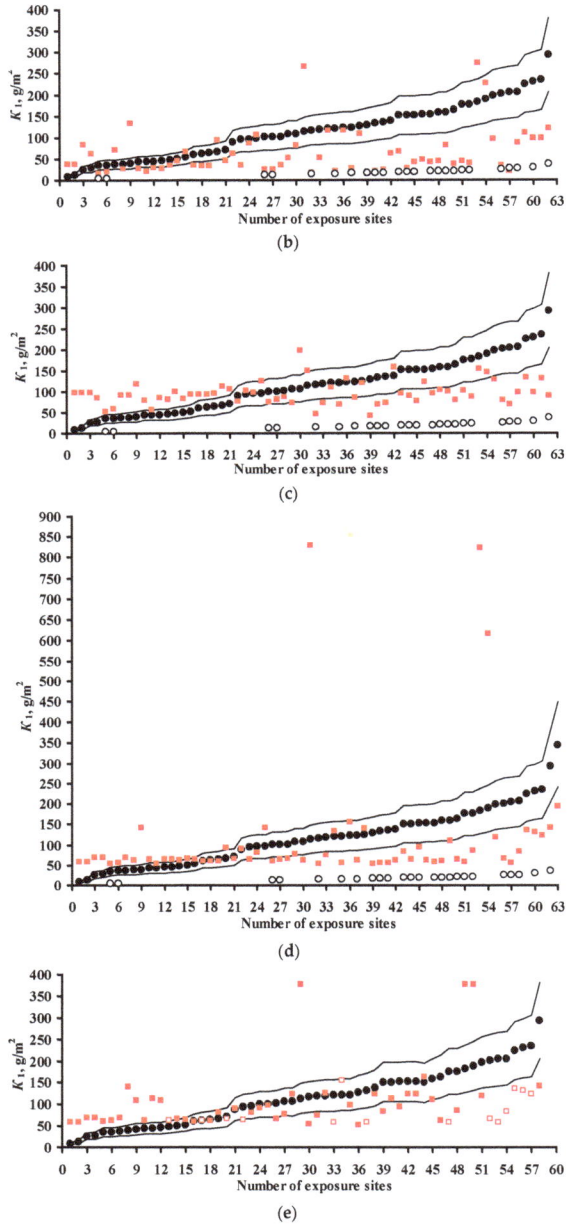

**Figure 4.** Carbon steel. MICAT program. $K_1$ predictions by the New DRF (**a**); Standard DRF (**b**); Unified DRF (**c**); linear model [20] (**d**); and linear model based on data from [20] (**e**). ●—experimental $K_1$ data; ■—$K_1$ predictions; □—the test locations in [20] which were not used (only for Figure 4e); ○—experimental $K_1$ data under the assumption that they were expressed in $g/m^2$ rather than in μm. Thin lines show the calculation error ($\pm$30%). The numbers of the exposure sites are given in accordance with Table 3.

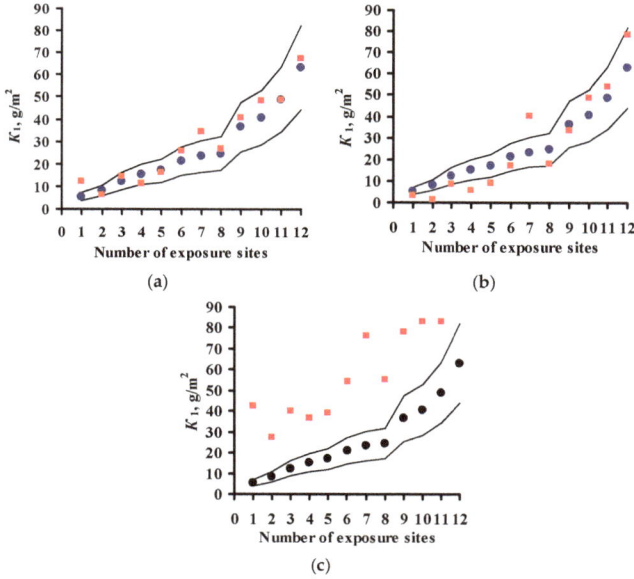

**Figure 5.** Carbon steel. RF program. $K_1$ predictions by the New DRF (**a**); Standard DRF (**b**); and Unified DRF (**c**). ●—experimental $K_1$ data; ■—$K_1$ predictions. Thin lines show the calculation error ($\pm 30\%$). The numbers of the exposure sites are given in accordance with Table 4.

If all DRFs give underestimated $K_1^{pr}$ values for the same locations, this may result from an inaccuracy of experimental data, i.e., corrosivity parameters and/or $K_1^{exp}$ values. We did not perform any preliminary screening of the test locations. Therefore, it is reasonable to estimate the reliability of $K_1^{exp}$ only in certain locations by comparing them with other locations. Starting from No. 26, $K_1^{pr}$ values are mostly either smaller or considerably smaller than $K_1^{exp}$. The locations with underestimated $K_1^{pr}$ that are common to all DRFs include: A4 (No. 5, No. 6), B1 (No. 28), B10 (No. 26), B11 (No. 41), E1 (No. 47, No. 48, No. 51), E4 (No. 43, 49, 50), EC1 (No. 45, No. 52, No. 56), CO3 (No. 40, 57), PE4 (No. 32, No. 39), M3 (No. 58, No. 60, No. 62). To perform the analysis, Table 6 was composed. It contains the test locations that, according to our estimates, have either questionable or reliable $K_1^{exp}$ values. It clearly demonstrates the unreliability of $K_1^{exp}$ in some test locations. For example, in the test locations PE4 and A4, with $RH = 33\%$–$51\%$ and $TOW = 0.003$–$0.114$ of the year at background $[SO_2]$, $K_1^{exp}$ are 4.5–16.5 µm (35.1–117 g/m$^2$), while under more corrosive conditions in E8 and M2 with $RH = 52\%$–$56\%$ and $TOW = 0.100$–$0.200$ of the year and $[SO_2] = 6.7$–$9.9$ µg/m$^3$, $K_1^{exp}$ values are also 3.3–15.2 µm (25.7–118.6 g/m$^2$). The impossibility of high $K_1$ values in PE4 and A4 is also confirmed by the 3D graph of the dependence of $K$ on $SO_2$ and $TOW$ in [20]. Alternatively, for example, in B1, CO3 and B11 with $RH = 75\%$–$77\%$ and $TOW = 0.172$–$0.484$ of the year and $[SO_2] = 1$–$1.7$ µg/m$^3$, $K_1^{exp} = 13.1$–$26.2$ µm (102.2–204.4 g/m$^2$), whereas in A2 and A3 with $RH = 72\%$–$76\%$ and $TOW = 0.482$–$0.665$ of the year and $[SO_2] = 1$–$10$ µg/m$^3$, $K_1^{exp}$ is as small as 5.6–16.1 µm (43.7–125.6 g/m$^2$). The $K_1$ values reported for locations with uncertain data are 2–4 times higher than the $K_1$ values in trusted locations. The reason for potentially overestimated $K_1^{exp}$ values being obtained is unknown. It may be due to non-standard sample treatment or to corrosion-related erosion. It can also be assumed that the researchers (performers) reported $K_1$ in g/m$^2$ rather than in µm. If this assumption is correct, then $K_1^{pr}$ values would better match $K_1^{exp}$ (Figure 4). Unfortunately, we cannot compare the questionable $K_1^{exp}$ values with the $K_1^{exp}$ values rejected in the study where an artificial neural network was used [20]. We believe that, of the $K_1^{exp}$ values listed, only the data for the test locations up to No. 26 in Figure 4 can be deemed reliable.

*Materials* **2017**, *10*, 422

**Table 6.** Atmosphere corrosivity parameters and first year corrosion losses of carbon steel in certain test locations under the MICAT project.

Locations with Uncertain Data

| Designation | No. | T, °C | RH, % | TOW, 1/a | Prec, mm/a | [SO₂], µg/m³ | $K_1^{exp}$ µm | $K_1^{exp}$ g/m² |
|---|---|---|---|---|---|---|---|---|
| PE4 | 32 | 16.4 | 37 | 0.003 | 17 | 1 | 15.0 | 117.0 |
| PE4 | 39 | 17.2 | 33 | 0.020 | 34 | 1 | 16.5 | 128.7 |
| A4 | 5 | 20.0 | 49 | 0.097 | 111 | 1 | 4.5 | 35.1 |
| A4 | 6 | 18.0 | 51 | 0.114 | 35 | 1 | 4.6 | 35.9 |
| M3 | 58 | 18.0 | 62 | 0.161 | 374 | 10.9 | 26.4 | 205.9 |
| M3 | 62 | 18.0 | 51 | 0.161 | 374 | 31.1 | 37.5 | 292.5 |
| M3 | 60 | 18.0 | 60 | 0.302 | 374 | 14.6 | 29.4 | 229.3 |
| E1 | 47 | 11.1 | 63 | 0.241 | 334 | 1.18 | 19.7 | 153.7 |
| E1 | 48 | 12.0 | 69 | 0.384 | 652 | 1.18 | 20.3 | 158.3 |
| E1 | 51 | 10.6 | 65 | 0.271 | 495 | 1.18 | 22.5 | 175.5 |
| E4 | 43 | 17.0 | 63 | 0.302 | 521 | 5.7 | 19.4 | 151.3 |
| E4 | 49 | 18.1 | 65 | 0.390 | 554 | 8.3 | 20.3 | 158.3 |
| E4 | 50 | 17.2 | 62 | 0.316 | 374 | 1.9 | 21.0 | 163.8 |
| B10 | 26 | 20.4 | 69 | 0.442 | 1440 | 1 | 12.9 | 100.6 |
| B1 | 28 | 21.2 | 75 | 0.484 | 996 | 1.67 | 13.1 | 102.2 |
| CO3 | 40 | 27.0 | 76 | 0.330 | 900 | 1 | 17.0 | 132.6 |
| CO3 | 57 | 27.0 | 76 | 0.330 | 900 | 1 | 26.2 | 204.4 |
| B11 | 41 | 25.9 | 77 | 0.172 | 1392 | 1 | 17.3 | 134.9 |
| EC1 | 56 | 24.8 | 75 | 0.354 | 564 | 2.1 | 25.8 | 201.2 |
| EC1 | 52 | 26.9 | 82 | 0.661 | 635 | 2.72 | 22.6 | 176.3 |

Locations with Trusted Data

| Designation | No. | T, °C | RH, % | TOW, 1/a | Prec, mm/a | [SO₂], µg/m³ | $K_1^{exp}$ µm | $K_1^{exp}$ g/m² |
|---|---|---|---|---|---|---|---|---|
| E8 | 3 | 8.8 | 52 | 0.100 | 738 | 9.1 | 3.3 | 25.7 |
| E8 | 4 | 6.9 | 52 | 0.100 | 624 | 8.9 | 3.6 | 28.1 |
| E8 | 7 | 7.8 | 52 | 0.100 | 681 | 9 | 4.8 | 37.4 |
| M2 | 29 | 21.0 | 56 | 0.200 | 1372 | 7.1 | 13.7 | 106.9 |
| M2 | 33 | 21.0 | 56 | 0.190 | 1352 | 6.7 | 15.2 | 118.6 |
| M2 | 22 | 21.0 | 56 | 0.212 | 1724 | 9.9 | 11.4 | 88.9 |
| E5 | 15 | 15.6 | 58 | 0.161 | 266 | 2.8 | 6.4 | 49.9 |
| E5 | 16 | 15.0 | 59 | 0.126 | 258 | 5.4 | 6.8 | 53.0 |
| M1 | 34 | 16.0 | 62 | 0.288 | 743 | 15.6 | 15.4 | 120.1 |
| M1 | 9 | 15.4 | 64 | 0.288 | 743 | 17.5 | 5.1 | 39.8 |
| M1 | 20 | 14.8 | 66 | 0.288 | 743 | 7.7 | 8.6 | 67.1 |
| A2 | 38 | 17.1 | 72 | 0.482 | 983 | 10.0 | 16.1 | 125.6 |
| A2 | 36 | 16.7 | 75 | 0.578 | 1729 | 10.0 | 15.7 | 122.5 |
| A2 | 25 | 17.0 | 74 | 0.555 | 1420 | 9 | 12.4 | 96.7 |
| A3 | 12 | 20.6 | 76 | 0.665 | 2158 | 1 | 5.7 | 44.5 |
| A3 | 13 | 20.9 | 74 | 0.631 | 2624 | 1 | 5.8 | 45.2 |
| A3 | 10 | 22.1 | 75 | 0.633 | 1720 | 1 | 5.6 | 43.7 |
| - | - | - | - | - | - | - | - | - |
| - | - | - | - | - | - | - | - | - |
| - | - | - | - | - | - | - | - | - |

For the RF program, the $K_1{}^{pr}$ values determined by the New DRF and the Standard DRF are pretty comparable with $K_1{}^{exp}$, but they are considerably higher for the Unified DRF (Figure 5).

The presented figures indicate that all DRFs which have the same parameters but different coefficients predict $K_1$ for same test locations with different degrees of reliability. That is, combinations of various coefficients in DRFs make it possible to obtain $K_1{}^{pr}$ results presented in Figures 3–5. In view of this, the analysis of DRFs in order to explain the principal differences of $K_1{}^{pr}$ from $K_1{}^{exp}$ for each DRF appears interesting.

### 3.3. Analysis of DRFs for Carbon Steel

The DRFs were analyzed by comparison of the coefficients in Equations (3), (5) and (9). Nonlinear DRFs can be represented in the form:

$$K_1 = A \times [SO_2]^{\alpha} \exp\{k_1 \times RH + k_2 \times (T-10) + k_3 \times Prec\}$$

or

$$K_1 = A \times [SO_2]^{\alpha} \times e^{k1 \cdot RH} \times e^{k2 \cdot (T-10)} \times e^{k3 \cdot Prec},$$

where $A \times e^{k1 \cdot RH} \times e^{k2 \cdot (T-10)} \times e^{k3 \cdot Prec} = K_{10}$.

The values of the coefficients used in Equations (3), (5) and (9) are presented in Table 7.

Table 7. Values of coefficients used in the nonlinear DRFs for carbon steel.

| DRF | A | | $\alpha$ | $k_1$ | $k_2$ | | $k_3$ |
| --- | --- | --- | --- | --- | --- | --- | --- |
| | µm | g/m$^2$ | | | $T \leq 10$ | $T > 10$ | |
| New | 0.99 | 7.7 | 0.47 | 0.024 | 0.095 | −0.095 | 0.00056 |
| Standard | 1.77 | 13.8 | 0.52 | 0.020 | 0.150 | −0.054 | - |
| Unified | 3.54 | 27.6 | 0.13 | 0.020 | 0.059 | −0.036 | - |

To compare the $\alpha$ values, $K_1{}^{\circ} = 63$ g/m$^2$ at $[SO_2] = 1$ µg/m$^3$ was used in Equation (8) for all DRFs. The $[SO_2]^{\alpha}$ plots for all the DRFs for all programs are presented in Figure 1. For the New DRF, the line $K = f(SO_2)$ was drawn approximately through the mean experimental points from all the test programs. Therefore, one should expect a uniform distribution of error $\delta$, e.g., in Figure 3a. For the Standard DRF, $\alpha = 0.52$ is somewhat overestimated, which may result in more overestimated $K_1$ values at high $[SO_2]$. However, in Figure 3b for CS1 (No. 76), CS3 (No. 73, 74, 77) and GER10 (No. 76), $K_1{}^{pr}$ overestimation is not observed, apparently due to effects from other coefficients in DRF. For Unified DRF $\alpha = 0.13$, which corresponds to a small range of changes in $K_1$ as a function of $SO_2$. Therefore, in Figures 3c and 4c, the $K_1{}^{pr}$ present a nearly horizontal band that is raised to the middle of the $K_1{}^{exp}$ range due to a higher value of $A = 3.54$ µm (27.6 g/m$^2$), Table 7. As a result, the Unified DRF cannot give low $K_1{}^{pr}$ values for rural atmospheres, Figures 3c and 5c, or high $K_1{}^{pr}$ values for industrial atmospheres, Figure 3c.

For the Linear DRF we present $K_1{}^{pr}$—$[SO_2]$ plots for *TOW* (fraction of a year) within the observed values: 0.043–0.876 for ISO CORRAG program; 0.5–1 based on the data in [19]; 0.17–0.62 from UN/ECE program; 0.003–1 from the MICAT project, and 0.002–0.8 based on the data [20] for the MICAT project, Figure 1. One can see that reliable $K_1{}^{pr}$ are possible in a limited range of *TOW* and $[SO_2]$. The $K_1{}^{pr}$ values are strongly overestimated at high values of these parameters (Figure 4c,d). That is, the Linear model has a limited applicability at combinations of *TOW* and $[SO_2]$ that occur under natural conditions. Furthermore, according to the Linear DRF, the range of $K_1{}^{pr}$ in clean atmosphere is 53–71 g/m$^2$, therefore the $K_1{}^{pr}$ values in clean atmosphere lower than 53 g/m$^2$ (Figures 3d and 4d,e) or above 71 g/m$^2$ cannot be obtained. Higher $K_1{}^{pr}$ values can only be obtained due to $[SO_2]$ contribution. The underestimated $K_1{}^{pr}$ values in comparison with $K_1{}^{exp}$ for the majority of test locations (Figure 3d) are apparently caused by the fact that the effects of other parameters, e.g., $T$, on corrosion are not taken into account.

Figure 6 compares $K = f(SO_2)$ for all the models with the graphical representation of the dependence reported in [20] (for [$SO_2$], mg/(m²·d) values were converted to μg/m³). The dependence in [20] is presented for a constant temperature, whereas the dependences given by DRFs are given for average values in the entire range of meteorological parameters in the test locations. Nevertheless, the comparison is of interest. Below 70 and 80 μg/m³, according to [20], $K$ has lower values than those determined by the New DRF and Standard DRF, respectively, while above these values, $K$ has higher values. According to the Unified DRF, $K$ has extremely low values at all [$SO_2$] values, whereas according to the Linear DRF (TOW from 0.03 to 1), the values at $TOW = 1$ are extremely high even at small [$SO_2$].

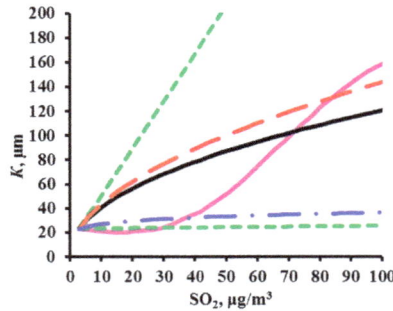

**Figure 6.** Comparison of $K = f(SO_2)$ plots for the DRF presented in [20]. – – – plot according to [20], – – by the New DRF; – – by the Standard DRF; –•– by the Unified DRF; – – – by the Linear DRF [20].

To perform a comparative estimate of $k_1$ and $k_2$, let us use the value $T_{lim} = 10\ °C$ accepted in the DRF, i.e., where the temperature dependence changes. Furthermore, it is necessary to know the $K_1$ value in clean atmosphere at $T_{lim}$ and at the $RH$ that is most common at this temperature. These data are unknown at the moment. Therefore, we'll assume that at $T_{lim} = 10\ °C$ and $RH = 75\%$, $K = 63\ g/m^2$. The dependences of $K$ on $T$ and $RH$ under these conditions and with consideration for the corresponding $k_1$ and $k_2$ for each DRF are presented in Figure 7.

The nearly coinciding $k_1$ values (0.020 for the Unified DRF and Standard DRF, and 0.024 for the New DRF, Table 8) result in an insignificant difference in the $RH$ effect on $K$ (Figure 7a).

**Table 8.** Values of coefficients used in the nonlinear DRFs for zinc.

| DRF | A | | $\alpha$ | $k_1$ | $k_2$ | | $k_3$ | B | |
| --- | --- | --- | --- | --- | --- | --- | --- | --- | --- |
| | μm | g/m² | | | $T \leq 10$ | $T > 10$ | | μg | g/m² |
| New | 0.0986 | 0.71 | 0.28 | 0.022 | 0.045 | −0.085 | 0.0001 | - | - |
| Standard | 0.0129 | 0.0929 | 0.44 | 0.046 | 0.038 | −0.071 | - | - | - |
| Unified | 0.188 | 1.35 | 0.22 | 0.018 | 0.062 | −0.021 | - | 0.00403 | 0.029 |

The temperature coefficient $k_2$ has a considerable effect on $K$. For the Unified *DRF*, the $k_2$ values of 0.059 (−0.036) for $T \leq 10\ °C$ ($T > 10\ °C$) create the lowest decrease in $K$ with a $T$ decrease (increase) in comparison with the other DRFs (Figure 7b). A consequence of such $k_2$ values can be demonstrated by examples. Due to the temperature effect alone, $K \sim 15\ g/m^2$ at $T = −12\ °C$ (Figure 7b) and $K \sim 45\ g/m^2$ at $T = 20\ °C$. The effects of other parameters and account for the $A$ value would result in even more strongly overestimated $K^{pr}$ values. For comparison: in Bilibino at $T = −12.2\ °C$ and $RH = 80\%$, $K_1^{exp} = 5.4\ g/m^2$ (Table 4) and $K^{pr} = 42\ g/m^2$ (Figure 5). In A3 test location, at $T = 20.6\ °C$ and $RH = 76\%$, $K_1^{exp} = 44.5\ g/m^2$ (Table 4), while due to $A$ and other parameters, $K_1^{pr} = 86.2\ g/m^2$, Figure 4c.

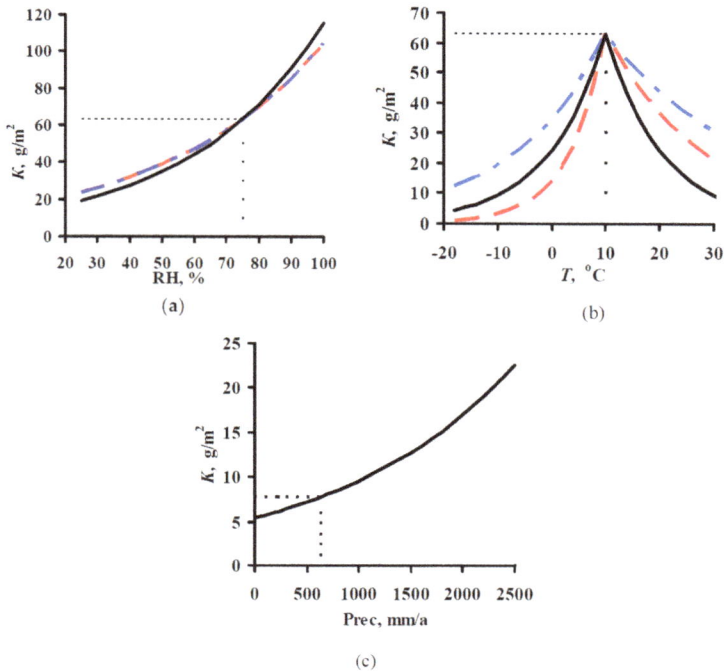

**Figure 7.** Variation of $K$ for carbon steel vs. relative humidity (**a**), temperature (**b**). and *Prec* (**c**) with account for the values of the DRF coefficients. ⎯⎯ by the New DRF; ⎯⎯ by the Standard DRF; ⎯•⎯ by the Unified DRF.

In the Standard DRF, the $k_2$ values are higher than in the Unified DRF: 0.150 and $-0.054$ for $T \leq 10\,°C$ and $T > 10\,°C$, respectively, so a greater $K$ decrease is observed, especially at $T \leq 10\,°C$, Figure 7b. At low $T$, the $K$ values are small, e.g., $K \sim 2\,g/m^2$ at $T = -12\,°C$. In $K_1{}^{Pr}$ calculations, the small $K$ are made higher due to $A$, and they are higher in polluted atmospheres due to higher $\alpha = 0.52$. As a result, $K^{Pr}$ are quite comparable with $K^{exp}$, Figure 3b. However, let us note that $K^{Pr}$ is considerably lower than $K^{exp}$ in many places. Perhaps, this is due to an abrupt decrease in $K$ in the range $T \leq 10\,°C$. This temperature range is mostly met in test locations under the UN/ECE program.

In the New DRF, $k_2$ has an intermediate value at $T \leq 10\,°C$ and the lowest value at $T > 10\,°C$, whereas $A$ has the lowest value. It is more difficult to estimate the $k_2$ value with similar $k_2$ values in the other DRFs, since the New DRF uses one more member, i.e., $e^{k3 \cdot Prec}$. The dependence of $K$ on *Prec* is presented in Figure 7c. The following arbitrary values were used to demonstrate the possible effect of *Prec* on $K$: $K = 7.8\,g/m^2$ at *Prec* = 632 mm/year. For example, in location PE5 (UN/ECE program) with *Prec* = 632 mm/year, $K = 7.8\,g/m^2$ at $T = 12.2\,°C$ and $RH = 67\%$. The maximum *Prec* was taken as 2500 mm/year, e.g., it is 2144 mm/year in NOR23 (UN/ECE program) and 2395 mm/year in B8 (MICAT project). It follows from the figure that, other conditions being equal, $K$ can increase from 5.4 to 22.6 $g/m^2$ just due to an increase in *Prec* from 0 to 2500 mm/year at $k_3 = 0.00056$ (Table 7).

Thus, it has been shown that the coefficients for each parameter used in the DRFs vary in rather a wide range. The most reliable $K_1{}^{Pr}$ can be reached if, in order to find the most suitable coefficients, the DRFs are based on the $K = f(SO_2)$ relationship obtained.

### 3.4. Predictions of $K_1$ Using Various DRFs for Zinc

The results on $K_1^{pr}$ for zinc for the UN/ECE program, MICAT project, and RF program are presented in Figures 8–10, respectively. In the UN/ECE program, the differences between the $K_1^{pr}$ and $K_1^{exp}$ values for zinc are more considerable than those for carbon steel. This may be due not only to the imperfection of the DRFs and the inaccuracy of the parameters and $K_1^{exp}$, but also to factors unaccounted for in DRFs that affect zinc. For all the DRFs, the $K_1^{pr}$ values match $K_1^{exp}$ to various extent; some of the latter exceed the error $\delta$ ($\pm 30\%$). Let us estimate the discrepancy between $K_1^{pr}$ and $K_1^{exp}$ for those $K_1^{pr}$ that exceed $\delta$. For the New DRF (Figure 8a) and the Standard DRF (Figure 8b), overestimated $K_1^{pr}$ values are observed for low and medium $K_1^{exp}$, while underestimated ones are observed for medium and high $K_1^{exp}$. In general, the deviations of $K_1^{pr}$ from $K_1^{exp}$ are symmetrical for these DRFs, but the scatter of $K_1^{pr}$ is greater for the Standard DRF. For Unified DRF (Figure 8c), $K_1^{pr}$ are mostly overestimated, considering that the $\Delta K^{[H+]} = 0.029 Rain[H^+]$ component was not taken into account for some test locations due to the lack of data on $[H^+]$. The $\Delta K^{[H+]}$ value can be significant, e.g., 2.35 g/m$^2$ in US39 or 5.13 g/m$^2$ in CS2.

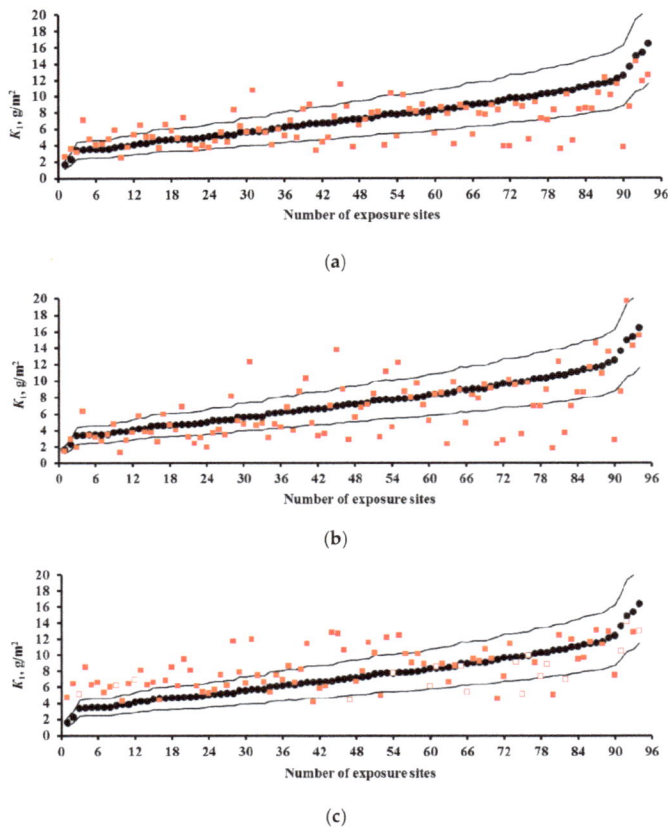

(a)

(b)

(c)

**Figure 8.** Zinc. UN/ECE program. $K_1$ predictions by the New DRF (**a**); Standard DRF (**b**); and Unified DRF (**c**). ●—experimental $K_1$ data; ■—$K_1$ predictions. □—$K_1$ predictions without taking $[H^+]$ into account (only for the Unified DRF). Thin lines show the calculation error ($\pm 30\%$). The numbers of the exposure sites are given in accordance with Table 2.

With regard to the MICAT project, the New and Unified DRFs (Figure 9a,c) give overestimated $K_1^{pr}$ at low $K_1^{exp}$, but the Standard DRF gives $K_1^{pr}$ values comparable to $K_1^{exp}$ (Figure 9b). Starting from test locations No. 33–No. 36, the $K_1^{pr}$ values for all the DRFs are underestimated or significantly underestimated. It is evident from Figure 2b that rather many test locations with small $[SO_2]$ have extremely high $K_1^{exp}$. This fact confirms the uncertainty of experimental data from these locations, as shown for carbon steel as well. The following test locations can be attributed to this category: A3 (No. 43, No. 44, No. 53), B10 (No. 50), B11 (No. 49), B12 (No. 57), CO2 (No. 55, No. 58, No. 60), CO3 (No. 54, No. 61), PE6 (No. 36, No. 38), and M3 (No. 35, No. 59). There is little sense in making $K_1$ predictions for these locations.

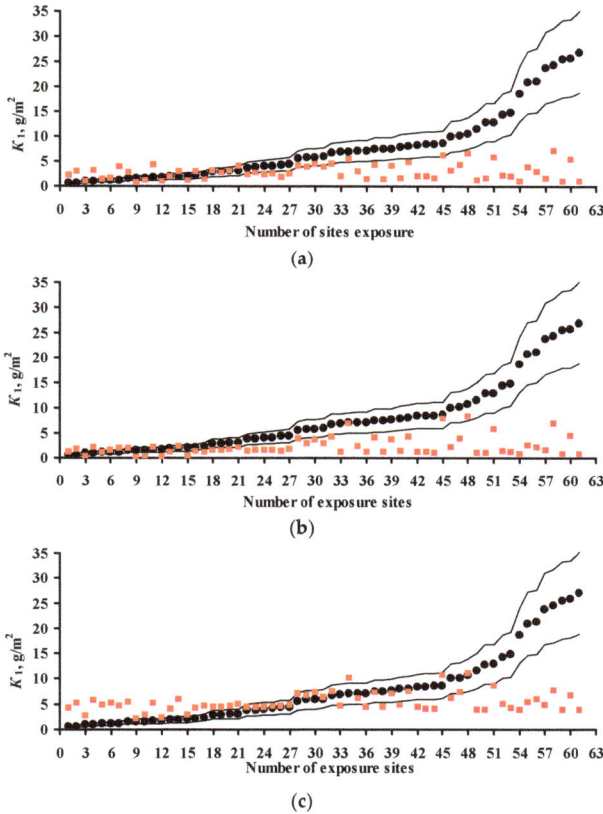

**Figure 9.** Zinc. MICAT program. $K_1$ predictions by the New DRF (**a**); Standard DRF (**b**); and Unified DRF (**c**). ●—experimental $K_1$ data; ■—$K_1$ predictions. Thin lines show the calculation error ($\pm$ 30%). The numbers of the exposure sites are given in accordance with Table 3.

For the RF program, the $K_1^{pr}$ values calculated by the New and Unified DRFs are more comparable to $K_1^{exp}$ than those determined using the Standard DRF (Figure 10).

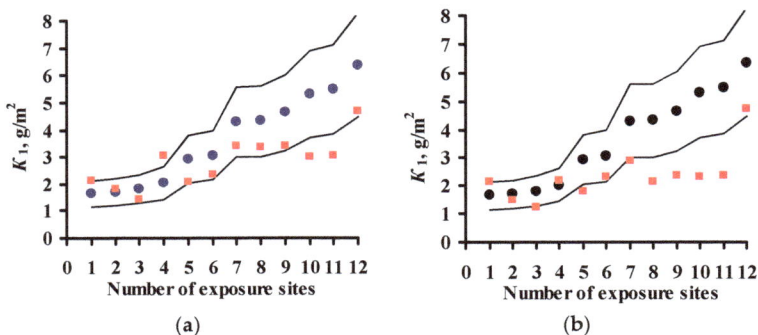

(a)

(b)

Figure 10. *Cont.*

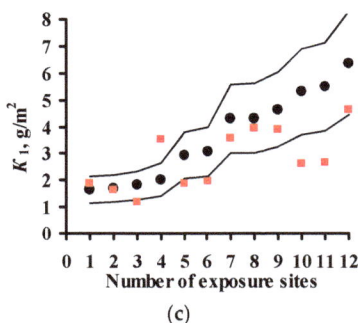

(c)

**Figure 10.** Zinc. RF program. $K_1$ predictions by the New DRF (**a**); Standard DRF (**b**); and Unified DRF (**c**). ●—experimental $K_1$ data; ■—$K_1$ predictions. Thin lines show the calculation error (±30%). The numbers of the exposure sites are given in accordance with Table 4.

### 3.5. Analysis of DRFs for Zinc

As for steel, DRFs were analyzed by comparison of their coefficients. The nonlinear DRFs for zinc can be represented in the form:

$$K_1 = A \times [SO_2]^\alpha \times \exp\{k_1 \times RH + k_2 \times (T-10) + k_3 \times Prec\} + B \times Rain \times [H^+]$$

or

$$K_1 = A \times [SO_2]^\alpha \times e^{k1 \cdot RH} \times e^{k2 \cdot (T-10)} \times e^{k3 \cdot Prec} + B \times Rain \times [H^+].$$

The values of the coefficients used in Equations (4), (6) and (10) are presented in Table 8.

To compare the $\alpha$ values, $K_1 = 4$ g/m² at $[SO_2] = 1$ µg/m³ was used for all DRFs. Let us note that the value $K_1 = 4$ g/m² was obtained during the estimation of $K = f(SO_2)$ for the development of the New DRF. The plots for all the programs are presented in Figure 2. For the New DRF, the line at $\alpha = 0.28$ mostly passes through the average experimental points. For the Standard DRF, $\alpha = 0.44$ is overestimated considerably, which may result in overestimated $K_1^{pr}$, especially at high $[SO_2]$. For the Unified DRF at $\alpha = 0.22$, the line passes, on average, slightly below the experimental points. The low $\alpha$ value, as for carbon steel, does not give a wide range of $K$ values as a function of $[SO_2]$, which may result in underestimated $K_1^{pr}$, especially at high $[SO_2]$.

Let us assume for a comparative estimate of $k_1$ and $k_2$ that $K = 4$ g/m² in a clean atmosphere at $T_{lim} = 10$ °C and $RH = 75\%$. Figure 11 demonstrates the plots of $K$ versus these parameters under these starting conditions. The Standard DRF ($k_1$ = 0.46) shows an abrupt variation in $K$ vs. *RH*. According to this relationship, at the same temperature, the $K$ value should be 0.5 g/m$^2$ at $RH$ = 30% and 12.6 g/m$^2$ at $RH$ = 100%. According to the New DRF and Unified DRF with $k_1$ = 0.22 and 0.18, respectively, the effect of *RH* is weaker, therefore $K$ = 1.5 and 1.8 g/m$^2$ at $RH$ = 30%, respectively, and $K$ = 6.9 and 6.4 g/m$^2$ at $RH$ = 100%, respectively.

The effect of temperature on $K$ is shown in Figure 11b. In the New DRF, $k_2$ = 0.045 at $T \leq 10\,^{\circ}$C has an intermediate value; at $T > 10\,^{\circ}$C, $k_2$ = −0.085 has the largest absolute value, which corresponds to an abrupt decrease in $K$ with an increase in temperature. In the Unified DRF, $k_2$ = −0.021 at $T > 10\,^{\circ}$C, i.e., an increase in temperature results in a slight decrease in $K$. As for the effect of $A$, this also contributes to higher $K_1^{\mathrm{pr}}$ values despite the small $\alpha$ value.

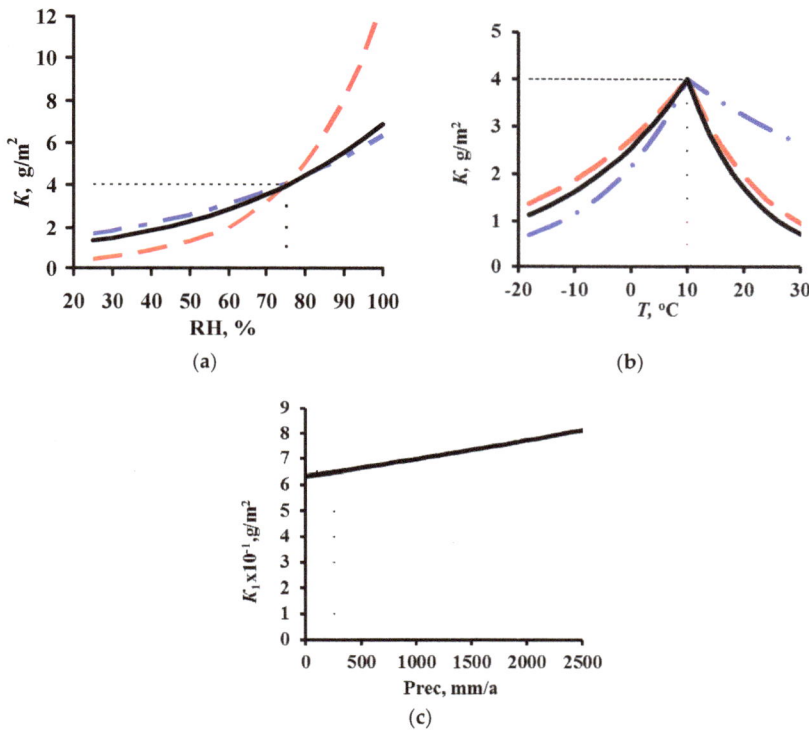

**Figure 11.** Variation of $K$ for zinc versus relative humidity (**a**), temperature (**b**) and *Prec* (**c**) with account for the values of the DRF coefficients. −− by the New DRF; −− by the Standard DRF; −•− by the Unified DRF.

In the Standard DRF, the value $A$ = 0.0929 (g/m$^2$), which is ~8 times smaller than in the New DRF, and a small $k_2$ = −0.71 at $T > 10\,^{\circ}$C were taken to compensate the $K_1^{\mathrm{pr}}$ overestimation due to the combination of high values, $\alpha$ = 0.44 and $k_1$ = 0.46. In the Unified DRF, the high $A$ value that is ~2 times higher than in the New DRF is not compensated by the combination of the low values, $\alpha$ = 0.22 and $k_2$ = −0.021 at $T > 10\,^{\circ}$C. Therefore, the $K_1^{\mathrm{pr}}$ values are mostly overestimated, Figures 8c and 9c for trusted test locations. However, small $K_1^{\mathrm{pr}}$ values were attained for low $T$ at $k_2$ = 0.62, Figure 10c.

The effect of *Prec* on $K$ at $k_3$ = 0.0001, which is taken into account only in the New DRF, given under the assumption that $K$ = 0.65 in a clean atmosphere at *Prec* (*Rain*) = 250 mm/year, $T$ = 15 $^{\circ}$C and

$RH = 60\%$ (e.g., location E5 in the MICAT project), is shown in Figure 11c. Upon an increase in *Prec* (*Rain*) from 250 to 2500 mm/year, $K$ can increase from 0.65 to 0.81 $g/m^2$.

As for carbon steel, the above analysis of coefficients in the DRFs for zinc confirms that the coefficients can be varied to obtain reliable $K_1^{pr}$ values. The New DRF based on $K = f(SO_2)$ gives the most reliable $K_1^{pr}$ values for zinc.

## 4. Estimation of Coefficients in DRFs for Carbon Steel and Zinc

Let us first note that the starting conditions that we took to demonstrate the effect of various atmosphere corrosivity parameters on $K$ of carbon steel and zinc (Figures 7 and 11) may not match the real values. However, the plots obtained give an idea on $K$ variations depending on the coefficients in the DRFs.

For continental test locations under all programs, the $K_1^{exp}$ values are within the following ranges: for carbon steel, from 6.3 (Oimyakon, RF program) to 577 $g/m^2$ (CS3, UN/ECE program); for zinc, from 0.65 (E5, MICAT project) to 16.41 $g/m^2$ (CS3, UN/ECE program). That is, the difference in the corrosion losses is at least ~10–35 fold, the specific densities of these metals being nearly equal. Higher $K_1^{pr}$ values for steel than for zinc are attained using different coefficients at the parameters in the DRFs.

In the New DRFs, $A$ is 7.7 and 0.71 $g/m^2$ for carbon steel and zinc, respectively, i.e., the difference is ~10-fold. Higher $K_1^{pr}$ values for steel than for zinc were obtained chiefly due to the contribution of $[SO_2]^\alpha$ at $\alpha = 0.47$ and 0.28, respectively. The values of $RH$ and *Prec* affect the corrosion of steel more strongly than they affect zinc corrosion. The coefficients for these parameters are: $k_1 = 0.024$ and 0.022; $k_3 = 0.00056$ and 0.0001 for steel and zinc, respectively. However, the temperature coefficients ($k_2 = 0.095$ and $-0.095$ for steel; $k_2 = 0.045$ and $-0.085$ for zinc) indicate that, with a deviation of $T$ from 10 °C, the corrosion process on steel is hindered to a greater extent than on zinc.

In the Standard DRF, $A$ is 1.77 and 0.0129 μm for carbon steel and zinc, respectively, i.e., the difference is ~137-fold. The $\alpha$ value for steel is somewhat higher than that for zinc, i.e., 0.52 and 0.44 respectively, which increases the difference of $K_1^{pr}$ for steel from that for zinc. As shown above, the difference should not be greater than 35-fold. This difference is compensated by the 2.3-fold higher effect of $RH$ on zinc corrosion than on steel corrosion ($k_1 = 0.046$ and 0.020 for zinc and steel, respectively). Furthermore, the temperature coefficient $k_2$ at $T \leq 10$ °C for steel is 3.95 times higher than that for zinc. This indicates that steel corrosion slows down abruptly in comparison with zinc as $T$ decreases below 10 °C. At $T > 10$ °C, the $k_2$ values for steel and zinc are comparable. Taking the values of the coefficients presented into account, the $K_1^{pr}$ values for steel are 15-fold higher, on average, than those for zinc at $T \leq 10$ °C, but ~60-fold at $T > 10$ °C. Of course, this is an approximate estimate of the coefficients used in the Standard DRF.

In the Unified DRF, $A$ is 3.54 and 0.188 μm for carbon steel and zinc, respectively, i.e., the difference is ~19-fold. The $\alpha$ value for steel is lower than that for zinc, i.e., 0.13 and 0.22 respectively, which decreases the difference of $K_1^{pr}$ of steel from that of zinc. Conversely, the $RH$ value affects steel corrosion somewhat more strongly than that of zinc ($k_1 = 0.020$ and 0.018 for steel and zinc, respectively). The $k_2$ values for steel and zinc are comparable in both temperature ranges. The $\Delta K^{[H+]}$ component was introduced only for zinc, which somewhat complicates the comparison of the coefficients in these DRFs.

All the presented DRFs are imperfect not only because of the possible inaccuracy of the mathematical expressions as such, but also due to the inaccuracy of the coefficients used in the DRFs. The $K_1^{pr}$ values obtained using the New DRF match $K_1^{exp}$ most accurately. However, while the $\alpha$ values that were assumed to be 0.47 and 0.28 for carbon steel and zinc, respectively, may be considered as accurate in a first approximation, the other coefficients need to be determined more accurately by studying the effect of each atmosphere corrosivity parameter on corrosion, with the other parameters being unchanged. Studies of this kind would allow each coefficient to be estimated and DRFs for reliable prediction of $K_1$ in atmospheres with various corrosivity to be created.

## 5. Conclusions

1.  $K = f(SO_2)$ plots of corrosion losses of carbon steel and zinc vs. sulfur dioxide concentration were obtained to match, to a first approximation, the mean meteorological parameters of atmosphere corrosivity.

2.  Based on the $K = f(SO_2)$ relationships obtained, with consideration for the nonlinear effect of temperature on corrosion, New DRFs for carbon steel and zinc in continental territories were developed.

3.  Based on the corrosivity parameters at test locations under the UN /ECE and RF programs and the MICAT project, predictions of first-year corrosion losses of carbon steel and zinc were given using the New DRF, Standard DRF, and Unified DRF, as well as the linear model for carbon steel obtained in [20] with the aid of an artificial neural network. The predicted corrosion losses are compared with the experimental data for each DRF. It was shown that the predictions provided by the New DRFs for the first-year match the experimental data most accurately.

4.  An analysis of the values of the coefficients used in the DRFs for the prediction of corrosion losses of carbon steel and zinc is presented. It is shown that more accurate DRFs can be developed based on quantitative estimations of the effects of each atmosphere corrosivity parameter on corrosion.

**Acknowledgments:** We are grateful to Manuel Morcillo for valuable comments.

**Author Contributions:** Y.M.P. performed the modelling and analysis and wrote the paper. A.I.M. contributed the discussion of the data.

**Conflicts of Interest:** The authors declare no conflict of interest.

## References

1.  McCuen, R.H.; Albrecht, P.; Cheng, J.G. A New Approach to Power-Model Regression of Corrosion Penetration Data. In *Corrosion Forms and Control for Infrastructure*; Chaker, V., Ed.; American Society for Testing and Materials: Philadelphia, PA, USA, 1992.

2.  Syed, S. Atmospheric corrosion of materials. *Emir. J. Eng. Res.* **2006**, *11*, 1–24.

3.  De la Fuente, D.; Castano, J.G.; Morcillo, M. Long-term atmospheric corrosion of zinc. *Corros. Sci.* **2007**, *49*, 1420–1436. [CrossRef]

4.  Landolfo, R.; Cascini, L.; Portioli, F. Modeling of metal structure corrosion damage: A state of the art report. *Sustainability* **2010**, *2*, 2163–2175. [CrossRef]

5.  Morcillo, M.; de la Fuente, D.; Diaz, I.; Cano, H. Atmospheric corrosion of mild steel (article review). *Rev. Metal.* **2011**, *47*, 426–444. [CrossRef]

6.  De la Fuente, D.; Diaz, I.; Simancas, J.; Chico, B.; Morcillo, M. Long-term atmospheric corrosion of mild steel. *Corros. Sci.* **2011**, *53*, 604–617. [CrossRef]

7.  Morcillo, M.; Chico, B.; Diaz, I.; Cano, H.; de la Fuente, D. Atmospheric corrosion data of weathering steels: A review. *Corros. Sci.* **2013**, *77*, 6–24. [CrossRef]

8.  Surnam, B.Y.R.; Chiu, C.W.; Xiao, H.P.; Liang, H. Long term atmospheric corrosion in Mauritius. *CEST* **2015**, *50*, 155–159. [CrossRef]

9.  Panchenko, Y.M.; Marshakov, A.I. Long-term prediction of metal corrosion losses in atmosphere using a power-linear function. *Corros. Sci.* **2016**, *109*, 217–229. [CrossRef]

10. *Corrosion of Metals and Alloys—Corrosivity of Atmospheres—Guiding Values for the Corrosivity Categories*; ISO 9224:2012(E); International Standards Organization: Geneva, Switzerlands, 2012.

11. Rosales, B.M.; Almeida, M.E.M.; Morcillo, M.; Uruchurtu, J.; Marrocos, M. *Corrosion y Proteccion de Metales en las Atmosferas de Iberoamerica*; Programma CYTED: Madrid, Spain, 1998; pp. 629–660.

12. Tidblad, J.; Kucera, V.; Mikhailov, A.A. *Statistical Analysis of 8 Year Materials Exposure and Acceptable Deterioration and Pollution Levels*; UN/ECE ICP on Effects on Materials; Swedish Corrosion Institute: Stockholm, Sweden, 1998; p. 49.

13. Tidblad, J.; Mikhailov, A.A.; Kucera, V. Unified Dose-Response Functions after 8 Years of Exposure. In *Quantification of Effects of Air Pollutants on Materials*; UN ECE Workshop Proceedings; Umweltbundesamt: Berlin, Germany, 1999; pp. 77–86.

14. Tidblad, J.; Kucera, V.; Mikhailov, A.A.; Henriksen, J.; Kreislova, K.; Yaites, T.; Stöckle, B.; Schreiner, M. UN ECE ICP Materials. Dose-response functions on dry and wet acid deposition effects after 8 years of exposure. *Water Air Soil Pollut.* **2001**, *130*, 1457–1462. [CrossRef]

15. Tidblad, J.; Mikhailov, A.A.; Kucera, V. *Acid Deposition Effects on Materials in Subtropical and Tropical Climates. Data Compilation and Temperate Climate Comparison*; KI Report 2000:8E; Swedish Corrosion Institute: Stockholm, Sweden, 2000; pp. 1–34.

16. Tidblad, J.; Mikhailov, A.A.; Kucera, V. Application of a Model for Prediction of Atmospheric Corrosion for Tropical Environments. In *Marine Corrosion in Tropical Environments*; Dean, S.W., Delgadillo, G.H., Bushman, J.B., Eds.; American Society for Testing and Materials: West Conshohocken, PA, USA, 2000; p. 18.

17. Tidblad, J.; Kucera, V.; Mikhailov, A.A.; Knotkova, D. Improvement of the ISO Classification System Based on Dose-Response Functions Describing the Corrosivity of Outdoor Atmospheres. In *Outdoor Atmospheric Corrosion, ASTM STP 1421*; Townsend, H.E., Ed.; American Society for Testing and Materials: West Conshohocken, PA, USA, 2002; p. 73.

18. *Corrosion of Metals and Alloys—Corrosivity of Atmospheres—Classification, Determination and Estimation*; ISO 9223:2012(E); International Standards Organization: Geneva, Switzerlands, 2012.

19. Cai, J.; Cottis, R.A.; Lyon, S.B. Phenomenological modelling of atmospheric corrosion using an artificial neural network. *Corros. Sci.* **1999**, *41*, 2001–2030. [CrossRef]

20. Pintos, S.; Queipo, N.V.; de Rincon, O.T.; Rincon, A.; Morcillo, M. Artificial neural network modeling of atmospheric corrosion in the MICAT project. *Corros. Sci.* **2000**, *42*, 35–52. [CrossRef]

21. Diaz, V.; Lopez, C. Discovering key meteorological variables in atmospheric corrosion through an artificial neural network model. *Corros. Sci.* **2007**, *49*, 949–962. [CrossRef]

22. Kenny, E.D.; Paredes, R.S.C.; de Lacerda, L.A.; Sica, Y.C.; de Souza, G.P.; Lazaris, J. Artificial neural network corrosion modeling for metals in an equatorial climate. *Corros. Sci.* **2009**, *51*, 2266–2278. [CrossRef]

23. Reddy, N.S. Neural Networks Model for Predicting Corrosion Depth in Steels. *Indian J. Adv. Chem. Sci.* **2014**, *2*, 204–207.

24. Knotkova, D.; Kreislova, K.; Dean, S.W. *ISOCORRAG International Atmospheric Exposure Program: Summary of Results*; ASTM Series 71; ASTM International: West Conshohocken, PA, USA, 2010.

25. Morcillo, M. Atmospheric corrosion in Ibero-America. The MICAT project. In *Atmospheric Corrosion*; Kirk, W.W., Lawson, H.H., Eds.; ASTM STP 1239; American Society for Testing and Materials: Philadelphia, PA, USA, 1995; pp. 257–275.

26. Panchenko, Y.M.; Shuvakhina, L.N.; Mikhailovsky, Y.N. Atmospheric corrosion of metals in Far Eastern regions. *Zashchita Metallov* **1982**, *18*, 575–582. (In Russian).

27. Knotkova, D.; Vlckova, J.; Honzak, J. Atmospheric Corrosion of Weathering Steels. In *Atmospheric Corrosion of Metals*; Dean, S.W., Jr., Rhea, E.C., Eds.; ASTM STP 767; American Society for Testing and Materials: Philadelphia, PA, USA, 1982; pp. 7–44.

28. Tidblad, J.; Mikhailov, A.A.; Kucera, V. Model for the prediction of the time of wetness from average annual data on relative air humidity and air temperature. *Prot. Met.* **2000**, *36*, 533–540. [CrossRef]

29. Feliu, S.; Morcillo, M.; Feliu, J.S. The prediction of atmospheric corrosion from meteorological and pollution parameters. I. Annual corrosion. *Corros. Sci.* **1993**, *34*, 403–422. [CrossRef]

30. Panchenko, Y.M.; Marshakov, A.I.; Nikolaeva, L.A.; Kovtanyuk, V.V.; Igonin, T.N.; Andryushchenko, T.A. Comparative estimation of long-term predictions of corrosion losses for carbon steel and zinc using various models for the Russian territory. *CEST* **2017**, *52*, 149–157. [CrossRef]

*Review*

# The Results of 45 Years of Atmospheric Corrosion Study in the Czech Republic

Katerina Kreislova * and Dagmar Knotkova

SVUOM Ltd., U Mestanskeho pivovaru 934, 170 00 Prague 7, Czech Republic; info@svuom.cz
* Correspondence: kreislova@svuom.cz; Tel.: +420-220-809-996

Academic Editor: Manuel Morcillo
Received: 27 February 2017; Accepted: 6 April 2017; Published: 7 April 2017

**Abstract:** Atmospheric corrosion poses a significant problem with regard to destruction of various materials, especially metals. Observations made over the past decades suggest that the world's climate is changing. Besides global warming, there are also changes in other parameters. For example, average annual precipitation increased by nearly 10% over the course of the 20th century. In Europe, the most significant change, from the atmospheric corrosion point of view, was an increase in $SO_2$ pollution in the 1970s through the 1980s and a subsequent decrease in this same industrial air pollution and an increase in other types of air pollution, which created a so-called multi-pollutant atmospheric environment. Exposed metals react to such changes immediately, even if corrosion attack started in high corrosive atmospheres. This paper presents a complex evaluation of the effect of air pollution and other environmental parameters and verification of dose/response equations for conditions in the Czech Republic.

**Keywords:** atmospheric corrosivity; atmospheric test exposure; yearly mass loss; long-term corrosion rate; structural metals

---

## 1. Introduction

Among several factors affecting the service life of metal structures, roofs, cladding, etc., atmospheric corrosion is recognized to be one of the major risks which impair the performance of constructions, resulting in huge economic and societal losses. Corrosion life prediction provides a key technology for the optimum selection of materials and/or coatings for such constructions. Corrosion scientists in many countries worldwide have carried out exposure tests to investigate the effects of the environment on corrosion rates. An evaluation period based on field tests usually takes 10–20 years.

In the Czech Republic (former Czechoslovakia) atmospheric corrosion has been studied since 1950 in the State research institute of material protection (SVUOM) [1–3]. Several exposure programs, both national and international, of structural metal atmospheric corrosion have been conducted at Czech atmospheric test sites since the 1970s, including on-site measurement of environmental data. As the Czech Republic is a relatively small country located in Central Europe, the most significant varying effect on the atmospheric corrosivity is imposed by air pollution, primarily that caused by industrial sources. In these repeated exposure programs, the standard flat samples of carbon steel, zinc, weathering steel, copper and aluminum were evaluated in intervals 1, 2, 3, 4, 5, 8, 10, 15, 20 and 25 years—Figure 1 [4]. The program data covering the period from 1985 to 2005, during which a significant reduction of $SO_2$ pollution occurred, show a practically immediate decrease in metals corrosion rate [5]. The exposure program carried between 2005 and 2015 was realized in a relatively steady multi-pollutant situation. In 2015, the last new exposure program was launched with planned withdrawals after 1, 2, 5 and 10 years.

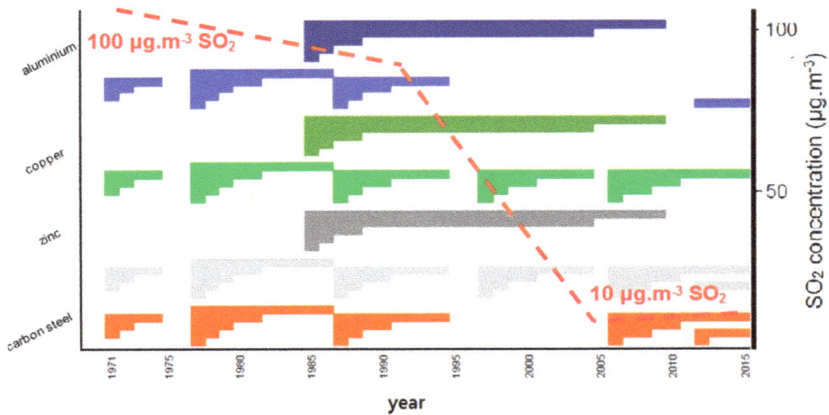

**Figure 1.** Scheme of exposure programs on atmospheric test sites during years 1970–2015.

## 2. Materials and Methods

### 2.1. Materials

During each exposure period the structural metals were exposed as $100 \times 150$ mm$^2$ flat panels on racks at an 45° to the horizontal, facing south:

(a)    Carbon steel 1.0338 according to EN 10130 (C < 0.08%, P < 0.03%, Mn < 0.40%, S < 0.03%) [6];

(b)    Weathering steel S 355JW Atmofix according to EN 10025-5 [7];

(c)    Zinc (98.5%);

(d)    Copper (99.8%);

(e)    Aluminum (99.99%).

Triplicate panels in each withdrawal were treated with a standard procedure—degreased, rinsed by water and dried before exposure.

### 2.2. Atmospheric Test Sites

Test sites are performed according to CSN 03 8110 [8] equivalent to ISO 8565 [9]. Over decades the network of atmospheric test sites changed. The test sites located in the current Slovak Republic have not been performed since 2000 (Hurbanovo, Lomnicky stit); some heavy industrial test site was closed too (Usti nad Labem, Czech Republic); in some specific exposure programs other test sites were involved (Ostrava, Telc, Horomerice, etc.), but there are 3 test sites that have been exploited for 45 years in the same locations (Figure 2):

(a)    Prague—Urban atmospheric environment;

(b)    Kopisty—Industrial atmospheric environment;

(c)    Kasperske Hory—Rural atmospheric environment.

**Figure 2.** Atmospheric test sites in the Czech Republic.

The environmental parameters significant for atmospheric corrosion have been measured at each test site during this period on a daily or monthly basis (temperature, relative humidity, precipitation, $SO_2$ pollution, $NO_x$ pollution, pH of precipitation, chemical composition of precipitation, etc.) according to CSN 03 8110, respectively ISO 8565 and ISO 9225 [10] and statistically treated as annual average values.

## 3. Results

### 3.1. Environmental Data

During 45-years the majority of climate parameters on these 3 test sites are stable with exception of yearly average temperature which increased (Figure 3a). During 1961–2010 average annual temperature increased for 0.7 °C and the annual amount of precipitation increased for 7% in North Bohemia region (test site Kopisty). In city Prague the thermal urban island effect is more evident. Estimated TOW is classified as $\tau_4$ for all test sites in the CR, e.g., for the Prague test site, the average TOW measured in period 1970–1980 was 3740 $h \cdot a^{-1}$.

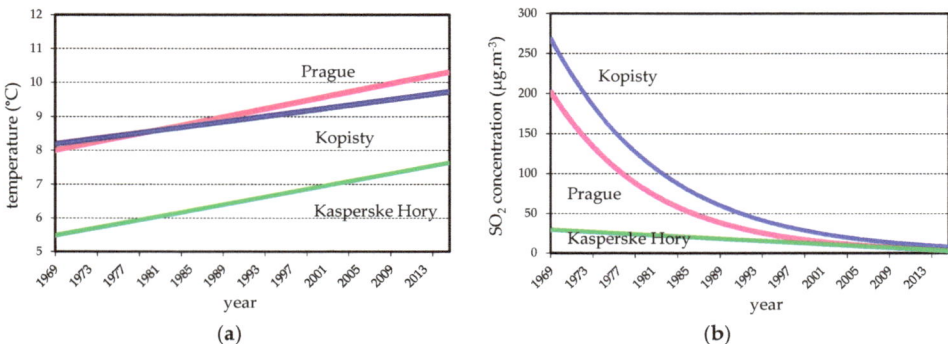

**Figure 3.** Trends in yearly average value of temperature (**a**) and $SO_2$ pollution (**b**).

When combined with climatic effects, air pollution can cause substantial deterioration of metallic materials. Sulphur dioxide and chlorides are the two dominant pollutants influencing atmospheric corrosion. The environmental parameters were relatively stable during the 45-year period, except $SO_2$ air pollution and pH of precipitation which followed the changes in air pollution of $SO_2$. The minor changes are observed in average annual temperature as opposed to very significant changes in average annual $SO_2$ pollution—Table 1 and Figure 3b [4].

**Table 1.** Selected average annual environmental parameters at test sites.

| Test Site | Year | T (°C) | RH (%) | $SO_2$ ($\mu g \cdot m^{-3}$) | $NO_x$ ($\mu g \cdot m^{-3}$) | Rain (mm) | pH of Precipitation |
|---|---|---|---|---|---|---|---|
| | 1970 | 8.6 | 76 | 92 | - | 490 | - |
| | 1980 | 7.4 | 77 | 99 | - | 653 | - |
| Prague | 1990 | 10.1 | 74 | 56 | 34 | 468 | 5.4 |
| | 2000 | 10.2 | 72 | 11 | 23 | 498 | 5.7 |
| | 2016 | 10.0 | 73 | 5 | 28 | 561 | 6.0 |
| | 1970 | 8.1 | 76 | 182 | - | 618 | - |
| | 1980 | 7.6 | 75 | 145 | - | 520 | - |
| Kopisty | 1990 | 9.7 | 71 | 67 | 32 | 398 | 4.9 |
| | 2000 | 10.2 | 76 | 16 | 28 | 486 | 4.7 |
| | 2016 | 9.9 | 76 | 18 | 23 | 567 | 6.0 |
| Kasperske | 1970 | 5.4 | 79 | 23 | - | 637 | - |
| | 1980 | 5.4 | 80 | 30 | - | 811 | - |
| Hory | 1990 | 7.0 | 78 | 26 | 9 | 722 | 4.9 |
| | 2000 | 7.8 | 76 | 8 | 9 | 798 | 5.4 |
| | 2016 | 8.7 | 74 | 11 | 24 | 841 | 6.0 |

In the years before 1990, the test sites Prague and Kopisty were affected by air pollution from local heating and industrial sources; i.e., mainly by $SO_2$. The first reports about the negative effect of the high concentration of $SO_2$ emission appeared in the then Czechoslovakia as early as the second half of the 1940s, a trend comparable to the other highly industrialized countries/regions. Sulphur dioxide $SO_2$ had the greatest influence on atmospheric corrosion rates in the 1970s to 1980s. Installation of desulphurization units at industrial sources, restructuring of industry and other large sources closing resulted in reduced $SO_2$ emission in the CR within a relatively short period of 1994 to 2004 [11]. Afterwards, reduction of $SO_2$ emission continued but not so intensively. Since 2010, a minimal difference in $SO_2$ air pollution among rural, urban and industrial test sites has been measured. This situation is observed at all atmospheric test sites, mainly at the industrial site Kopisty and at the urban site Prague. Together with this acidic air pollution decrease the pH value of precipitation increase slightly from 5.0 to 6.0.

Measurement of nitrogen oxides ($NO_x$) air pollution started only later, namely in the middle of 1980s. The effect of $NO_x$ on atmospheric corrosion are still only assumed above 30 $\mu g \cdot m^{-3}$ [12] and long-term measured data from Prague and Kopisty show that average values oscillate around this value. The ratio $SO_2/NO_x$ is ca. 0.5 since 2000. $NO_x$ reacts with air to form strong corrosive gaseous nitrogen acid $HNO_3$ but its average concentrations measured at the Prague and Kopisty test sites in the years 2003 to 2015 were 0.82 $\mu g \cdot m^{-3}$ and 0.57 $\mu g \cdot m^{-3}$, respectively.

Prior to the 1950s, atmospheric corrosion caused by airborne salinity was limited to coastal areas. However, in the 1970s the widespread use of de-icing materials on roadways led to a serious corrosion problem in the snowbelt regions/countries. Field tests have shown that approx. 40 wt. % of NaCl de-icing salt applied to roadways is carried by the air to be deposited within 100 m of treated road and may accumulate on their side and become part of dust. $Cl^-$ deposition was measured in 1980s at the Prague test site and average annual value was 4.8 $mg \cdot m^{-2} \cdot d^{-1}$. The chloride deposition rate has been measured at the atmospheric test sites Prague and Kopisty according to EN ISO 9225 since 2016. The chloride deposition at the Prague atmospheric test site was approximately twice (ca. 3.0 $mg \cdot m^{-2} \cdot d^{-1}$) the values identified at the industrial atmospheric test site Kopisty (ca. 1.3 $mg \cdot m^{-2} \cdot d^{-1}$). These values are at the background level $S_0$ according to ISO 9223.

## 3.2. Corrosion Data

After exposure, all panels were visually examined before determining gravimetric corrosion mass losses by removing corrosion product layers according to ISO 8407 [13] for each metal. Weight loss was calculated by area and time.

In 45 years many repeated one-year exposure of structural metals had been performed (see Figure 1). As $SO_2$ pollution level changed during this period the one-year corrosion loss decreased too. There was some varying in corrosion data due to not stable climate conditions (warm or cold summer, warm or cold winter, rainy season, etc.). The trend analysis based on statistic treatment of measured values undertaken within a formal regression analysis of one-year corrosion loss of carbon steel, zinc and copper panels are shown in Figure 4 (bolt line) along with the trend of $SO_2$ decrease (dashed line) in the Prague and Kopisty test sites; there are fewer corrosion data available for Kasperske Hory test site throughout this entire period. These trend analyses based on statistic treatment of measured values show that the atmospheric corrosion of carbon steel and copper is more strongly affected by $SO_2$ than the atmospheric corrosion of zinc.

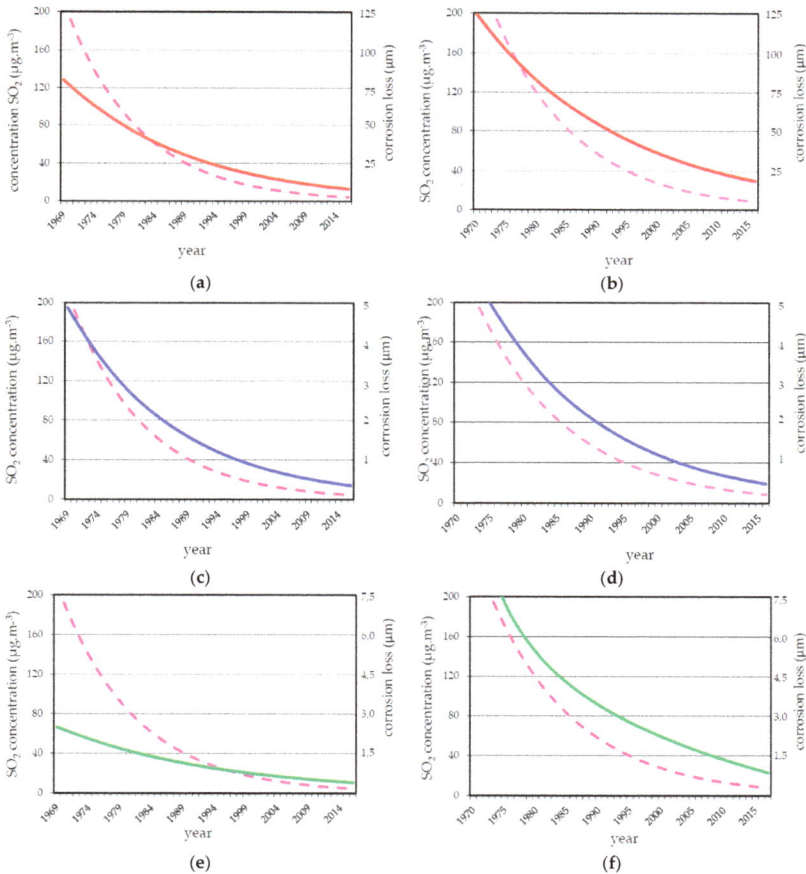

**Figure 4.** Trend analysis in annual corrosion losses (bolt line) and $SO_2$ pollution (dashed line) at Prague (**a,c,e**) and Kopisty (**b,d,f**) test sites in different exposure periods for carbon steel (**a,b**); zinc (**c,d**); and copper (**e,f**).

These databases can be used to verify models of atmospheric corrosion. The functions that allow calculating corrosion loss based on environmental data are derived to predict the corrosion behavior of metals in the future.

## 4. Discussion

Corrosion losses $r_{corr}$ estimated after one year of exposure on carbon steel, zinc and copper are used to classified corrosivity at the test sites in accordance with standard ISO 9223 [14]. Local climate is defined as conditions prevailing within the radius of an object/locality up to 1000 m. Local climate and the pollutant content provide a basis for the determination of an atmospheric corrosivity category.

The corrosion growth rate has always been a function of time. For long- term exposure predictions this ISO 9224 [15] equation can be used:

$$D = r_{corr} \cdot t^b$$

where t is the exposure time in years, $r_{corr}$ is the attack experienced in the first year in mm or μm, b is the metal-environment specific time exponent which is different for each metal.

The changes in average air temperature and mainly evident changes in amount of precipitation occurred since 2000 more frequently may affect the short-time, one-year corrosion data but they are not yet evident for long-term corrosion behavior of structure metals. But for application of long-term corrosion model it is necessary to use $r_{corr}$ as average value for some last exposures or from period with relatively average climate parameters. For the statistical treatment of this extensive database of 45 years of atmospheric corrosion field exposure measured values and the trend analyses resulting therefrom, three periods may be created: two periods with relatively stable environmental conditions (the years 1970–1994 and the years 2004–2016) and a transmission period 1994–2004 with quickly changing air pollution level. The average data for these periods are given in Table 2.

**Table 2.** The comparison of estimated and predicted long-term corrosion rate.

| Test Site | Period | Environmental Data | | | Corrosion Rate ($\mu m \cdot a^{-1}$) | | | | | |
|---|---|---|---|---|---|---|---|---|---|---|
| | | T (°C) | RV (%) | $SO_2$ ($\mu g \cdot m^{-3}$) | Carbon Steel | | Zinc | | Copper | |
| | | | | | Real | Cal | Real | Cal | Real | Cal |
| Prague | 1970–1994 | 8.8 | 77 | 87.6 | 16.1 | 15.1 | 2.82 | 2.30 | 1.30 | 0.86 |
| | 2004–2015 | 10.1 | 73 | 7.3 | 2.6 | 4.0 | 0.13 | 0.45 | 0.22 | 0.23 |
| Kopisty | 1970–1994 | 8.7 | 74 | 119.2 | 21.4 | 25.0 | 3.66 | 2.66 | 2.26 | 2.35 |
| | 2004–2015 | 9.3 | 78 | 14.2 | 9.3 | 8.6 | 0.33 | 0.89 | 0.51 | 0.73 |

A number of damage functions or dose-response equation, which are compared to the atmospheric corrosion of metals using environmental parameters applicable in Europe, were derived in several studies:

(a)  ISOCORRAG-ISO 9223 and ISO 9224 [16];
(b)  UN ECE ICP Materials [17];
(c)  EU project EVK4-CT 2001-00044 MULTI-ASSESS [18].

All of these dose-response equations are derived from field exposure results obtained in the years 1986–2004 where the $SO_2$ level was relative high at the urban test sites and very high at the industrial test sites. The data from all exposure programs performed in the Czech Republic (or former Czechoslovakia) were used to verify these prediction models. The best fitting was established for ISO 9224 equations [19]. With regard to long-term corrosion data, comparison of the results of the Czech national programs performed in the years 1970–1990 and 2005–2015 was completed for Prague and Kopisty test sites—Figure 5.

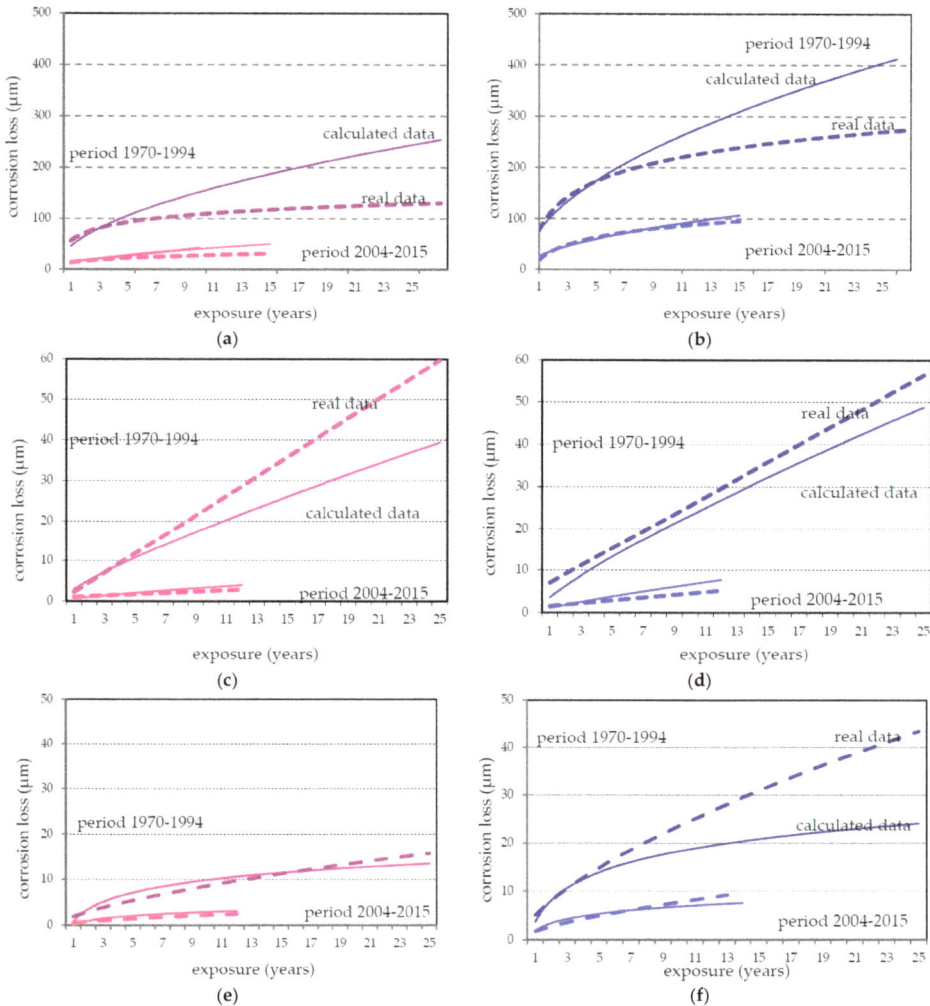

**Figure 5.** Trend analysis in long-term corrosion losses at Prague (**a**,**c**,**e**) and Kopisty (**b**,**d**,**f**) test sites in different exposure periods for carbon steel (**a**,**b**); zinc (**c**,**d**); and copper (**e**,**f**).

A survey of long-term exposed materials (weathering steel structures, zinc coating on galvanized structures, copper roofs, aluminum products) was also included in these data [20].

The differences between predicted (calculated) values and real (estimated) corrosion loss was caused the fact that dose-response functions were derived from database which does not include these corrosion values from high pollution years. The ISO prediction model for one-year atmospheric corrosion corresponds relatively well with real (determined) values but in exposures with $SO_2 > 90$ mg·m$^{-2}$·d$^{-1}$ calculated values are significantly different from determined corrosion loss; in case of the long-term prediction model the fitting of real (determined) to calculated values is better for atmospheres with lower $SO_2$ pollution. The first results show it would be better to divide the dose-response functions according to limit pollution values to obtain more accurate prediction models. To enhance atmospheric corrosion prediction, interaction between all air pollutants should be considered in current multi-pollutant environmental condition, especially for zinc and copper.

Changing environmental conditions also affected significant structural material of weathering steel. This effect has been evident since the first year of exposure, but a decreasing corrosion rate is more important for long-term data. The protective effect of patina is more evident for atmospheres with high $SO_2$. Only 8 years of data are available from the Czech test sites for this material with regard to the current environmental situation. Comparison of the long-term trends of the atmospheric corrosion of carbon and weathering steel in industrial locality Kopisty shows that the corrosion rate of weathering steel decreases less than that of carbon steel (Figure 6). The corrosion loss of weathering steel, respectively the protective effect of patina, is more significantly affected by climate parameters in environments without salinity.

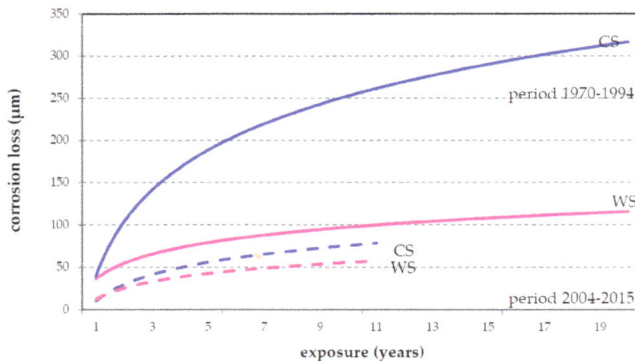

**Figure 6.** Comparison of corrosion loss of carbon (CS) steel and weathering steel (WS) at Kopisty.

The corrosion rates shown in ISO 9223 and ISO 9224 were estimated as uniform corrosion, but aluminum corrosion manifests itself rather as pitting corrosion. The rate of the deepening of pits formed in atmospheric environments decreases with time. Studies showed that the deepening rate of pits follows the equation [21]:

$$d = k \cdot t^{1/3}$$

where $d$ is the depth of the pit, $t$ is the time, and $k$ is a constant depending on the alloy and the service conditions (nature of the alloy, temperature, atmospheric corrosivity category, etc.).

Maximum pit depth is a better indicator of potential damage, but this characteristic cannot be evaluated after the first year of exposure. Evaluation of long-term exposed aluminum materials shows that the largest increasing of pit depth occurred within 5 years. Average rate of the long-term pitting corrosion of aluminum exposed in CR atmospheric environment is 1.5 $\mu m \cdot a^{-1}$ (Figure 7). This value corresponds to published data. From experimentally evaluated data, constant $k_{ave}$ for CR atmospheres was estimated as 18 for the average rate of pitting corrosion. The same approach is defined for maximum pitting corrosion rate and constant $k_{max}$ is 50.

*Materials* **2017**, *10*, 394

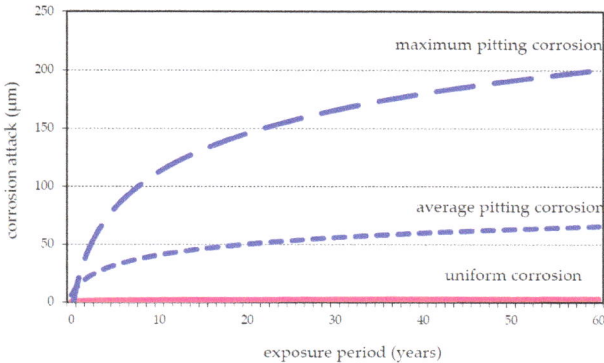

**Figure 7.** Uniform and pitting corrosion of long-term exposed aluminum in CR.

## 5. Conclusions

The short- and mainly long-term atmospheric corrosion tests of structural metals provide significant information for improving prediction models of the atmospheric corrosion. Data mining, applying to corrosion, represents the selection tools available to quantify a response that a given metal will give in a given environment and the effect that environment changes will have on that metal. A large database of environmental and corrosion data available for the Czech test sites allows us to follow the trends and effects of air pollution changes on the atmospheric corrosion of structural metals.

The main changes in atmospheric corrosion rate were caused by $SO_2$ deposition. The obtained results confirm the significant dependence of atmospheric corrosion of all exposed metals on $SO_2$ air pollution and show that carbon steel atmospheric corrosion depends on $SO_2$ pollution level much more than zinc and copper atmospheric corrosion. The reduction of $SO_2$ has an impact on both short- and long-term corrosion rate. The exposure programs continue as one-year and long-term programs with extended measurement of the other types of pollution because, in past decades, the impact of $SO_2$ has markedly prevailed over any other pollutants.

**Acknowledgments:** This research work and costs to publishing them were supported by program MPO—IP 9/2017.

**Conflicts of Interest:** The authors declare no conflict of interest.

## References

1. Barton, K.; Knotkova, D.; Spanily, J. Errechnung der Geschwindigkeit der atmospharischen Korrosion nach meteorologischen und Luftverunreinigungswerten. In *Vortrage des Wissenschaftlichen Kollquims Bearbeitung meteorologischer Werte in Zusammenhang mit der Erforschung der Atmospharischen Korrosion*; CSSR: Kladno, Czech Republic, 1973; pp. 36–52.
2. Barton, K. *Protection against Atmospheric Corrosion*; ISBN 0-471-01349-8; J. Wiley & Sons: Hoboken, NJ, USA, 1976.
3. Knotkova, D.; Bosek, B.; Vlckova, J. Corrosion Aggressively of model regions of Czechoslovakia. In *Corrosion in Natural Environments*; ASTM STP 558; ASTM: West Conshohocken, PA, USA, 1974; pp. 52–74.
4. Kreislova, K.; Geiplova, H.; Majtas, D. Long-term study of structural metals' atmospheric corrosion in the Czech Republic. In Proceedings of the EUROCORR 2016, Montpellier, France, 11–15 September 2016.
5. Kreislova, K.; Knotkova, D. Corrosion behaviour of structural metals in respect to long-term changes in the atmospheric environment. In Proceedings of the EUROCORR 2011, Stockholm, Sweden, 5–8 September 2011.
6. *Cold Rolled Low Carbon Steel Flat Products for Cold Forming—Technical Delivery Conditions*; British Standards Institution: London, UK, 2006; EN 10130.

7.   *Hot Rolled Products of Structural Steels—Part 5: Technical Delivery Conditions for Structural Steels with Improved Atmospheric Corrosion Resistance*; EN 10025-5; British Standards Institution: London, UK, 2004.

8.   *Protection against Corrosion—Atmospheric Test Stations—General Requirements (Czech Technical Standard)*; CSN 03 8110; Czechoslovak Standardization Institute: Prague, Czech Republic, 1978.

9.   *Metals and Alloys—Atmospheric Corrosion Testing—General Requirements*; ISO 8565; International Organization for Standardization: Geneva, Switzerland, 2011.

10.  *Corrosion of Metals and Alloys—Corrosivity of Atmospheres—Measurement of Environmental Parameters Affecting Corrosivity of Atmospheres*; ISO 9225; International Organization for Standardization: Geneva, Switzerland, 2012.

11.  Kreislova, K.; Knotkova, D.; Tidblad, J.; Henriksen, J. Trends in corrosivity of atmosphere and material deterioration in Europe region in period 1987–2001. In Proceedings of the Abstracts of Conference Acid Rain 2005, Prague, Czech Republic, 12–17 June 2005; p. 697.

12.  Arroyave, C.; Morcillo, M. The effect of nitrogen oxides in atmospheric corrosion of metals. *Corros. Sci.* **1995**, *37*, 293–305. [CrossRef]

13.  *Corrosion of Metals and Alloys—Removal of Corrosion Products from Corrosion Test Specimens*; ISO 8407; International Organization for Standardization: Geneva, Switzerland, 2009.

14.  *Corrosion of Metals and Alloys—Corrosivity of Atmospheres—Classification, Determination and Estimation*; ISO 9223; International Organization for Standardization: Geneva, Switzerland, 2012.

15.  *Corrosion of Metals and Alloys—Corrosivity of Atmospheres—Guiding Values for Corrosivity Categories*; ISO 9224; International Organization for Standardization: Geneva, Switzerland, 2012.

16.  Knotkova, D.; Kreislova, K.; Dean, W.S. *ISOCORRAG—International Atmospheric Exposure Program: Summary of Results*; ASTM Data Serie 71; ASTM International: West Conshohocken, PA, USA, 2010; ISBN: 978-0-8031-7011-7.

17.  Kucera, V.; Tidblad, J.; Kreislova, K.; Knotkova, D.; Faller, M.; Reiss, D.; Snethlage, R.; Yates, T.; Henriksen, J.; Schreiner, M.; et al. UN/ECE ICP Materials dose-response functions for the multi-pollutant situation. *Water Air Soil Pollut. Focus* **2007**, *7*, 249–258. [CrossRef]

18.  Watt, J.; Tidblad, J.; Kucera, V.; Hamilton, R. (Eds.) *The Effect of Air Pollution on Cultural Heritage*; Springer: Berlin, Germany, 2009; ISBN: 978-0-387-84892-1.

19.  Geiplova, H.; Kreislova, K.; Mindos, L.; Turek, L. Evaluation of long-term exposed structures and their maintenance. Corrosion and Surface Treatment in Industry. *Mater. Sci. Forum* **2016**, *844*, 79–82. [CrossRef]

20.  Hatch, J.E. (Ed.) *Aluminium—Properties and Physical Metallurgy*; ASM: Geauga County, OH, USA, 1984; pp. 242–264.

21.  Vargel, C. *Corrosion of Aluminium*; Elsevier Science: Oxford, UK, 2004; ISBN: 0-08-044495-4.

*materials*

MDPI

*Review*

# Marine Atmospheric Corrosion of Carbon Steel: A Review

Jenifer Alcántara, Daniel de la Fuente, Belén Chico, Joaquín Simancas, Iván Díaz and Manuel Morcillo *

National Centre for Metallurgical Research (CENIM/CSIC), Avda. Gregorio del Amo n° 8, 28040 Madrid, Spain; j.alcantara@cenim.csic.es (J.A.); delafuente@cenim.csic.es (D.d.l.F.); bchico@cenim.csic.es (B.C.); jsimancas@cenim.csic.es (J.S.); ivan.diaz@cenim.csic.es (I.D.)
* Correspondence: morcillo@cenim.csic.es; Tel.: +34-915-538-900

Academic Editor: Yong-Cheng Lin
Received: 9 March 2017; Accepted: 7 April 2017; Published: 13 April 2017

**Abstract:** The atmospheric corrosion of carbon steel is an extensive topic that has been studied over the years by many researchers. However, until relatively recently, surprisingly little attention has been paid to the action of marine chlorides. Corrosion in coastal regions is a particularly relevant issue due the latter's great importance to human society. About half of the world's population lives in coastal regions and the industrialisation of developing countries tends to concentrate production plants close to the sea. Until the start of the 21st century, research on the basic mechanisms of rust formation in $Cl^-$-rich atmospheres was limited to just a small number of studies. However, in recent years, scientific understanding of marine atmospheric corrosion has advanced greatly, and in the authors' opinion a sufficient body of knowledge has been built up in published scientific papers to warrant an up-to-date review of the current state-of-the-art and to assess what issues still need to be addressed. That is the purpose of the present review. After a preliminary section devoted to basic concepts on atmospheric corrosion, the marine atmosphere, and experimentation on marine atmospheric corrosion, the paper addresses key aspects such as the most significant corrosion products, the characteristics of the rust layers formed, and the mechanisms of steel corrosion in marine atmospheres. Special attention is then paid to important matters such as coastal-industrial atmospheres and long-term behaviour of carbon steel exposed to marine atmospheres. The work ends with a section dedicated to issues pending, noting a series of questions in relation with which greater research efforts would seem to be necessary.

**Keywords:** atmospheric corrosion; marine environment; carbon steel

---

## 1. Introduction

Steel is the most commonly employed metallic material in open-air structures, being used to make a wide range of equipment and metallic structures due to its low cost and good mechanical strength. Much of the steel that is manufactured is exposed to outdoor conditions, often in highly polluted atmospheres where corrosion is much more severe than in clean rural environments.

The atmospheric corrosion (AC) of carbon steel (CS) is an extensive topic that has been studied by many researchers. Useful books and chapters have been published by a number of authors [1–8].

Since the 1920's much time and effort has been devoted to studying the corrosion of metals in natural atmospheres. As a result, the importance of various meteorological and pollution parameters on metallic corrosion in now fairly well known. The effect of sulfur dioxide ($SO_2$) on AC has been widely studied, but until relatively recently researchers have paid surprisingly little attention to the action of marine chlorides in AC, despite it being well known that airborne salt in coastal regions promotes a marked increase in AC rates compared to clean atmospheres.

The issue of corrosion in coastal regions is particularly relevant in view of the latter's great importance to human society. About half of the world's population lives in coastal regions and the industrialisation of developing countries tends to concentrate production plants close to the sea.

The first rigorous study on the salinity of marine atmospheres and its effect on metallic corrosion was carried out in Nigeria by Ambler and Bain [9] and dates from 1955. For many years it was simply accepted that marine chlorides dissolved in the aqueous adlayer considerably raised the conductivity of the electrolyte on the metal surface and tended to destroy any passivating films. In 1973 Barton noted that the mechanism governing the effects of chloride ions ($Cl^-$) in AC had not been completely explained, and that the higher corrosion rate of steel in marine atmospheres could also be due to other causes, such as: (a) the hygroscopic nature of $Cl^-$ species (sodium chloride (NaCl), calcium chloride ($CaCl_2$), magnesium chloride ($MgCl_2$)), which promotes the electrochemical corrosion process by favouring the formation of electrolytes at relatively low relative humidity (RH); and (b) the solubility of the corrosion products. Thus, in the case of iron, which does not form stable basic chlorides, the action of chlorides is more pronounced than with other metals (zinc, copper, etc.) whose basic salts are only slightly soluble [4].

In the year 2000, Nishimura et al. noted that with the exception of a few studies, research on the basic mechanisms of rust formation in $Cl^-$-rich marine atmospheres had been rather scarce [10]. Since then, scientific knowledge of marine atmospheric corrosion (MAC) has advanced greatly, perhaps as a result of the need to develop new weathering steels (WS) with greater MAC resistance than conventional WS, whose main limitation is precisely their low corrosion resistance in this type of environment [11]. This hypothesis seems to be confirmed by the high proportion of MAC studies that consider this type of materials.

Therefore, this is a relatively young scientific field and there continue to be great gaps in its comprehension [12]. Nevertheless, in the authors' opinion a considerable body of knowledge has been built up in a large number of published scientific papers, and it is now time to make an up-to-date review of the current state-of-the-art and to assess what issues still need to be addressed. That is the purpose of the present review.

## 2. Basic Concepts

The AC of metals is an electrochemical process which is the sum of individual processes that take place when an aqueous adlayer forms on the metal. This electrolyte can be either an extremely thin moisture layer (just a few monolayers) or an aqueous film of hundreds of microns in thickness (when the metal is perceptibly wet). Aqueous precipitation (rain, fog, etc.) and humidity condensation due to temperature changes (dew), capillary condensation when the surfaces are covered with corrosion products or with deposits of solid particles, and chemical condensation due to the hygroscopic properties of certain polluting substances deposited on the metallic surface, are the main promoters of metallic corrosion in the atmosphere [2]. Recent studies on the wetting of metal surfaces in order to understand the process controlling AC, as well as the effect of RH on steel corrosion in the presence of sea salt aerosols (NaCl and $MgCl_2$) can be found in references [13–16].

The magnitude of AC is basically controlled by the length of time that the surface is wet, though it ultimately depends on a series of factors such as RH, temperature, exposure conditions, atmospheric pollution, metal composition, rust properties, etc. [5,17]. The AC process involves simultaneous oxidation and reduction reactions which can be accompanied by other chemical reactions in which the corrosion products may take part.

The anodic reaction, consisting of the oxidation of the metal, can be given as:

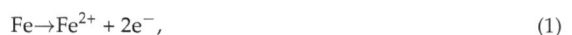

$$Fe \rightarrow Fe^{2+} + 2e^-, \tag{1}$$

Oxygen ($O_2$), which is highly soluble in the aqueous layer, is a possible electron acceptor. Oxygen reduction in neutral or basic media takes place according to the reaction:

$$O_2 + 2H_2O + 4e^- \rightarrow 4OH^-, \tag{2}$$

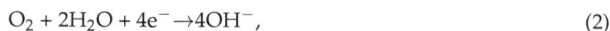

The hydroxide ions migrate to anodic areas, forming ferrous hydroxide [$Fe(OH)_2$] as the initial corrosion product.

Oxygen diffusion through the aqueous adlayer is usually a corrosion rate-controlling factor. The corrosion rate reaches a maximum value for intermediate thicknesses of the aqueous adlayer on the metal surface. The joining up of individual droplets to form relatively thick electrolyte layers somewhat reduces the rate of attack, as it hampers the arrival of oxygen. On the other hand, an excessive decrease in the moisture layer thickness halts the corrosion process, due to the high ohmic resistance of very thin layers where the ionisation and dissolution reactions of the metal are obstructed. Fast drying and repeated wetting of the surface leads to stronger corrosion effects. During drying periods, the convective currents caused by evaporation of the electrolyte lead to a decrease in the effective thickness of the diffusion layer, with the consequent rise in the transportation rate of cathodic depolariser, thus making the corrosion rate a cathodically controlled process. The electrolyte is self-stirring during evaporation [3,18].

Another factor that substantially determines the intensity of the corrosive phenomenon is the chemical composition of the atmosphere (air pollution by gases, acid vapours or seawater aerosols). $SO_2$ and $NaCl$ are the most common corrosive agents in the atmosphere. Nitrogen oxides ($NO_x$) are another important source of atmospheric pollution.

## 2.1. Sulfur Dioxide

The effect of $SO_2$ on AC has been studied by many authors [7]. $SO_2$ is often found in the atmosphere in concentrations that vary considerably depending on the type of industries in the region, the presence of power plants, time of year, etc. $SO_2$ is much more aggressive to steel when its concentration exceeds $0.1 \ mg \cdot m^{-3}$, a level that is easily reached in many towns, especially in winter. Fortunately, the $SO_2$ concentration in urban air has decreased greatly in recent years due to efforts to reduce pollution [19].

Rozenfeld [3] has shown that $SO_2$ is also an active cathodic depolarising agent due to its susceptibility to be reduced on metals. $SO_2$ is some 2600 times more soluble in water than oxygen, so even if the $SO_2$ gas content in the atmosphere is very small, its concentration in the electrolyte and its effect can be similar to that of oxygen, which is the depolarising agent par excellence. Thus, above a certain acidity level in polluted atmospheres, $SO_2$ can act as an oxidising agent and greatly accelerate the cathodic process.

Rainwater can absorb $SO_2$ from the atmosphere as it falls, giving rise to what is known as acid rain. For this reason the pH of rainwater collected downwind of highly industrialised regions of Europe sometimes presents clearly acid values, as in Norway [20], where average daily and monthly measurements of down to pH 2.9 have been recorded. In such situations, the cathodic reaction of hydrogen evolution can be relevant.

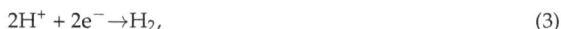

$$2H^+ + 2e^- \rightarrow H_2, \tag{3}$$

Kucera [21] distinguishes between the rinsing effect of rainwater, which tends to wash away pollutants that accumulate on the metallic surface, and the harmful effect of acid precipitation. In terms of corrosion, Kucera suggests a predominance of the rinsing effect in appreciably polluted areas, whereas in rural areas rainwater with a circumstantially low pH may worsen the situation.

$SO_2$ gives rise to the formation and propagation of sulfate "nests", according to reactions (4,5), which start to appear at isolated points on the surface but whose number increases until all the surface is coated with a rust film (Figure 1a) [22].

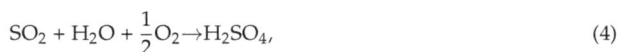

$$SO_2 + H_2O + \frac{1}{2}O_2 \rightarrow H_2SO_4, \tag{4}$$

$$2H_2SO_4 + 2Fe + O_2 \rightarrow 2H_2O + 2FeSO_4, \tag{5}$$

Hydrolysis of the ferrous sulfate formed in these nests controls their propagation (reactions 6,7).

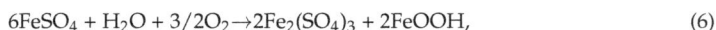

$$6FeSO_4 + H_2O + 3/2O_2 \rightarrow 2Fe_2(SO_4)_3 + 2FeOOH, \tag{6}$$

$$Fe_2(SO_4)_3 + 4H_2O \rightarrow 2FeOOH + 3H_2SO_4, \tag{7}$$

Osmotic pressure may cause the nests to burst, thus raising the corrosion rate [22].

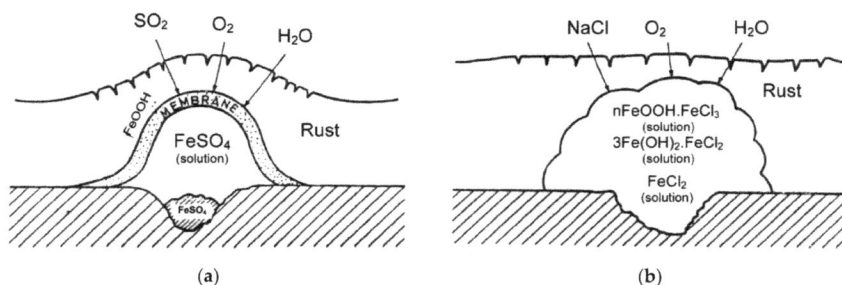

Figure 1. Schematic representation of a sulphate nest (a) and a chloride agglomeration (b) [22].

## 2.2. Saltwater Aerosols

The deposition of salt particles on a metallic surface accelerates its corrosion, especially, as in the case of chlorides, if they can give rise to soluble corrosion products rather than the only slightly soluble products formed in pure water.

$Cl^-$ ions are abundant in marine atmospheres, where the fundamental source of mineralisation consists of saltwater particles that are carried along by air masses as they pass over seas, oceans and salt lakes [3]. According to Ambler and Bain [9], only salt particles and droplets of more than 10 μm cause corrosion when deposited on a metallic surface. Given that such particles remain in the atmosphere for a short time, usually corrosion completely loses its marine character just a few kilometres inland.

For salt to accelerate corrosion the metallic surface needs to be wet. The RH level that marks the point at which salt starts to absorb water from the atmosphere (hygroscopicity) seems to be critical from the point of view of corrosion.

As has been noted above, the effect of $Cl^-$ ions on CS corrosion mechanisms has been much less widely studied than the effect of $SO_2$. A high $Cl^-$ concentration in the aqueous adlayer on the metal and high moisture retention in very deteriorated areas of the rust give rise to the formation of ferrous chloride ($FeCl_2$), which hydrolyses the water:

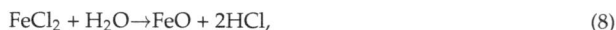

$$FeCl_2 + H_2O \rightarrow FeO + 2HCl, \tag{8}$$

Notably raising the acidity of the electrolyte. In this situation the cathodic reaction (3) becomes important, accelerating the corrosion process. The anolyte on the steel surface and in the pits that have formed becomes saturated (or close to saturation) with the highly acidic $FeCl_2$ solution. Both the metallic cations and hydrogen ions require neutralisation, which occurs by the entry of $Cl^-$ ions, but this leads to an increase in the $Cl^-$ concentration which intensifies metal dissolution, giving rise in turn to the entry of more $Cl^-$, which further intensifies the corrosion process. This attack mechanism is fed by the corrosion products themselves (feedback mechanism), and it is sometimes referred to as "autocatalytic" [23].

Unlike $SO_2$ pollution, $Cl^-$ pollution does not cause the formation of nests but $Cl^-$ agglomerates. The literature also sometimes mentions the formation of "chloride nests" [24], but the osmotic pressure of $FeCl_2$ or NaCl does not influence corrosive activity, which instead is determined by other causes such as the ability of ferrous and ferric chlorides to form complexes ($nFeOOH \cdot FeCl_3$) or a solution of $FeCl_3$ in FeOOH in the form of a gel. No amorphous oxide/hydroxide membrane is originated (Figure 1b) [22].

*2.3. Hydrogen Chloride Vapours*

Askey et al. [25] published an interesting study of iron corrosion by atmospheric hydrogen chloride (HCl), in which they suggest a direct reaction between CS and HCl. HCl reacts directly with the metal to produce soluble $FeCl_2$, which is then oxidised to FeOOH, releasing HCl.

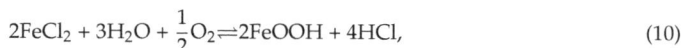

$$Fe + 2HCl + \frac{1}{2}O_2 \rightleftharpoons FeCl_2 + H_2O, \tag{9}$$

$$2FeCl_2 + 3H_2O + \frac{1}{2}O_2 \rightleftharpoons 2FeOOH + 4HCl, \tag{10}$$

It should be noted that this is a reaction cycle in which rereleased HCl reacts with iron to form fresh $FeCl_2$. Once started, therefore, the cycle is independent of incoming HCl. Corrosion continues until the $Cl^-$ ions are removed (possibly by washing away of $FeCl_2$) after which fresh incoming HCl reacts with the metal surface and reinitiates the cycle. The cycle proposed above is analogous to the acid regeneration cycle proposed by Schikorr [26] for the action of $SO_2$ on iron.

## 3. Experimentation on Marine Atmospheric Corrosion

Most studies of MAC have involved the performance of field exposure tests. Specimens of appropriate dimensions are mounted on racks using porcelain or plastic insulating clips. The exposure angle is generally 45° to the horizontal in Europe or 30° to the horizontal in the United States. It is general practice to have the panel racks facing south in the northern hemisphere or north in the southern hemisphere. However, when panels are mounted in coastal locations it is desirable to have the racks facing the shore. Further to these general requirements it is advised to follow the appropriate specific standards that have been published [27,28].

Atmospheric exposure tests usually involve the use of flat specimens of metals and alloys, although wire specimens are sometimes also used, such as in the ISOCORRAG International Atmospheric Exposure Program [29]. It is increasingly common to use both flat and wire specimens to evaluate the aggressivity (corrosivity) of atmospheres [30]. It is also interesting to note the quick response and high sensitivity of the wire-on-bolt technique, using specimens originally devised by Bell Telephone Laboratories [31] and subsequently developed by the Canadian company Alcan International [32]. This technique consists of the atmospheric exposure for just three months of metallic wires wound firmly around bolts of another metal. The functioning of the galvanic couple depends among other factors on the atmosphere where it is exposed. Doyle and Godard report that aluminium wire wound around an iron bolt is highly sensitive to the marine atmosphere [33].

The specimens are exposed to the atmosphere for a given time and subsequently analysed in the laboratory by gravimetric techniques to determine the corrosion losses experienced. Mass-loss data allows structural integrity to be estimated after a given number of years of service. Structural engineers typically use mass-loss data to overbuild a structure, allowing for a given mass loss over the predicted lifetime.

Over the course of time different accelerated tests have been developed to simulate AC in the laboratory. The disadvantage of accelerated corrosion tests is that the results obtained do not always coincide with those found in real atmospheric exposure tests. For instance, the laboratory simulation of marine atmosphere exposure has long been carried out by constant exposure to salt fog. However, the classic salt fog test has a bad reputation and is unanimously considered to offer poor reproducibility and correlation with atmospheric exposure. The application of intermittent salt spray, on the other hand, is a much better approximation to marine and coastal conditions [34], and the cyclic salt fog test, along with the use of alternative saline solutions to NaCl, provides much better correlations [35].

As has been noted above, the existence and duration of wetting and drying stages play an important role in AC mechanisms. The need for wet/dry cycles to simulate AC is now well established and any accelerated laboratory test for this purpose needs to take this aspect into account. Today's standard accelerated tests for the simulation of atmospheric exposure are all based on wet/dry cycles. Although some standards set out conditions for the performance of such tests [36], they tend to be carried out in many different ways, and the results of one researcher are not always comparable to those of another. Despite this difficulty, analysis of the data obtained by researchers throughout the world in the most varied of experimental conditions has led to great advances in the knowledge of MAC. As will be mentioned below, it would nevertheless be desirable to standardise a universal wet/dry cyclic test that could be followed by all researchers so that the results obtained would be comparable.

Considerable efforts have been made to obtain reliable estimates of AC without the limitations of gravimetric tests, especially in terms of their enormous duration. Very good results have been obtained with electrochemical cells [37]. Electrochemical techniques, and in particular impedance measurements, have been widely used by many researchers in numerous studies related with AC.

In particular, attention is drawn to the important atmospheric rusting cycle mechanism proposed by Stratmann [38]. In an electrochemical study of phase transitions in rust layers, Stratmann showed that when a pre-rusted iron sample was wetted, iron dissolution was not immediately balanced by a reaction with oxygen, but rather by the reduction of the preexisting rust (lepidocrocite) with later reoxidation of the reduced species. Thus, Stratmann [38] proposed dividing the AC mechanism of pure iron into the following three stages (Figure 2): (a) wetting of the dry surface; (b) wet surface; and (c) drying-out of the surface.

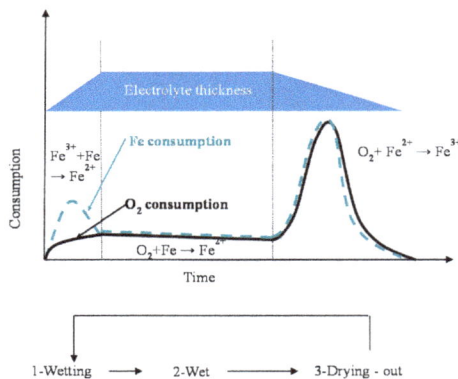

**Figure 2.** Schematic representation according to Stratmann [38] of atmospheric rusting cyclic mechanism.

Nishimura et al. [10,39] carried out a laboratory study of the electrochemical behaviour of rust formed on CS in wet/dry cycles in solutions containing $Cl^-$ ions, simulating exposure in marine atmospheres. They observed that akaganeite formation was the cause which enormously accelerated the AC process in this type of atmosphere, being electrochemically reduced and consumed during the wetting of the metallic surface, in contrast to the important role played by lepidocrocite in steel corrosion in $Cl^-$-free atmospheres [38].

Over the last decade, great advances have been made in the understanding of AC mechanisms. Many of these advances have been due to the research groups of Legrand [40] and Dillmann [41]. Basic research carried out in this field has been related with a greater knowledge of the electrochemical reactivity of the ferric phases that constitute atmospheric rust and the coupling or decoupling of anodic and cathodic reactions.

A fine characterisation of corrosion product layers identifying the oxide phases on a metal surface yields valuable information on the evolution of the corrosion process in a given atmosphere. Not only is it important to identify the different oxides but also to ascertain the fraction of each corrosion product and its distribution in the rust layer in order to gain a better understanding of the corrosion process. The distribution of the phases can drastically influence local corrosion mechanisms. Elemental composition can be determined by energy dispersive X-ray (EDX) analysis and electron probe microanalysis (EPMA), and several macroscopic techniques such as X-ray diffraction (XRD), infrared spectroscopy (IRS) and Mössbauer spectroscopy (MS) are commonly used for corrosion products characterisation. However, it may be very important to obtain a fine and local determination of the structure of corrosion products in order to understand corrosion mechanisms. In such cases the local structure of the corrosion layers must be characterised with the help of microprobes: Raman microspectroscopy (μRS), X-ray microdiffraction (μXRD), X-ray absorption spectroscopy (XAS), etc. The specificities of each analysis method strongly influence the type of phase identified [11].

XRD is one of the most commonly used techniques for identifying the rust composition and the structure of different components in corrosion products. One of the limitations of XRD is the separate identification of magnetite and maghemite. Both oxides have a cubic structure and nearly identical lattice parameters at room temperature, making them nearly indistinguishable by XRD. However, their magnetic and electric properties are quite different, thereby allowing MS to identify each. According to Cook [42], corrosion research is one area in which MS has become a required analytical technique. This is in part due to the need to identify and quantify the nanophase iron oxides that are nearly transparent to most other spectroscopic techniques.

Rust composition studies using μRS usually demand a high laser power for the excitation of spectra because some of the most common iron oxides and oxyhydroxides are poor light scatterers. Sample degradation frequently occurs under intense sample illumination and may lead to the misinterpretation of spectra. Low laser power minimises the risks of spectral changes due to sample degradation [43]. Moreover, in some cases, particularly when the phases are less crystallised, it is difficult to discriminate one phase from another only by μRS because the Raman shift is very close.

According to Monnier et al., the use of complementary analytical techniques (μXRD, XAS and MS) is needed to obtain accurate Raman phase characterisation. Each technique provides complementary information. μXRD is more sensitive to crystallised phases while μRS presents a higher spatial resolution and allows the detection and location of crystallised phases (goethite, lepidocrocite, maghemite, akaganeite) from less crystallised ones (feroxyhyte, ferrihydrite). Discrimination of maghemite, feroxyhyte and ferrihydrite could be partially solved by the use of XAS [44].

Knowledge of the rust layer structure is another aspect widely studied by researchers. The techniques traditionally used are optical microscopy (OM), polarised light microscopy, scanning electron microscopy (SEM) and transmission electron microscopy (TEM)/electron diffraction (ED). In order to characterise corrosion product structures in various scales, Kimura et al. [45] use several analytical approaches that are sensitive to three structural-correlation lengths: long-range order (LRO) (>50 nm), middle-range order (MRO) (~1–50 nm), and short-range order (SRO) (<1 nm) and Konishi et al. employ X-ray absorption fine structure analysis (XAFS) methods, including extended XAFS and X-ray absorption near-edge structure (XANES), for characterisation of rust layers formed on Fe, Fe-Ni and Fe-Cr alloys exposed to $Cl^-$-rich environments [46].

Complementary analyses on the porosity of rust layers have also been conducted by several researchers. Dillmann et al. use various techniques such as small-angle X-ray scattering (SAXS), Brunauer-Emmett-Teller (BET) and mercury intrusion porosimetry (MIP) [47]. Attention is also drawn to the studies of Ishikawa et al., where the specific surface area of the pores was calculated by fitting the BET equation to $N_2$ adsorption isotherms [48].

Over the past few decades, the new analytical techniques developed to study properties of solid surfaces, such as chemical composition, oxidation state, morphology, structure, etc. have continued to increase and improve in terms of resolution and sensitivity. The more recent analytical techniques are both surface-sensitive and able to provide information under in-situ conditions. According to Leygraf et al. [8], it is anticipated that the number and variety of in-situ techniques for probing surfaces will continue to increase.

## 4. The Marine Atmosphere

From the point of view of MAC, the marine atmosphere is characterised by the presence of marine aerosol. $Cl^-$ ions are abundant in marine atmospheres, where the fundamental source of mineralisation consists of saltwater particles that are carried along by air masses as they pass over seas, oceans, and salt lakes. Marine salts are mainly NaCl, but quite appreciable amounts of potassium, magnesium and calcium ions are also found in rainfall.

### 4.1. Atmospheric Salinity

Atmospheric salinity is a parameter related with the amount of marine aerosol present in the atmosphere at a certain geographic point. Marine aerosols, consisting of wet aerosols, partially wet aerosols and non-equilibrium aerosols depending on the atmospheric humidity, are carried along by the wind and can come into contact with metallic structures and greatly accelerate the corrosion process. The sizes of the three types of aerosols and the resultant dry aerosols were estimated by Cole et al. [49].

Salinity in marine atmospheres varies within very broad limits (<5 to >300 mg $Cl^-/m^2 \cdot d$) [50]. While extremely high values have been recorded close to surf, salinity at other points on the shoreline near calmer waters is no more than moderate. The concentration of marine aerosol decreases with altitude [51,52]. Meira et al. find that this relationship can be represented by an exponential decrease function which is influenced by the wind regime [52].

An increase in wind speed, even on the same coast, does not always lead to an increase in salinity, as the final result is dependent on the wind direction. In fact, an increase in wind speed can even reduce the degree of pollution by purifying the exposure site of pollutant. This will naturally depend on the situation of the exposure site in relation with the sea, and on the direction and type of winds blowing at a given time.

Marine aerosol is comprised of fine particles suspended in the air (jet drops, film drops, brine drops and sea-salt particles), solid or liquid, whose sizes vary from a few angstroms to several hundred microns in diameter [53]. Marine aerosol particles are usually classified by size into two classes: coarse particles, with an equivalent aerodynamic diameter of >2 μm; and small particles, with a diameter of <2 μm. Fine particles are in turn subdivided into Aitken nuclei (<0.05 μm) and particles formed by accumulation (with diameters of between 0.05 and 2 μm). In coastal locations (<2 km from the seashore) the most common aerosols deposited are in the coarse size range: 2–100 μm in diameter [54,55].

Large marine aerosol particles (diameter >10 μm) remain for only a short time in the atmosphere; the larger the particle size, the shorter the time. On the other hand, particles of a diameter of <10 μm may travel hundreds of kilometres in the air without sedimenting [9,56].

Li and Hihara [55] in a study of natural salt particle deposition on CS for only 30 min in a severe marine site in Hawaii, found that most airborne salt particles had diameters ranging from approximately 2–10 μm, and varied in composition ranging from almost pure NaCl or KCl to mixtures of NaCl, KCl, $CaCl_2$ and $MgCl_2$. These differences in composition may depend on whether the seawater droplets dehydrate, crystallise and fragment while airborne, or if they are deposited as liquid droplets before crystallising.

## 4.2. Production of Marine Aerosol

Cole et al. described marine aerosol formation, chemistry, reaction with atmospheric gases, transport, deposition onto surfaces, and reaction with surface oxides [57].

The wind, which stirs up and carries along seawater particles, is the force responsible for the salinity present in marine atmospheres. Oceanic air is rich in marine aerosols resulting from the evaporation of drops of seawater, mechanically transported by the wind. The origin, concentration and vertical distribution of marine aerosol over the surface of the sea has been studied by Blanchard and Woodcock [51].

The first step in the production of aerosol particles is the breaking of waves [51,58,59]. The turbulence that accompanies this phenomenon introduces air bubbles into the water which subsequently burst and launch sea salt particles into the atmosphere. On the high seas the breaking of waves depends on the speed of the wind blowing over them. In the coastal surf zone, waves can break without the need for simultaneous wind action, and the amount of aerosol generated is largely dependent on the type of sea floor (uniformity, slope, etc.) and the width of the surf zone.

Aerosol levels at the seashore depend both on the aerosol that is generated out at sea, which is carried to the coast by marine winds, and that which is generated in the surf zone close to the shoreline [60–62]. Of the two, the latter seems to be the main contributor to the $Cl^-$ levels measured in the lower layers of the atmosphere in coastal areas [61,63].

A relationship has been seen between salinity and wave height [64]. The graph in Figure 3 shows the variation of atmospheric salinity with the average spectral wave height. As can be seen in the figure, monthly average spectral wave height values of 1.5–2.0 m are sufficient to produce high monthly average salinity values of 100–200 mg $Cl^-/m^2 \cdot d$.

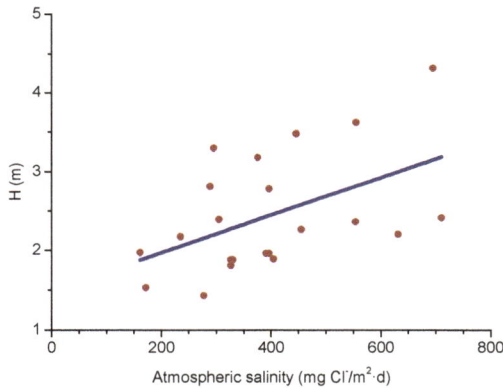

**Figure 3.** Variation of monthly average salinity with monthly average spectral wave height values (H). The regression line shows the general trend [64].

Several studies in the literature have attempted to relate aerosol levels measured at high sea and on the coast with wind speed. Potential and exponential type functions express the considerable effect of this variable on marine aerosol production (especially when the wind speed exceeds some 3–5 m/s) [51,65–67]. In Figure 4, Morcillo et al. note that the wind only needs to blow short time at speeds above 3 m/s in directions with high entrainment of marine aerosol (they call them "saline winds") for atmospheric salinity to reach important values [66].

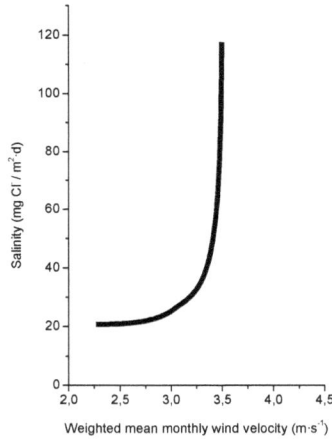

**Figure 4.** Relation between atmosphere salinity and weighted mean monthly wind velocity for marine winds [66].

### 4.3. Entrainment of Marine Aerosol Inland

Aerosol particles can be entrained inland by marine winds (winds proceeding from the sea), settling after a certain time and after travelling a certain distance. The wind regime directly influences aerosol production and transportation, and is significantly affected by geostrophic winds, large-scale atmospheric stability, and the difference between diurnal land and sea temperatures, which varies according to the season of the year. It is also dependent on the latitude, ruggedness of the coastline, and undulation of the land surface [51,58]. A reduction in the size and mass of aerosol particles due to drying of the droplets can considerably increase the entrainment distance.

In a recent study by Alcántara et al. [64], it was seen that the variation in salinity with the distance from the shoreline (Figure 5) clearly showed an exponential relationship.

$$Y = 78288.23 \exp(-X/91.34) + 108.40, \qquad (11)$$

Being Y the atmospheric salinity expressed as mg $Cl^-/m^2 \cdot d$ and X the distance from the shore in meters (m).

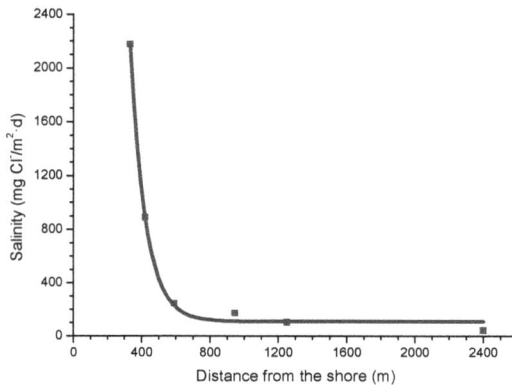

**Figure 5.** Variation in atmospheric salinity with distance from the shore [64].

Comparison of the atmospheric salinity at an exposure site and the average wind speed often fails to yield a clear relationship between both parameters. Calculation of the average wind speed takes into account the speeds recorded in all wind directions, and not only marine winds, which it would be reasonable to suppose are those that govern the presence of marine aerosol masses in coastal regions. In this sense it would be interesting to know whether the atmospheric salinity at the site is related with the run of marine winds, which is the sum of adding together the marine wind speed in each direction multiplied by the time it has been blowing [64]. Figure 6 has been prepared accordingly and clearly explains the decrease in the salinity value obtained in the second three-month period by a decline in the run of all marine winds, especially the most frequent (north-easterly, NE). Therefore, more than the average wind speed in the study area, the total run of marine winds is the parameter that has the greatest influence on the atmospheric salinity of the test site.

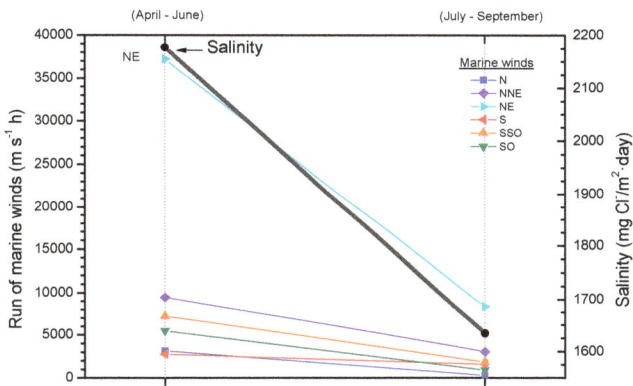

**Figure 6.** Salinity data and run of different marine winds in two three-month periods [64].

Nevertheless, the topography of the land and the general wind regime of the zone can also lead continental winds to influence salinity values. This has been shown in a study by the Academy of Sciences in Russia [65,68] involving long-term studies at Murmansk and Vladivostok, which concluded that chloride entrainment in both areas was dependent on both the average speed of total winds (marine + continental) and the product of the wind speed by its duration (wind power).

Further studies on this effect would be of great interest as a fuller knowledge would allow, for instance, the estimation of atmospheric salinities simply by analysing information on winds available in existing meteorological databases. The inclusion of salinity values in the numerous published damage functions between corrosion and environmental factors would also make it possible to estimate MAC at a specific site from meteorological data without the need to carry out lengthy and expensive natural corrosion tests.

### 4.4. Effect of Salinity on Steel Corrosion

For salt to accelerate corrosion the metallic surface must be wet. Preston and Sanyal [69] showed that corrosion of an iron surface under a deposit of NaCl particles starts to be seen at 70% RH, and is notably accelerated at higher RH. However, Evans and Taylor also note that sea salt particles cause corrosion at a lower RH than NaCl particles, due to the fact that sea salt contains very hygroscopic magnesium salts [70].

#### 4.4.1. Steel Corrosion versus Salinity

In studies of MAC a direct relationship is generally established between corrosion and the saline content of the atmosphere. Ambler and Bain were the first to demonstrate this relationship [9].

In Figure 7, corresponding to the studies of Ambler and Bain [9], it is clearly seen how steel corrosion already experiences a notable acceleration at low atmospheric salinities, increasing from 10 to 100 mg $Cl^-/m^2 \cdot d$, as has also been reported by many other researchers. This effect has subsequently been addressed in other papers [64,71]. Figure 8 shows the variation in the CS corrosion rate with atmospheric salinity over a broad spectrum of airborne salt concentrations [64].

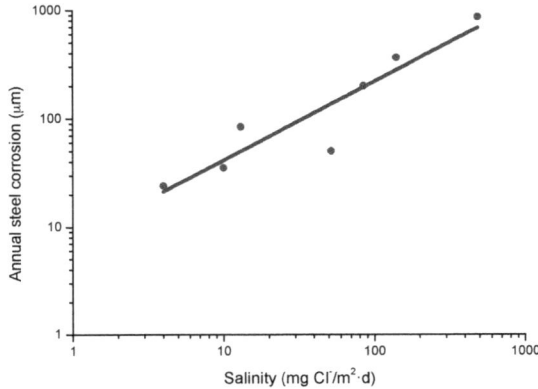

**Figure 7.** Annual steel corrosion versus salinity according to Ambler and Bain [9].

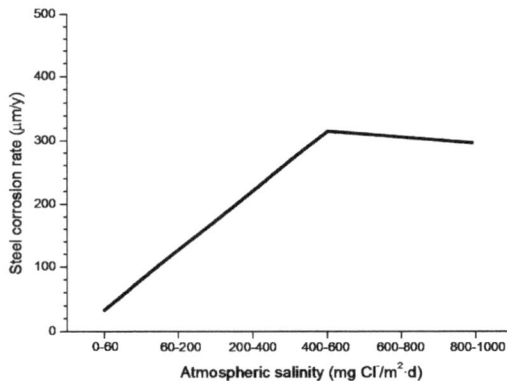

**Figure 8.** Variation in the corrosion rate of mild steel with salinity over a broad spectrum of atmospheric salinities. The graph shows a trend. Information obtained in an exhaustive bibliographic search [64].

For salinities of less than 600 mg $Cl^-/m^2 \cdot d$, a linear relationship trend between both parameters can be deduced, with the CS corrosion rate increasing considerably as the atmospheric salinity rises. For salinities above this value the corrosion rate seems to be stabilised.

Only a small number of MAC studies have been carried out at sites with very high atmospheric salinities. Morcillo et al. observed less steel corrosion at sites with a high $Cl^-$ deposition rate (1905 mg/$m^2 \cdot d$) than at other very nearest site with lower atmospheric salinity values (824 mg/$m^2 \cdot d$) [12]. The explanation of this fact lies in the lower oxygen solubility in the aqueous layer on the metallic surface at a very high $Cl^-$ concentration. Oxygen is a fundamental element for the cathodic process of metallic corrosion. This finding is not an isolated occurrence. Pascual Marqui [72] explains this effect in terms of competitive adsorption: at high $Cl^-$ concentrations the adsorbed $O_2$ concentration on the metal surface is lower, in contrast to the adsorption of $Cl^-$ ions. Espada et al. [73] also observed this effect with salt fogs at high NaCl concentrations. In another study by Hache [74] it was experimentally seen in immersion tests that both steel corrosion and dissolved oxygen decreased when the saline solution concentration exceeded a threshold of 10 g NaCl/L.

The literature contains very little steel corrosion data corresponding to salinities above 600 mg $Cl^-/m^2 \cdot d$. It would be important to have more information from very severe marine atmospheres in order to rigorously confirm these observations.

### 4.4.2. Steel Corrosion versus Distance from the Shore

The influence of the distance from the sea is one of the most important aspects of MAC in coastal areas. Empirically, it is known that the effect of marine atmospheres basically runs to a few hundred metres from the shoreline and decays rapidly further inland.

The complexity of the phenomena associated with MAC makes it difficult to devise a model that can cover all possible scenarios. However, for areas closest to the shoreline (~400 to 600 m), published data shows that the decrease in the corrosion rate with the distance from the sea is fairly well represented by a simple exponential relationship [60].

$$C = C_0 \exp(-\beta X) + A, \tag{12}$$

where C is the corrosion rate, $C_0$ is the corrosion rate at the shoreline; $\beta$ is a constant; X is the distance inland from the shoreline; and A is the corrosion rate at zero salinity.

Bearing in mind the aforementioned relationship between steel corrosion and atmospheric salinity, the variation in corrosion with the distance from the shore (Figure 9) should be an exponential function similar to that observed in Figure 5 between this variable and atmospheric salinity.

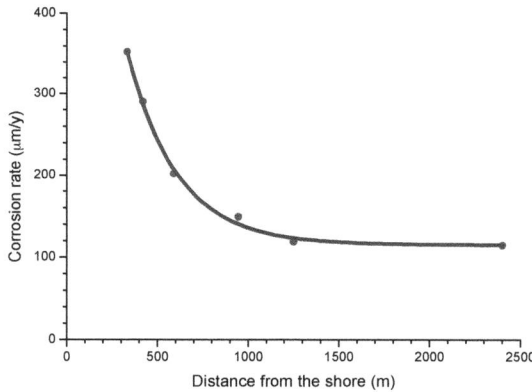

**Figure 9.** Variation in the corrosion rate of mild steel with distance from the shore [64].

### 4.5. Measurement of Atmospheric Salinity

Airborne salinity is the amount of marine aerosol present in a given marine atmosphere, and a value that is commonly measured in corrosion studies. Strekalov carried out an important review of this matter [75].

In MAC studies chlorides are usually captured by the wet candle method [9] or the dry cloth (or gauze) method [76,77], both of which are set out in ISO standard 9225 [78]. The dry cloth method was developed in the former Soviet Union and is also widely used in Asia. Both methods offer great benefits from the point of view of corrosion studies as they are suitable for long-term measurements (usually one month in duration) and the fact that their data refers to the amount of salt deposited per unit of surface area (generally expressed as mg $Cl^-/m^2 \cdot d$), which is a more relevant indicator for the corrosion process than the saline content per unit of air volume. Foran et al. [79] suggest the possibility of measuring atmospheric salinity simply by determining the amount of chlorides dissolved in the rainwater collected in pluviometers.

As is noted in ISO 9223 standard [30], the results obtained by applying these various methods are not always directly comparable or convertible. In fact, ISO 9225 standard [78] provides a number of conversion factors. Corvo et al. [80] find that the following relationship:

$$[Cl^-]_{wc} \ (mg/m^2 \cdot d) = -54.5 + 1.6 \ [Cl^-]_{dc} \ (mg/m^2 \cdot d), \tag{13}$$

where:
$[Cl^-]_{wc}$ = salinity determined by the wet candle method
$[Cl^-]_{dc}$ = salinity determined by the dry cloth method
Is only valid for salinity values of a considerable magnitude.

Figure 10 shows the relationship between the $Cl^-$ deposition rates measured using both methods at two sites in Japan [81]. It is seen that the wet candle method is more sensitive to the presence of NaCl, capturing a greater amount of aerosol than the gauze method for NaCl levels of more than 5 $mg/m^2 \cdot d$.

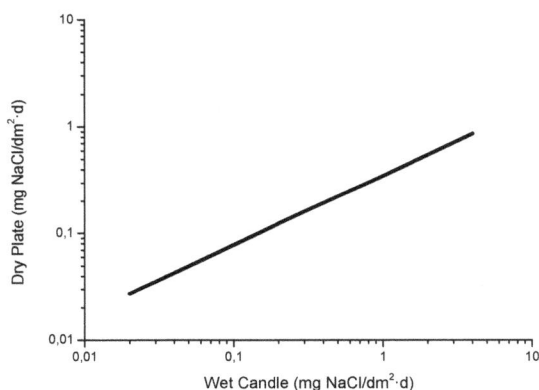

**Figure 10.** Relationship between salinity values reported by wet candle and dry plate methods [81].

In Australia, the INGALV Corrosion Mapping System [82] appears to calculate the $Cl^-$ deposition rate at any location primarily on the basis of its proximity to the coast. However, it may be misleading to rely on simple subjective appreciations in the hope of correlating environment and pollution. For instance, points relatively close to the shoreline may in fact have lower $Cl^-$ levels that a simple glance at the area and its surroundings might seem to suggest.

Finally, the Civil Research Institute (CRI) of the Ministry of Construction in Japan developed the CRI method in order to allow the absorption of a larger amount of salt than the relatively limited gauze method (Japanese standard JIS-Z-2381 [83]). This method uses a large capacity salt collector [84].

### 4.6. Salt Lake Atmospheres

Very little research work has focused on steel corrosion in salt lake environments, though several papers have recently been published in relation with Qinghai salt lake in north-west China [85,86]. Qinghai salt lake possesses an extremely high $Cl^-$ ion concentration, with an average of 34 wt % $Cl^-$ ions in the salt lake water, ten times that of seawater. It is important to note that the Mg content in the salt water of this lake is very high, much higher than other cations in the salt lake water and in seawater, and must be taken into account that the critical RH for $MgCl_2$ is 35%, much lower than the 75% corresponding to NaCl.

According to the authors, as the steel surface went though wet/dry cycles the alien magnesium cations got a good chance to participate in the corrosion reactions by replacing the ferrous ions and forming Mg-containing intermediates.

*4.7. Deicing Salts*

Although typically associated with marine environments, NaCl is actually more prevalent in the environment from the use of road deicing salt. Extensive use of deicing salts for snow removal, generally NaCl with small amounts of $CaCl_2$ and $MgCl_2$, began in the early 1960s. Heavy use of deicing salt, as much as 20 tons per lane mile per year, is common throughout regions of the snow belt in the northern states of the US. The widespread use of salt has been associated with a significant amount of damage to the environment and highway structures. Road spray, dirt and salts are carried by the air blast created by heavy traffic and quickly contaminate horizontal specimens. The prolonged wet period caused by deposits, chlorides and sulfates in close contact with the steel tends to accelerate poultice corrosion [42,87].

It is well known that deicing salts often cause corrosion problems and produce thick and flaky rust on steel bridges. This kind of rust is strongly dependent upon the local environment and topography around the bridges, where the RH is usually high, the air circulation is poor, and the steels are exposed to wetness for long times. In addition, chlorides accumulate in the rusts on girders that receive less washing from rainfall. Cook et al. [42] have evaluated several WS bridges in the USA exposed to road deicing salt and showing signs of significant corrosion and exfoliated rust. Rust samples have been collected from steel girders directly above roadways that are regularly deiced during winter. In these locations total thickness losses of about 1.5 mm have been measured on the girders over a period of 20 years.

Takebe et al. [88] estimated the amount of $Cl^-$ from deicing salts on WS used for bridges and developed a method to evaluate the amount of salt present on bridge girders due to deicing salts. The sampling method is described in [89].

In a review of publications on this matter in the USA and Japan, Hara et al. [90] reported that $Cl^-$ concentrations exceeding approximately 0.2–0.3 wt % in the rusts accelerated the increase in rust thickness and led to the development of extremely thick rust layers. A countermeasure to this problem is to periodically wash adhered salts from the girders. According to Hara et al. [90], periodic washing with pressurised tap water, delivering 2–4 MPa (high pressure washing) at the outlet nozzle, effectively suppresses the growth of rust particles by $Cl^-$ ions and the development of thick rust layers, and may be useful as a suppression technology for deicing salts.

## 5. Atmospheric Corrosion Products

Atmospheric corrosion products of iron, referred to as rust, comprise various types of oxides, hydroxides, oxyhydroxides and miscellaneous crystalline and amorphous substances (chlorides, sulfates, nitrates, carbonates, etc.) that form as a result of the reaction between iron and the atmosphere [91] (Table 1).

In marine environments other rust products not listed in Table 1 may also appear, in some cases quite significantly. These include ferrous and ferric chlorides ($FeCl_2$ and $FeCl_3$), ferrous-ferric chloride [$Fe_4Cl_2(OH)_7$], etc., which are highly stable and therefore easily leachable from the corrosion product layers during atmospheric exposure. Table 2 shows the iron corrosion species that contain chlorine in their composition. Gilberg and Seeley [92] have investigated the context in which $Cl^-$ ions can be found within iron corrosion products. Thus, they note that $FeCl_3$ and FeOCl are unstable to hydrolysis, being converted to akaganeite.

After short-term atmospheric exposure, oxyhydroxides (lepidocrocite, goethite, and akaganeite) and oxides (magnetite and maghemite) are the main crystalline products comprising the rust layers. The composition of the rust layer depends on the conditions in the aqueous adlayer and thus varies according to the type of atmosphere.

**Table 1.** Iron corrosion species according Cornell and Schwertmann [91].

| Type | Name | Formula |
|---|---|---|
| Oxides | Magnetite | $Fe_3O_4$ |
| | Maghemite | $\gamma\text{-}Fe_2O_3$ |
| | Hematite | $\alpha\text{-}Fe_2O_3$ |
| Hydroxides | - | $Fe(OH)_2$ |
| | Bernalite | $Fe(OH)_3$ |
| | Green rusts | $Fe_x^{III} Fe_y^{II} (OH)_{3x+2y-z} (A^-)_z$ where $A^- = Cl^-; \frac{1}{2}SO_4{}^{2-}$ |
| | Ferrydrite | $Fe_5O_8H \cdot H_2O$ |
| Oxyhydroxides | Goethite | $\alpha\text{-}FeOOH$ |
| | Lepidocrocite | $\gamma\text{-}FeOOH$ |
| | Akaganeite | $\beta\text{-}FeOOH$ |
| | Feroxyhite | $\delta\text{-}FeOOH$ |
| | Schwertmannite | $Fe_{16}O_{16}(OH)_y(SO_4)_z \cdot nH_2O$ |

**Table 2.** Iron corrosion species containing chloride [92].

| Name | Formula |
|---|---|
| Ferrous chloride (lawrencite) | $FeCl_2$ |
| Ferric chloride (molysite) | $FeCl_3$ |
| Ferric oxychloride | $FeOCl$ |
| Ferrous hydroxychloride | $\beta\text{-}Fe_2(OH)_3Cl$ |
| Green rusts | GR1 (GR Cl) |
| $\beta$-oxihydroxide (akaganeite) | $\beta\text{-}FeOOH$ |

One matter that has not yet been completely clarified is the content and composition of the amorphous phase of rust. Authors often try to study the structure of the corrosion products by quantitative powder XRD. The amorphous phase represents the difference between the sum of all the crystallised phase portions and 100%. According to Dillmann et al. [47], because powder XRD quantitative measurements are not very precise (about 10–20 relative percent error), measurements of the amorphous part of rust provided by this method need to be considered with great caution. If the techniques normally used to identify the different crystalline phases of rust (XRD, Fourier transform infrared (FTIR), MS, Raman spectroscopy (RS)) often have difficulty discriminating one phase from another, in the case of less crystallised (amorphous) rust phases such as feroxyhyte ($\delta$-FeOOH), ferrihydrite, etc. this difficulty is further exacerbated. Monnier et al. [93] and Neff et al. [94] suggest combining the use of different complementary techniques in order to obtain improved characterisation, e.g., using $\mu$XRD, X-ray absorption under synchrotron radiation, $\mu$RS, etc.

It is unanimously accepted that lepidocrocite ($\gamma$-FeOOH) is the primary crystalline corrosion product formed in the atmosphere. In marine atmospheres, where the surface electrolyte contains chlorides, akaganeite ($\beta$-FeOOH) is also formed. As the exposure time increases and the rust layer becomes thicker, the active lepidocrocite is partially transformed into goethite ($\alpha$-FeOOH) and spinel (magnetite ($Fe_3O_4$)/maghemite ($\gamma$-$Fe_2O_3$)). An increase in the airborne $Cl^-$ deposition rate is accompanied by a drop in the lepidocrocite content of the rust and a rise in the goethite, akaganeite and spinel contents [95], as will be seen later.

Bernal et al. [96] in 1959 identified the conditions for the formation of the different iron oxides and hydroxides, noting the importance of the physicochemical conditions of the aqueous adlayer on the oxidation products of $Fe(OH)_2$: goethite, feroxyhite, green rusts, spinels, etc. They also noted that the common feature of the group of iron oxides and hydroxides was that they were composed of different stackings of close-packed oxygen/hydroxyl sheets, with various arrangements of the iron ions in the octahedral or tetrahedral interstices.

## 5.1. Most Significant Corrosion Products in Steel Corrosion in Marine Atmospheres

The following section focuses on the most significant corrosion products of steel when exposed in $Cl^-$-rich atmospheres, describing their formation mechanisms, structure, etc.

### 5.1.1. Green Rust 1 (GR1 or GR($Cl^-$))

Green rusts (GR) are unstable intermediate products, very often amorphous, which occasionally emerge in the presence of anions such as $Cl^-$, $SO_4^{2-}$, etc., and replace $OH^-$ ions in processes involving the ferrous-ferric transformation of hydroxides, oxides and oxyhydroxides in poorly aerated environments. Their name is derived from their bluish-green colour [97]. Two broad groups of GR have been distinguished. One contains primarily monovalent anions such as $OH^-$ and $Cl^-$ and is designated GR1, while the other contains mainly divalent ions such as $SO_4^{2-}$ and is designated GR2 [98].

Green rusts rarely exhibit a well-defined stoichiometry and their composition depends on the particular environmental conditions. The formula of GR1 sometimes reported in the literature is $[3Fe(OH)_2 \cdot Fe(OH)_2Cl \cdot nH_2O]$, containing an equal number of $Cl^-$ and $Fe^{3+}$ ions, while GR2 conforms to the formula $[2Fe(OH)_3 \cdot 4Fe(OH)_2 \cdot FeSO_4 \cdot nH_2O]$ [99].

The crystal structures of GRs are assumed to be similar to that of the mineral pyroaurite [100], $Mg_6^{II}Fe_2^{III}(OH)_{16}CO_3 \cdot 4H_2O$. According to Refait et al. [99] a structural model derived from the pyroaurite structure can be reasonably proposed for GR1. The Fe atoms of the hydroxide layers are randomly distributed among the octahedral positions. The interlayers are mainly composed of $Cl^-$ ions and $O_2$ atoms belonging to the water molecules connecting two $OH^-$ ions of adjacent hydroxide layers.

GR1 is usually prepared by aerial oxidation of $Fe(OH)_2$ suspensions in the presence of a slight excess of dissolved $FeCl_2$. Thus, in slightly basic and $Cl^-$-containing aqueous media, GR1 should be obtained as a corrosion product of iron and steels either by oxidation of an initial $Fe(OH)_2$ layer or by direct precipitation in the simultaneous presence of $Fe^{2+}$ and $Fe^{3+}$ dissolved species.

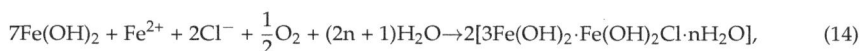

$$7Fe(OH)_2 + Fe^{2+} + 2Cl^- + \frac{1}{2}O_2 + (2n+1)H_2O \rightarrow 2[3Fe(OH)_2 \cdot Fe(OH)_2Cl \cdot nH_2O], \qquad (14)$$

GR1 found in $Cl^-$-containing aqueous media occurs during the corrosion of steels before the formation of the end products such as lepidocrocite, goethite, akaganeite and magnetite, as its formation is more favoured [99].

### 5.1.2. Akaganeite (($\beta$-FeOOH) or $\beta$-FeO(OH,$Cl^-$))

Akaganeite is the rust phase of capital importance in the MAC process of steel, and thus is discussed here in the greatest detail. Akaganeite is one of the polymorphs of ferric oxyhydroxides (-FeOOH). Its formation requires halogen ions to stabilise its crystalline structure. Since it always contains $Cl^-$ ions, this compound is not strictly speaking an oxyhydroxide. Stahl et al. have determined its chemical formula as $FeO_{0.833}(OH)_{1.167}Cl_{0.167}$ [101].

Watson et al. [102] observed that the crystal possessed a regular porous structure and suggested that the subcrystals might not be solid rods but tubes which, though externally still square prisms, contained a circular central channel or tunnel running the whole length of the subcrystal. The tunnels in the akaganeite structure, with a diameter of 0.21–0.24 nm, are stabilised by $Cl^-$ ions, and $Cl^-$ levels ranging from 2 to 7 mol % have been reported. A minimum amount of $Cl^-$, 0.25–0.50 mmol/mol seems essential to stabilise the crystalline structure of akaganeite [91]. According to Keller [103], akaganeite has been shown to contain up to 5 wt % $Cl^-$ ions in marine atmospheres. At ambient temperature these tunnels are full of water and $Cl^-$ [91]. The impossibility of leaching the $Cl^-$ by washing confirms that at least part of the $Cl^-$ ions are found in the crystalline lattice, as noted by Rezel et al. [104] and Ståhl et al. [101].

Gallagher [105] describes akaganeite as a fascinating substance that precipitates as unusual cigar-shaped crystals with a tetragonal unit cell, although this has given rise to much controversy.

The structural refinement by XRD (Rietveld) carried out by Post and Buchwald confirms that the unit cell is monoclinic, having eight formula units per unit cell [106]. The $Fe^{3+}$ ions were each surrounded octahedrally by six $OH^-$ ions.

The crystals are very small and the crystallographic structure is isostructural with hollandite ($BaMn_8O_{16}$) characterised by the presence of tunnels parallel to the C-axis of the lattice. The size distribution of the crystals is fairly narrow and their length is only exceptionally greater than 500 nm. Due to its special structure (presence of tunnels) akaganeite is less dense than other oxyhydroxides like lepidocrocite or goethite [107,108]. In this respect, Shiotani et al. note that akaganeite has a relatively larger volume in relation to the initial iron [109].

Akaganeite displays two basic morphologies: somatoids (spindle-shaped crystals) and rods (cigar-shaped crystals). The former type is the usual morphology of akaganeite when it forms in laboratory conditions by hydrolysis of acid $FeCl_3$ solutions at 25–100 °C [107,108]. The latter type, according to the authors' experience, is the usual morphology of the akaganeite crystals that form in atmospheric conditions. Researchers have assigned SEM morphologies to akaganeite without an unequivocal characterisation of this oxyhydroxide. Morcillo et al. were able to do this using the SEM/µRS technique [110,111], observing aggregates of akaganeite crystals, a sponge-type morphology, constituted by a lattice of elongated cylinder- or tube-shaped crystals typical of the rod morphology (cigar-shaped crystals) of this oxyhydroxide.

With regard to akaganeite formation mechanisms when steel is exposed to a $Cl^-$-rich marine atmosphere, the following may be noted. The formation of akaganeite is preceded by the accumulation of $Cl^-$ ions in the aqueous adlayer giving rise to the formation of $FeCl_2$, which hydrolyses water according to:

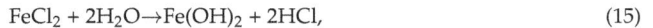

$$FeCl_2 + 2H_2O \rightarrow Fe(OH)_2 + 2HCl, \tag{15}$$

At the steel/corrosion products interface, where $Cl^-$ ions accumulate, high $Cl^-$ concentrations and acidic conditions with pH values between 4 and 6 give rise to the formation of ferrous hydroxychloride ($\beta$-$Fe_2(OH)_3Cl$), a very slow process requiring the transformation of metastable precursors [112,113]. Remazeilles and Refait concluded that large amounts of dissolved Fe(II) species and high $Cl^-$ concentrations are both necessary for akaganeite formation [114]. The oxidation process of $\beta$-$Fe_2(OH)_3Cl$ which leads to akaganeite formation passes through different steps via the formation of intermediate GR1 ($Fe_3{}^{II}Fe^{III}(OH)_8Cl^-nH_2O$) [99,113,115]. In all, the whole oxidation process leading to akaganeite can be summarised as follows [99,112–115]:

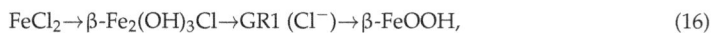

$$FeCl_2 \rightarrow \beta\text{-}Fe_2(OH)_3Cl \rightarrow GR1\ (Cl^-) \rightarrow \beta\text{-}FeOOH, \tag{16}$$

Thus requiring a relatively long time.

5.1.3. Magnetite ($Fe_3O_4$)/Maghemite ($\gamma$-$Fe_2O_3$)

The structure of magnetite is that of an inverse spinel. Magnetite has a face-centred cubic unit cell based on thirty-two $O^{2-}$ ions which are regularly cubic close packed. There are eight formula units per unit cell [91]. Magnetite differs from most other iron oxides in that it contains both divalent and trivalent iron. Its formula is written as $Y[XY]O_4$, where $X = Fe^{II}$, $Y = Fe^{III}$ and the brackets denote octahedral sites. Eight tetrahedral sites are distributed between $Fe^{II}$ and $Fe^{III}$, i.e., the trivalent ions occupy both tetrahedral and octahedral sites. The structure consists of octahedral and mixed tetrahedral/octahedral layers [91]. However, magnetite, if it were the normal spinel structure, would have eight tetrahedral sites occupied by eight $Fe^{2+}$ ions and sixteen octahedral sites occupied by sixteen $Fe^{3+}$ ions [116].

In stoichiometric magnetite $Fe^{II}/Fe^{III} = 0.5$, however, magnetite is frequently non-stoichiometric, in which case it has a cation-deficient $Fe^{III}$ sub-lattice. Magnetite is also said to have a defect structure with a narrow composition range, the Fe:O ratio of which varies from 0.750 to 0.744 [117]. Thus, magnetite usually presents vacancies, preferably on octahedral sites, which form to maintain the electroneutrality of the crystal when $H_2O$ or $OH^-$ molecules enter the network, as well as ferrous and ferric ions sharing their valence electrons.

Maghemite has a similar structure to magnetite, but differs in that all or most Fe is in the trivalent state. Cation vacancies compensate for the oxidation of $Fe^{II}$. Maghemite also has a cubic unit cell, each cell contains thirty-two $O^{2-}$ ions, twenty-one and one-third $Fe^{III}$ ions and two and a third vacancies. Eight cations occupy tetrahedral sites and the remaining cations are randomly distributed over the octahedral sites. The vacancies are confined to the octahedral sites [91]. Maghemite is also a defect structure with the Fe:O ratio in the range of 0.67–0.72 [117].

XRD presents an important limitation when it comes to differentiating the magnetite phase from the maghemite phase, as both show practically identical diffractograms (similar crystalline structures) and are very hard to differentiate when mixed with large amounts of other phases (lepidocrocite, goethite and akaganeite), as occurs in the corrosion products formed on steel when exposed to marine atmospheres. Both phases are associated to the diffraction angle at 35° [118]. Both phases are usually detected in the inner part of the rust adhering to the steel surface, where oxygen depletion can occur [119,120].

Spinel phase (magnetite and/or maghemite) may form by oxidation of $Fe(OH)_2$ or intermediate ferrous-ferric species such as green rust [119]. It may also be formed by lepidocrocite reduction in the presence of a limited oxygen supply [120] according to:

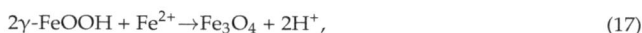

$$2\gamma\text{-FeOOH} + Fe^{2+} \rightarrow Fe_3O_4 + 2H^+, \tag{17}$$

With a broader view, Ishikawa et al. [121] and Tanaka et al. [122] found that the formation of magnetite particles was caused by the reaction of dissolved ferric species of oxyhydroxides with ferrous species in the solution, in the following order: akaganeite > lepidocrocite >> goethite. The formation of magnetite can be represented as the following reaction:

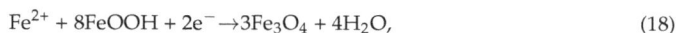

$$Fe^{2+} + 8FeOOH + 2e^- \rightarrow 3Fe_3O_4 + 4H_2O, \tag{18}$$

Remazeilles and Refait [114], Nishimura et al. [10] and Lair et al. [40] all observe that the electrochemical reduction of oxyhydroxides leads to spinel phase formation.

As Hiller noted some time ago, the rust formed in marine atmospheres contains more magnetite than that formed in $Cl^-$-free atmospheres [123]. In severe marine atmospheres the spinel phase can be the main rust constituent, as was found by Jeffrey and Melchers [124] and by Haces et al. [125].

There is often uncertainty as to which of the two phases, magnetite or maghemite, is present in AC products, and indeed both species could be present depending on the local formation conditions and the corrosion mechanisms involved in the process. This lack of definition may also be intimately related with the analytical techniques used for their determination. Many researchers have reported the presence of magnetite in AC products on the basis of XRD data, but much of this data is suspect since the XRD patterns of magnetite and maghemite are very similar. The same happens when the ED method is used [126]. However, Graham and Cohen [127] do show convincing evidence on the basis of MS that magnetite is a component of corrosion products on several samples. However, Leidheiser and Music [128] and Chico et al. [129], also using this technique, found no evidence of magnetite. Likewise, Oh et al. [130], using MS and RS, find a high magnetic maghemite content in the exposure of CS at 250 m from the seashore. In contrast, Nishimura et al. [10], using X-ray photoelectron spectroscopy (XPS) and TEM, find high magnetite contents.

Antony et al. [131] reported that FTIR is also not very appropriate to precisely identify magnetite, and Monnier et al. [93] also note that MS has difficulty in discriminating phases of the same oxidation state that have similar local environments, particularly in the case of complex mixes, as is the case of magnetite and maghemite. Thus, it seems that the specific nature of each analysis method strongly influences the type of phase identified.

The identification of rust amorphous phases as well as the classification of the type of spinel formed (magnetite or maghemite) are two issues where more research effort is needed.

### 5.2. Other Characteristics of the Steel Atmospheric Corrosion Products

#### 5.2.1. Towards a Greater Knowledge of the Structure of Iron Oxides

As Bernal et al. [96] suggested in 1959, the common feature of the group of iron oxides and hydroxides is that they are composed of different stackings of close-packed oxygen/hydroxyl sheets, with various arrangements of the iron ions in the octahedral or tetrahedral interstices, and their mutual transformations are topotactic by rearrangement of the atoms. These authors interpreted in a rational crystallochemical way the transformations involving the compounds $Fe(OH)_2$, $\delta$-FeOOH, FeO, $\gamma$-$Fe_2O_3$, $\alpha$-FeOOH, $\alpha$-$Fe_2O_3$ and $Fe_3O_4$. Only some of these transformations were not topotactic and seemed to have dissimilar structures, with renucleation being necessary for the transformation process. This is the case, for instance, with $\beta$-FeOOH$\rightarrow\alpha$-$Fe_2O_3$ [96].

According to Matsubara et al. [132], it is important to know the fundamental structures of the components of iron corrosion products in order to understand the characteristic features of various types of corrosion products. The ideal crystallographic structures of three ferric oxyhydroxides—lepidocrocite, goethite and akaganeite—are described using $FeO_6$ octahedral units (Figure 11). Furthermore, the structure of a $Fe(OH)_2$ is composed of layers of $FeO_6$ octahedra intercalated with hydroxyl $OH^-$. There are also several kinds of GR containing ferric and ferrous ions which have a layered structure as $Fe(OH)_2$. In the structure of GR, the fractions of ferric and ferrous ions in layers of $FeO_6$ octahedra are variable and different anions and water molecules are intercalated between the layers. Although there are other iron oxide structures including hydroxides, they are fundamentally described in a similar way.

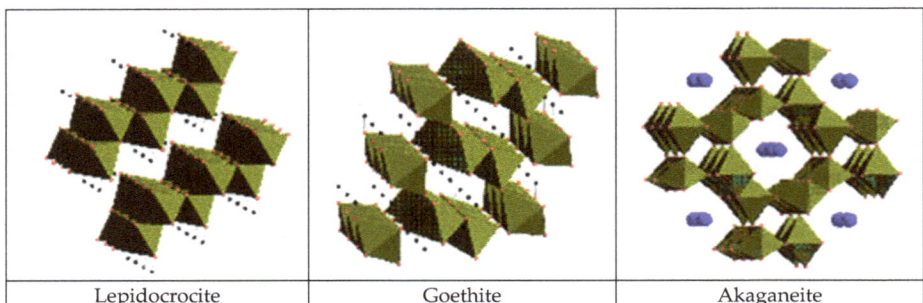

| Lepidocrocite | Goethite | Akaganeite |

**Figure 11.** Ideal crystallographic structures of lepidocrocite ($\gamma$-FeOOH), goethite ($\alpha$-FeOOH) and akaganeite ($\beta$-FeOOH). The structures are described using $FeO_6$ octahedral units. Small circles: hydrogen, medium circles: chlorine [132,133].

MS has been used to identify components in corrosion products and to analyse their fine structures. Other analytical methods, such as EPMA, TEM, FTIR and RS are often used for analysing corrosion products formed on the surface of steel. The results obtained by these methods provide information on the composition, morphology and structure of corrosion products. However, structural information on corrosion products obtained by these methods is limited [132].

As was seen in Section 3, in order to characterise corrosion product structures in various scales, Kimura et al. [45] have used several analytical approaches that are sensitive to three structural-correlation lengths (LRO, MRO and SRO). Thus, conventional XRD techniques can detect detailed structural information in terms of LRO. However, this technique yields broad peaks when the grain size is smaller than ~50 nm, as is often found in corrosion products. Contrarily, XFAS is useful for in-situ observation of SRO. In XAFS, oscillatory modulation near an X-ray absorption edge of a specific element of a specimen provides information in terms of the local structure around an atom (Fe, Cl, etc.) in the rust layer or determines the distance between the centred atom and the neighbouring ligands, the number of ligands, and the stereographic arrangement of ligands. However, some reservations should be made regarding information about linkage of the $FeO_6$ octahedral unit structure, because XAFS data is obtained only from near neighbour atomic arrangements. This strongly suggests the great importance of middle-range ordering (MRO) for characterising corrosion products. This has been achieved by a combination of X-ray scattering (AXS) and reverse Monte Carlo simulation (RMC), which visualises the atomic configurations [45].

### 5.2.2. Morphology

The rust formed on steel when exposed to the atmosphere is usually a complex mixture of several phases. Moreover, each of these phases can take on a wide variety of morphologies depending on their growth conditions. Thus, the diversity of rust morphologies formed on CS exposed to marine atmospheres is enormous, with a great variety of shapes and sizes of the crystalline aggregates that reflect to a large extent the different growth conditions: chemical characteristics of the aqueous adlayers formed by humidity condensation, rainfall, etc., temperature, wet/dry cycle characteristics, etc.

Cornell and Schwertmann dedicated an entire chapter of their well known and well referenced book "The iron oxides, structure, properties, occurrence and uses" to iron oxide crystal morphology and size, mainly concerned with synthetic iron oxides. Table 3 shows the principal habits (morphologies) of iron oxides according to Cornell and Schwertmann [91].

**Table 3.** Principal habits (morphologies) of iron oxides according to Cornell and Schwertmann [91].

| Iron Oxide | Morphology |
|---|---|
| Lepidocrocite ($\gamma$-FeOOH) | Laths |
| Goethite ($\alpha$-FeOOH) | Acicular |
| Akaganeite ($\beta$-FeOOH) | Rods, somatoids |
| Feroxyhyte ($\delta$-FeOOH) | Plates |
| Magnetite ($Fe_3O_4$) | Octohedra |
| Maghemite ($\gamma$-$Fe_2O_3$) | Laths or cubes |
| Hematite ($\alpha$-$Fe_2O_3$) | Hexagonal plates, rhombohedra |
| Ferrihydrite ($Fe_5HO_8 \cdot 4H_2O$) | Spheres |

It should however be noted that the morphologies of synthetic iron oxides produced in laboratory conditions may be very different to those obtained during CS corrosion in the atmosphere, as pointed out by Waseda and Suzuki in the preface to an interesting book on "Characterisation of corrosion products on steel surfaces" [134]. As they note, the morphology of AC products is often not describable in terms of typical iron oxide structures but is much more complicated; the component phases in rust formed on steel in outdoor exposure show imperfections in their structures and real component structures appear to diverge from an ideal crystallographic structure of typical iron oxides.

For some time now, articles published on AC studies usually include SEM views of rust formations and in some cases even attribute certain morphologies to specific rust phases without an analytical characterisation. An exception can be seen in the pioneering work of Raman et al. [135]. These researchers attempted to indirectly identify the morphologies observed by SEM by comparison with the morphologies of standard rust phases grown in the laboratory and identified by XRD and IRS.

Very recently, the research group of Morcillo et al. has progressed in this field using the powerful SEM/$\mu$RS spectroscopic technique to perform a more direct and rigorous characterisation of the different morphologies that can be displayed by the main rust phases (lepidocrocite, goethite, akaganeite and magnetite) formed on CS specimens exposed to marine atmospheres for a certain time [110,111,136,137].

Without seeking to be exhaustive, there follows a tentative classification of the different types of morphology observed by the authors in the rust formed on steel exposed to marine atmospheres [137]:

(a) Globular: hemispheric-shaped aggregated formations like small mounds.
(b) Acicular: aggregates with a similar appearance to needles, hairs, or threads.
(c) Laminar: this can appear in a wide range of different formations in which laminas grow perpendicularly to the surface: bar shape, worm nest shape, bird's nest shape, flower petal shape, feather shape, etc.
(d) Tubular: formations in which the crystalline aggregates are constituted by prisms, tubes, or rods, etc.
(e) Toroidal
(f) Geode-type: unusual or singular oolitic or globular morphology constituted by fish-egg-like spherical formations.

Figure 12 presents typical characteristic morphologies of the four rust phases normally present among the corrosion products formed in marine atmospheres: lepidocrocite, goethite, akaganeite and magnetite.

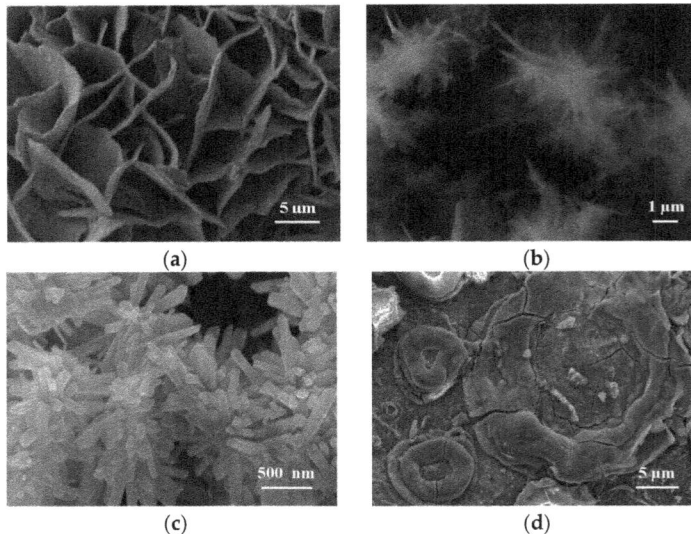

**Figure 12.** SEM view of laminar lepidocrocite (**a**); acicular goethite (**b**); tubular akaganeite (**c**) and toroidal magnetite (**d**) formations [137].

### 5.2.3. Grain Size (Granulometry)

As shown in Figure 13, from Kimura et al. [45], which shows a schematic illustration of corrosion on an iron surface in the atmosphere, the formation process of solid particles can be visualised by three steps: (a) nucleation; (b) growth; and (c) ageing [138].

(a) Nucleation corresponds to the first step of precursor condensation and solid formation. On an atomic scale, iron forms cations that are coordinated by six water molecules $[Fe(H_2O)_6]^{2+}$ [139]. In a neutral solution, metal cations react with $OH^-$, $O_2$, and $H_2O$ resulting in the formation of hydroxo cations $[Fe(OH)_X(H_2O)_{6-X}]^{(3-X)+}$.
(b) Then the growth process follows, where $Fe(O,OH)_6$ octahedra units as cations or smaller sized growing nuclei accumulate to form larger particles. On a colloidal scale, polymerisation of

these Fe(O,OH)$_6$ octahedra leads to the formation of fine particles of hydroxides, oxyhydroxides or oxides.

(c)     These particles grow into grains or layers through a long period of ageing processes affected by repeated wet and dry cycles. Reaction conditions (concentration, acidity, temperature, nature of anions, etc.) have a strong influence on the structural or morphological changes of poly-octahedra during corrosion. Coagulation and adhesion processes ensue to generate corrosion products, which undergo ageing processes leading the system to stability. During ageing the particles may undergo modifications such as increases in size, changes in crystal type, changes in morphology, etc. [140]. Thus, according to Ishikawa et al. [141], steel rusts can be regarded as agglomerates of colloidal nanoparticles of ferric oxyhydroxides (goethite, akaganeite and lepidocrocite), spinels (magnetite/maghemite), and poorly recrystallised iron oxides (amorphous substances). Voids of different sizes form between the fine particles in the rust layer.

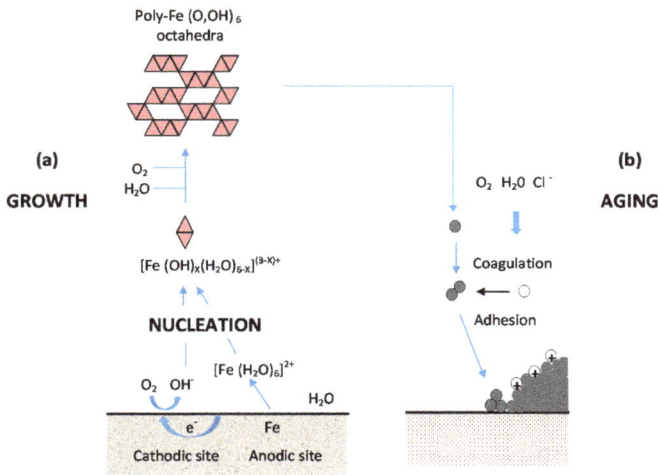

**Figure 13.** Scheme of iron corrosion in the atmosphere according Kimura et al. [45]: (**a**) reactions in the initial wet cycle; (**b**) reactions during repetition of wet/dry cycles for a long period. Triangle pairs represent Fe(O,OH)$_6$ octahedra.

Resulting from these complicated processes, corrosion products are generally classified as coarse or fine grains, both of which are composed of crystallites and inter-crystallites. The structure in the former are similar to those of ideal crystals, while in the latter the linkage of Fe(O,OH)$_6$ octahedra is disordered, due to the existence of defects and/or different sizes of Fe(O,OH)$_6$.

A practical laboratory method for determining the grain size of rusts formed on steel during atmospheric exposure, known as the "tape method" [142], consists of adhering a $2 \times 2$ cm$^2$ piece of adhesive tape to the outermost surface of the rust layer, pressing firmly and evenly on the surface, and lifting off to examine the size and density of rust particles. The morphology takes the form of grains or particles, agglomerates of grains, flakes, and even exfoliations (layers or laminates) [143] (Figure 14). The texture of rust is seen to vary according to the atmospheric aggressivity (Figure 15). A more heterogeneous surface appearance and coarser granulometry is found in more aggressive atmospheres (industrial and marine) [144]. In marine atmospheres, the granulometries are coarser and become more accentuated with airborne salinity and exposure time (Figure 16). In the marine atmosphere with the highest Cl$^-$ deposition rate (665 mg/m$^2$·d) the formation of coarse flakes and exfoliations is seen [64]. These results confirm the observations of Ishikawa et al. [144,145].

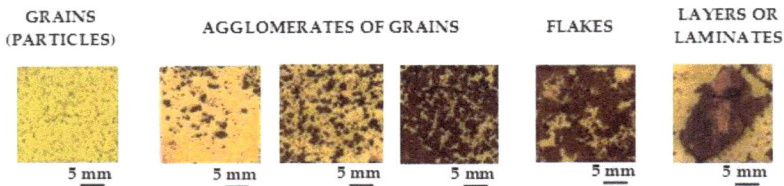

**Figure 14.** Type of rust morphologies formed on carbon steel exposed to marine atmospheres [143].

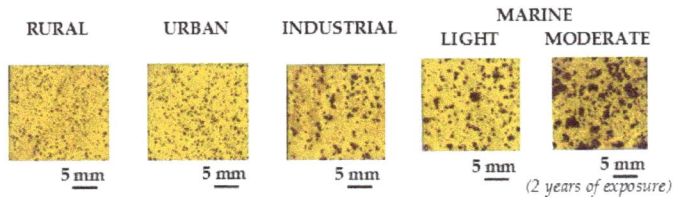

**Figure 15.** Granulometries of outermost rusts formed on skyward- facing side of carbon steel exposed for 5 years at different type of atmospheres.

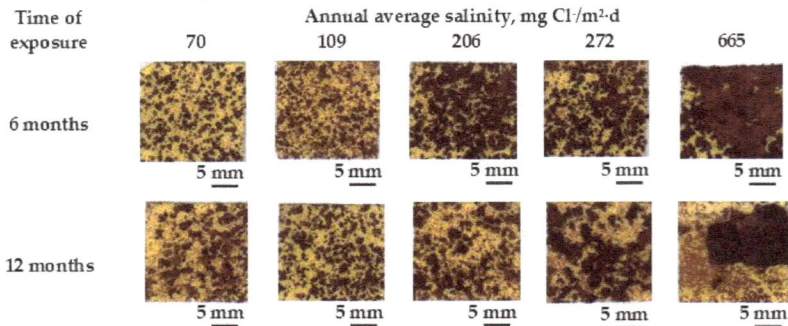

**Figure 16.** Granulometries of outermost rusts formed on skyward- facing side of carbon steel exposed for 6 and 12 months at marine atmospheres of different aggressivity [64].

The compactness of the rust layers depends on the morphology of the rust particles; smaller particles form more compact and less permeable layers. However, as Ishikawa et al. note [144], particle size analysis of rust is not easy because of the heterogeneous morphology and strong aggregation of rust particles. Ishikawa et al. [146] use the $N_2$ adsorption method to estimate the particle size of rust formed on steel exposed to various situations. It was revealed that the specific surface area (SSA) obtained by $N_2$ adsorption decreased with increasing airborne salinity (Figure 17). This finding shows that rust particles grow with an increase in airborne salinity, and that less compact rust layers with low corrosion resistance are formed in $Cl^-$ environments such as coastal areas. In contrast, in a low salinity environment fine rust particles assemble to form densely packed rust layers with high corrosion resistance. Ishikawa et al. attribute the high SSA obtained by $H_2O$ adsorption on the rusts generated on the coast to the tunnels of akaganeite crystals, accessible to $H_2O$ but not to $N_2$ [146].

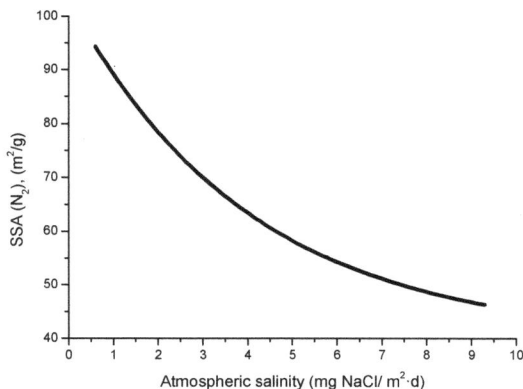

**Figure 17.** Relationships (trend curve) between SSA(N2) of rusted carbon steel and atmospheric NaCl contents [146].

Ishikawa et al., examining the texture of rusts, note that the rusts formed at coastal sites were aggomlerates of large particles and had larger pores than rusts formed at rural and urban sites. This finding suggests that NaCl promotes rust particle growth, resulting in the formation of larger pores as voids between larger particles in the rust layer and facilitating further corrosion.

## 6. The Rust Layer

When the thin layer of corrosion products has grown to cover the whole surface, further growth requires reactive species from the aqueous adlayer to be transported inwards through the rust layer while metal ions are transported outwards. In addition to this, electrons must be transported from anodic to cathodic sites on the surface, so that those produced in the anodic reaction can be consumed in the cathodic reaction. As long as the metal substrate is covered only by a thin oxide film, electron transportation through the film is generally not a rate-limiting step. However, when the corrosion products grow in thickness, electron transportation may become rate-limiting [8].

This section considers the different physical and chemical properties of corrosion product layers. It starts by addressing the organoleptical properties of rust layers, such as their colour and texture, before going on to consider other properties more related with their protective capacity: stratification, stabilisation, adhesion, thickness, and porosity and their evaluation using different indices.

### 6.1. Organoleptical Properties

#### 6.1.1. Colour

CS exposed to the atmosphere develops ochre-coloured rust which becomes dull brown as the exposure time increases. Lighter rust colours are seen in atmospheres with greater salinity (more corrosive) and darker rusts in less aggressive atmospheres [64]. In marine atmospheres, the colour of rust varies not only with the salinity of the atmosphere, but also according to the steel type, exposure time, etc.

Some time ago, LaQue [147] exposed different steels for 6 months to the marine atmosphere of Kure Beach (250 m from the shoreline) and found that the colouring of 84% of the tested steels was within the range seen in Figure 18, which shows the relationship between the rust colour ratings as developed early in the test and the corrosion resistance of the steels after long-term exposure.

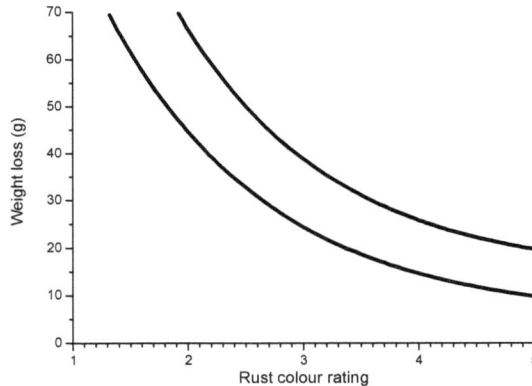

**Figure 18.** Band showing the relationships between rust colour rating after 6 months of exposure and corrosion of steels (4 × 6 in. specimens) after 7.5 years of exposure in marine atmosphere at Kure beach [147].Rust color rating: 1 (lightest)–5(darkest).

### 6.1.2. Texture

In Section 5.2.3 reference was made to the granulometry of corrosion products. In part, this property is closely related with the texture of the outer surface of the rust layer. Sense of touch is used to determine aspects of texture such as smoothness, unevenness and roughness. Ph. Doctoral Thesis of I. Díaz [148] and H. Cano [149] reported one to three-year exposure of a variety of CS in different types of atmospheres, where differences in texture were observed in the rust layers formed. Patinas with smoother textures (more homogenous appearance and finer granulometry) were found in rural and urban atmospheres, while rougher textures with a more heterogeneous appearance and a coarser granulometry were observed in industrial and marine atmospheres, all the more so the higher the corrosivity of the atmosphere (higher Cl⁻ deposition rate) and the longer the exposure time.

### 6.2. Properties More Related with the Protective Capacity of Rust Layers

#### 6.2.1. Stratification of Rust Layers

There is controversy about the stratification of the rust layer in different sublayers on unalloyed CS [97]. According to Díaz et al. [150] rust is always stratified irrespective of the steel composition, be it WS or plain CS. In their investigations these authors have found the presence of two sublayers in all rust films: an uncoloured (dark grey) inner layer and an orangey-brown-coloured outer layer (Figure 19). Thus, the dual nature of the rust layer is not an exclusive characteristic of WS since plain CS with less AC resistance also generates a stratified rust.

According to Suzuki [151], rust layers usually present considerable porosity, spallation, and cracking. Cracked and non-protective oxide layers allow corrosive species easy access to the metallic substrate, and is the typical situation in atmospheres of high aggressivity. However, compact oxide layers formed in atmospheres of low aggressivity favour the protection of the metallic substrate. The higher the Cl⁻ deposition rate in marine atmospheres, the greater the degree of flaking observed, with loosely adherent flaky rust favouring rust film breakdown (detachment, spalling) and the initiation of fresh attack. As time elapses, the number and size of defects may decrease due to compaction, agglomeration, etc. of the rust layer, thereby lowering the corrosion rate [152,153].

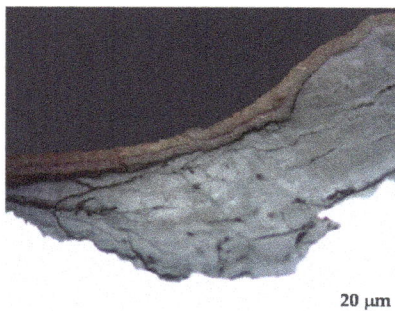

20 µm

**Figure 19.** Dual structure of a consolidated rust layer formed during one year on carbon steel exposed al unsheltered conditions in a marine atmosphere with low chloride deposit (21 mg $Cl^-/m^2 \cdot d$). Optical micrograph obtained by polarized light. The outer orange-coloured layer is mainly lepidocrocite while the inner greyish layer is mainly goethite and magnetite [150].

### 6.2.2. Stabilisation of Rust Layers and Steady-State Corrosion Rate

Bibliographic information on this aspect is highly erratic and variable. The gradual development of a corrosion layer takes several years before steady-state conditions are obtained, though the exact time taken to reach a steady state of AC will obviously depend on the environmental conditions of the atmosphere where the steel is exposed.

Morcillo et al. have determined the stabilisation times of rust layers formed on WS [87], considering the steady state corrosion rate to be the rate corresponding to the year from which corrosion slows by $\leq$10%. Previously it was confirmed that the corrosion rate (y) plotted against the exposure time (x) fitted an exponential decrease equation:

$$y = A_1 \exp(-x/t_1) + y_0, \tag{19}$$

where

y = corrosion rate, µm/y

x = time, years

$1/t_1$ = decrease constant

$y_0$ = steady state corrosion rate, µm/y

$A_1 + y_0$ = corrosion rate at x = 0, µm/y

The rust layer stabilisation time decreases as the corrosivity category (ISO 9223) [30] of the atmosphere rises. The stabilisation time depends, among other factors, on the exposure time, the existence of wet/dry cycles, the corrosivity of the atmosphere, and in short on the volume of corrosion products formed. However, a shorter stabilisation time does not imply a greater protective capability of the rust. In this respect, stabilisation of the rust layer occurs faster in marine atmospheres, due to their greater corrosivity, but the protective value of this rust is lower than that of rusts formed in less aggressive atmospheres (rural, urban, etc.) where stabilisation times are longer. The steady-state corrosion rate increases in line with the corrosivity of the atmosphere in both rural, urban, industrial, and marine atmospheres (Figure 20) [87].

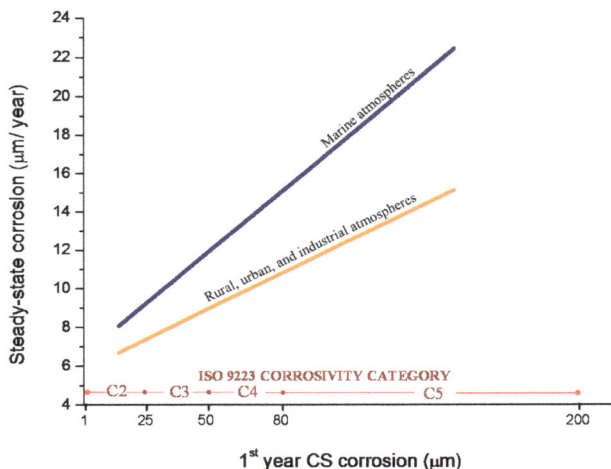

**Figure 20.** Relationship between steady-state corrosion rate of carbon steel and atmospheric corrosivity category according ISO 9223 for different type of atmospheres [87].

### 6.2.3. Adhesion

According to Honzak [154] it is possible to differentiate three layers in rust: surface rust that is easily removable (e.g., by light scraping), an intermediate rust layer that can be "burst off" by bending the specimen, and a very adherent layer on the metal surface which cannot be removed by scraping but is removable by abrasive cleaning or chemical methods (e.g., pickling).

Not all steel corrosion products become incorporated in the rust layer. Some examples [124] include:

(a) Brown stains as a consequence of run-off processes
(b) Leaching of soluble components of the rust layer (iron chlorides in marine atmosphere) by rainwater, and
(c) Rust lost through abrasion and erosion, the latter particularly in high wind areas.

It has long been known that not all the corroded metal becomes part of the measurable rust product [155], but there have been very few attempts to quantify this part. One exception is the work of García et al. [156–158], who classified adherent rust as: (i) removable by scraping the steel surface with a metallic brush; and (ii) removable by hitting the steel with a hammer. According to these authors, the protective properties of rust on CS in a given corrosive environment depends on the characteristics of the adherent rust, i.e., that which is bonded to the metal surface, but a full overview of the corrosion process also requires the characterisation of non-adherent rust, i.e., that which is loosely bound to the metal surface, and that which is lost during the corrosion process. The authors found that the amount of corroded iron that is converted into adherent rust on steels exposed to $Cl^-$ in wet/dry cycles ranged from 0.55 to 0.90, the amount of corrode iron converted into non-adherent rust ranged between 0.3 and 0.18, and the amount of iron that is lost ranged between 0.2 and 0.38.

### 6.2.4. Thickness and Internal Structure

The thickness of the rust layer increases with time of exposure and the aggressivity of the atmosphere. A direct (linear) relationship is found between the rust layer thickness and the substrate corrosion rate [159] (Figure 21).

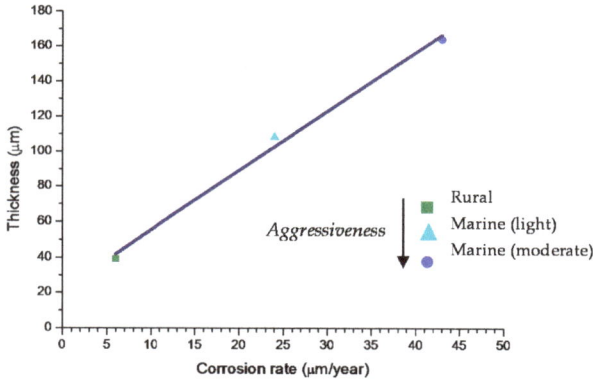

**Figure 21.** Corrosion rate versus rust thickness for low carbon steel exposed during two years in atmospheres of different aggressivity [159].

The thickness of the rust layer formed in marine atmospheres is not usually uniform, being thicker in some areas than in others, and the attack profile of the underlying steel generally shows the abundant formation of pits of variable depths [64]. Asami and Kikuchi [160] published an interesting study on in-depth distribution of rust components (determined by TEM/ED) on steel exposed to the atmosphere for 17 years under a bridge in a coastal-industrial region of Japan. They saw the aforementioned thick and thin areas within the rust layers and found that akaganeite was preferentially located in the thick areas and was scarce in the thin areas of the rust layers. The existence of akaganeite inside the pits formed on CS exposed for three months in a marine atmosphere with a high $Cl^-$ deposition rate (1136 mg/m$^2$·d) has been confirmed by RS in unpublished results obtained by the authors of this work (Figure 22).

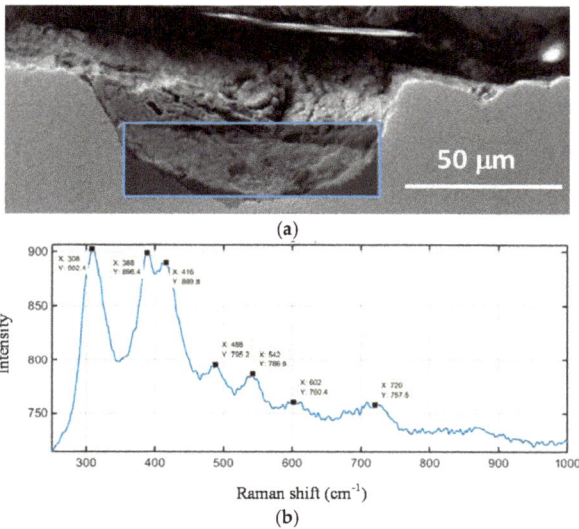

**Figure 22.** The rust present inside the pits formed on carbon steel substrates when exposed to severe marine atmospheres is almost entirely composed of akaganeite: (**a**) cross section affecting a deep pit; (**b**) Raman spectrum of rust inside the pit.

In not highly aggressive marine atmospheres, consistent (consolidated), adherent, and continuous rust layers present a two-sublayer organisation, as has been seen above (Figure 19). However, the exposure of CS in very aggressive marine atmospheres can in certain circumstances lead to the formation of heterogeneous and anomalous thick rust layers. High times of wetness of the metallic surface and an atmosphere with a high $Cl^-$ deposition rate lead to the formation of this type of rust. These thick rust layers tend to become detached from the steel substrate (exfoliated), leaving it uncovered and without protection and thus accelerating the metallic corrosion process [129]. The rust exfoliation phenomenon can only take place if such anomalous thick rust layers are formed, as has also been observed in studies carried out by other researchers [42,161]. In studies by the authors on CS corrosion in $Cl^-$-rich atmospheres, the average $Cl^-$ deposition rate needed to exceed a critical threshold of close to 300 mg/m$^2$·d for exfoliation to take place; the annual steel corrosion at that atmospheric salinity was higher than 100 μm [129].

The exfoliated rust layers are composed of multiple rust strata, this can clearly be seen in the cross section of Figures 23 and 24. Observation by optical microscopy shows that in general the thick rust layer contains one or more strata of compact rust, exhibiting a greyish colouring and a metallic shine, whose number varied according to the area of the rust layer observed. With regard to the rust exfoliation mechanism, it is recommended to consult recent publications by the authors [129,153,162].

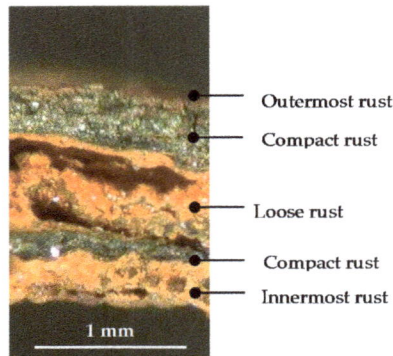

**Figure 23.** Optical micrograph of exfoliated (multilayered) rust formed on mild steel after one year in a very aggressive marine atmosphere (390 mg Cl$^-$/m$^2$·d) [129]. The characteristic of the different rust sublayer within the rust multilayer are described in Figure 24 [162].

| | DENOMINATION | CHARACTERISTICS |
|---|---|---|
| | Outermost rust | Red-brown coloured. Its surface appearance, texture and chemical composition are all typical of the outermost rust formed on mild steel in marine atmospheric |
| | Compact rust | Fragile compact rust layer exhibiting a greyish colouring and a metallic shine. Its internal structure is highly porous with abundant fine linearly arrayed voids |
| | Loose rust interlayer | Orange-red coloured. Highly porous loose rust layer |
| | Compact rust | |
| | Innermost rust | Heterogeneous pattern with orange and brown patches |

**Figure 24.** Schematic illustration of different rust sublayers in exfoliated rust (Figure 23). The denomination and characteristic of each rust layer is given [162].

6.2.5. Porosity

Voids of different sizes are formed among the rust particles in the corrosion layer, whose compactness depends on the rust particle size.

Important parameters for the protective ability of rust layers include their thickness, porosity and Specific Surface Area (SSA). According to Dillmann et al. [47] these parameters directly influence the amount of oxyhydroxides that will be in contact with the electrolyte and will be reduced. Thus, these characteristics of the rust layer directly influence the AC mechanisms.

The pore structure of the rust layer is clearly related with steel corrosion because various molecules and ions such as $O_2$, $H_2O$, and $Cl^-$ diffuse through the rust layer in the corrosion process [146,163]. Despite this fact, few studies of rust pore structures have been reported. The pore size of atmospheric rust is in the range of up to 15 nm, and the highest peak always appears below 5 nm [151].

Dillmann et al. [47] characterised the overall porosity, pore diameter and SSA of pores in rust layers using different complementary methods: MIP, BET and SAXS. However, as the authors note, these methods do not provide quantitative data on the three-dimensional distribution of pores in the rust layer, their tortuosity or their connectivity; three other parameters about which information is desirable for AC modelling [41].

Ishikawa et al. [48,141,146] have published abundant information on this subject, using the adsorption method to obtain more complete information than the aforementioned techniques on a wide range of rust layers. The SSA of rust layers is evaluated by fitting the BET equation [164] to the adsorption isotherms of nitrogen ($S_N$) and water molecules ($S_W$) and using the cross-sectional area of nitrogen and water molecules (0.162 and 0.108 $nm^2$, respectively) [165]. Among the results obtained using this technique, attention is drawn to the following which are considered particularly relevant:

(a)  Estimation of rust particle size [141]

Ishikawa et al. apply adsorption of nitrogen molecules to estimate the rust particle size, successfully demonstrating that the SSA of the rust layer is a valid means of assessing its protective nature. They saw that the pores formed between larger particles were more accessible to $O_2$, $H_2O$ and $SO_4^{2-}$ and $Cl^-$ ions, which are important substances in the AC of steel.

Compact rust layers with a high $S_N$ or a small particle size exhibit high corrosion resistance (Figure 25). The corrosion resistance of rust layers is a result of pore filling by the adsorption and capillary condensation of water [141], according to the proposed scheme shown in Figure 26.

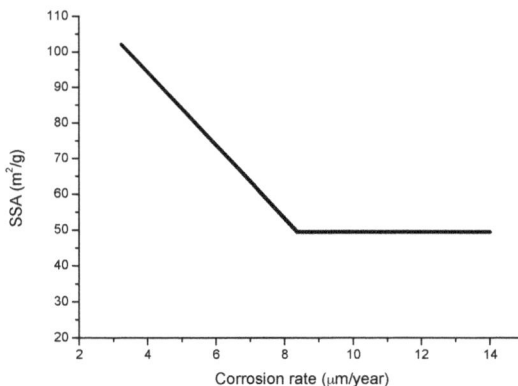

**Figure 25.** Trend plot of specific surface area (SSA) against corrosion rate for the rusts formed by exposing carbon steels at different bridges in Japan for 17 years [141].

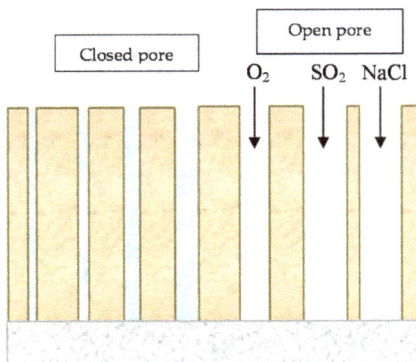

**Figure 26.** Schematic representation of pore filling by adsorption and capillary condensation of water, according to Ishikawa et al. [141].

(b)    Pore size distribution of rusts formed on exposed steels [146]

Figure 27 depicts the pore size distribution trend curve of rusts formed on steels, as calculated by the Dollimore-Heal (D-H) method [165] from $N_2$ adsorption isotherms. The curve rises steeply at pore diameters (D) of <~5 nm and still increases at D = ~2 nm, suggesting that these rusts contained micropores with a D < 2 nm.

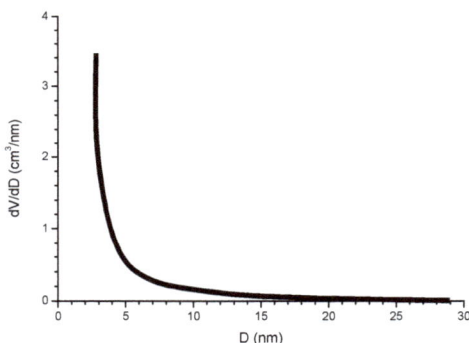

**Figure 27.** Trend plot of pore size distributions of the rust formed on steels exposed to coastal conditions [48]. V: adsorbed amount of $N_2$; D: pore diameter.

(c)    Influence of particle size and pore size distribution on rusts formed in marine atmospheres

Rusts formed in saline environments, such as marine or coastal regions or districts where deicing salts are used, show larger particle sizes than rusts formed in rural and urban areas, resulting in the formation of larger pores which act as voids between larger particles in the rust layer and facilitate further corrosion.

The $Cl^-$ ion promotes the growth of rust particles to yield less compact rust layers composed of larger particles, which leads to a high corrosion rate of steels in the saline environment [48,146].

(i)    Effect of chloride deposition rate on rust particle size [146]

Figure 28 shows the adsorption isotherms of nitrogen ($S_N$) and water ($S_W$) molecules on rusts obtained on CS exposed to atmospheres with different $Cl^-$ deposition rates and times of wetness (TOW). $S_N$ and $S_W$ depend strongly on the $Cl^-$ deposition rate and the TOW duration.

Thus, the formation and growth of rust particles is influenced by both parameters. These results confirm that the development of rust particles in aqueous solutions and the particle growth of rusts is promoted by $Cl^-$ ions.

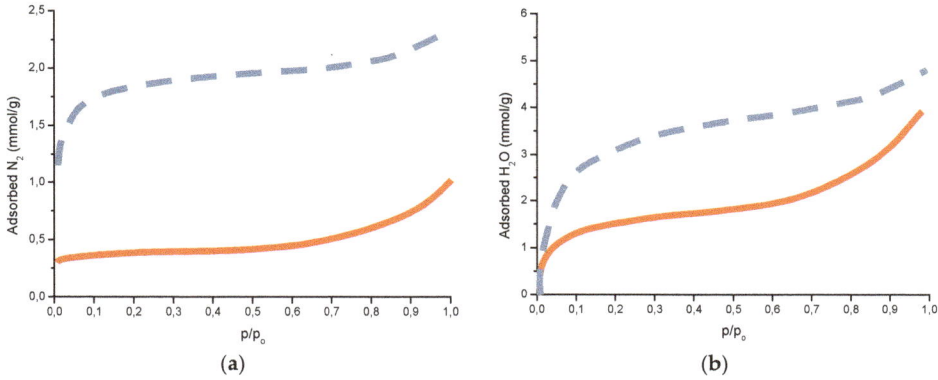

**Figure 28.** Adsorption isotherms of nitrogen (**a**) and water (**b**) on the rusts generated by exposing a carbon steel for 3 months at two atmospheres with different NaCl deposition rates [146]. ─ ─ 5.6 mg NaCl/$m^2$; wet period: 56%; ────── 10.7 mg NaCl /$m^2$; wet period: 82%; p/p$_o$ is the relative pressure of water or relative humidity (RH).

(ii)     Effect of chloride deposition rate on rust pore size [146]

The SSA (determined by $N_2$ adsorption) (Figure 17) decreases as the NaCl content of the atmosphere increases (the micropores volume (MPV) shows a similar tendency), indicating that the rusts formed at coastal sites are agglomerates of large particles which give rise to large pores, consistent with the fast corrosion rate at the coast. This finding again suggests that NaCl promotes rust particle growth.

6.2.6. Indices to Evaluate the Protective Capacity of Rust Layers

Yashamita and Misawa [166] found that the mass ratio of goethite to lepidocrocite contents ($\alpha/\gamma$) in the rust layer formed on WS exposed in industrial and rural environments, determined by XRD, is a function of the exposure time, as is shown in Figure 29a. The $\alpha/\gamma$ ratio increases proportionally with exposure time due to the long-term phase transformation. The relationship between $\alpha/\gamma$ and the corrosion rate is shown in Figure 29b. It can be said that the corrosion rate decreases as the $\alpha/\gamma$ increases, and that $\alpha/\gamma > 2$ is a necessary condition for the final protective rust layer. $\alpha/\gamma$ can be considered a useful index to evaluate the protective ability of the rust layer.

Kamimura et al. [167] and Hara et al. [142] later noted that this tendency for the $\alpha/\gamma$ mass ratio to increase with exposure time was not seen in marine atmospheres where the corrosion species akaganeite ($\beta$) and magnetite (M) also form on steel. In contrast, they saw that the relationship between goethite and the total mass of lepidocrocite, akaganeite and magnetite ($\alpha/\gamma^*$) was related to the corrosion rate, even in coastal environments with more than 0.2 mg/$dm^2$·d of airborne sea salt particles. When the $\alpha/\gamma^*$ ratio exceeded a certain critical threshold, steel corrosion was less than 10 µm/year.

Despite the fact that magnetite is a conducting phase, Dillmann et al. [47] consider it to be protective because of its relatively good stability, suggesting a new protective ability index ($\alpha^*/\gamma^*$):

$$\alpha^*/\gamma^* = (\alpha + M)/(\gamma + \beta), \tag{20}$$

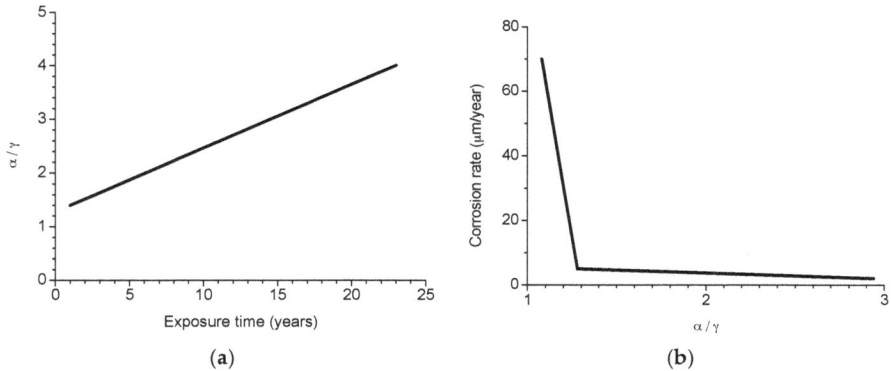

**Figure 29.** Relation between $\alpha/\gamma$ of the rust layer formed on weathering steel and exposure time (**a**) and corrosion rate (**b**). According to Yamashita and Misawa [166].

## 7. Mechanisms of Steel in MAC

### 7.1. First Researches

As has been noted in the introduction to this review, it is rather surprising that despite the great practical importance of this issue it has only recently that it has started to attract the interest of corrosion scientists.

It is well known that the presence of atmospheric pollutants (natural or anthropogenic) notably accelerates the AC process of CS. The two most common pollutants, which have drawn the majority of research efforts, are $SO_2$ and marine chlorides. In principle most of the attention has been focused on $SO_2$, and considerable progress has been made in this respect [4]. However, as Nishimura et al. [10,39] point out, it was not until the final decade of the 20th century that major research started to be carried out on the fundamental mechanisms of rust formation in $Cl^-$-rich marine atmospheres. Until then very little research was undertaken in this field, the most notable being the work of Keller [103], Feitknecht [24], Henriksen [168] and Misawa [169–171].

Keller in 1948 [103] reported three basic chlorides obtained by partial precipitation from $FeCl_2$ solution in various concentrations, noting that these three basic chlorides were presumably the precursors of GR1. Feitknecht [24] reported the existence of chloride accumulations (nests) containing $FeCl_2$ in the rust layers formed on steel exposed in coastal areas and their role in stimulating AC. Henriksen [168], using autoradiography, noted that the AC of CS in marine atmospheres starts at weak spots in the oxide film. $Na^+$ and $Cl^-$ in the aqueous adlayer migrate to these weak spots and once $Cl^-$ has been adsorbed and corrosion has begun, $Na^+$ migrates to the cathodic areas. Misawa et al. [169–171], who had carried out important basic research on the formation mechanisms of the different AC products of iron, noted in their work the formation of an intermediate compound, GR1, in marine atmospheres, also in accordance with Keller [103].

At around the same time, Barton [4], in his important book on AC, pointed out three causes which explained the high corrosion rates of steel exposed to marine atmospheres: (a) the increase in the ionic conductivity of the aqueous adlayer due to the presence of ionising substances (chlorides); (b) the hygroscopic nature of the $Cl^-$-containing corrosion products formed; and (c) the solubility of the latter, unlike the stable basic chlorides that form in the case of other metals (Cu, Zn, etc.), indicating that the mechanism which governs the effects of $Cl^-$ ions in AC had not been completely explained.

In addition to the above, two other causes of enormous importance should be mentioned: (a) he strong cathodic depolarising role of $Cl^-$ ions, accelerating the cathodic process by tens and hundreds of times, as formulated in 1961 by Rozenfeld [18]; and (b) the catalytic role of chlorides [23,172]. With regard to the latter effect, it is noted that the anodic reaction generates cations by dissolution and $H^+$ by hydrolysis of the dissolved cations. Both $Fe^{2+}$ and $H^+$ require neutralisation, which is accomplished by the ingress of chlorides. The locally higher $Cl^-$ concentration enhances local metal dissolution, then draws more $Cl^-$ and enhances dissolution even further [23]. Migration of $Cl^-$ to the corroding substrate is facilitated by its high permeability in the rust layer. As noted in [23], this is a feedback mechanism, sometimes referred to as autocatalytic. The high concentrations of $Cl^-$ in the inner rust layer will facilitate the formation of akaganeite, as was also previously noted by Keller [103] and Misawa [169].

According to Misawa et al. [169], when the aqueous adlayer on the metal surface is neutral or slightly acidic, $Fe(OH)_2$ cannot be formed, but various Fe(II) hydroxo-complexes may be formed, depending on the existing anion in the aqueous solution. Fe(II) hydroxo-complexes thus formed are oxidised by dissolved oxygen, resulting in lepidocrocite through an intermediate GR.

$$\text{Fe(II) hydroxo-complexes} \rightarrow GR \rightarrow \gamma\text{-FeOOH}, \tag{21}$$

In marine atmospheres, the intermediate generally has a GR1 structure that forms in the presence of $Cl^-$ ion [96,107,108]. GR1 is converted to black magnetite by slow oxidation in solution and this reaction is considered to correspond to the formation of magnetite in the underlying rust layer where the oxygen supply is limited. Also, akaganeite can be obtained by the dry oxidation of solid $\beta\text{-Fe}_2(OH)_3Cl$ precipitated from the slightly acidic solution in the presence of $Cl^-$ ions.

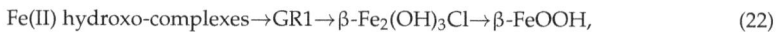

$$\text{Fe(II) hydroxo-complexes} \rightarrow GR1 \rightarrow \beta\text{-Fe}_2(OH)_3Cl \rightarrow \beta\text{-FeOOH}, \tag{22}$$

Worch et al. reported in 1983 [173] that GR1 is often seen on iron and steel exposed to marine environments and that chloride may also be involved catalytically in the formation of akaganeite. Akaganeite is produced only in the presence of sufficient concentrations of $Cl^-$ [107,108].

Askey et al. [25] suggest a cyclical rust formation process, similar to the acid regeneration cycle proposed by Schikorr [26,174] for the action of $SO_2$ in iron, by which the accumulation of $Cl^-$ ions in the underlying steel gives rise to the formation of $FeCl_2$, which hydrolyses water according to

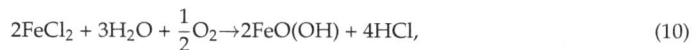

$$2FeCl_2 + 3H_2O + \frac{1}{2}O_2 \rightarrow 2FeO(OH) + 4HCl, \tag{10}$$

Releasing HCl. It should be noted that this represents a reaction cycle in which rereleased HCl will react with iron to form fresh $FeCl_2$. Once started, therefore, the cycle will be independent of incoming HCl. Corrosion will continue until the $Cl^-$ ions are removed (possibly by the washing away of $FeCl_2$).

### 7.2. The Fundamental Role of Akaganeite in the Atmospheric Corrosion Process of Steel in Marine Atmospheres

A fundamental advance in relation with the role played by akaganeite in the AC process of steel in marine atmospheres was made by Nishimura et al. in studies carried out in the last decade of the 20th century [10,39]. Nishimura et al. in 1995 used a wet and dry corrosion test to study the relationship between steel corrosion resistance and NaCl concentration, analysing the corrosion products by in-situ XRD [39]. The steel corrosion rate increased as the NaCl concentration rose, and a very strong increase in the akaganeite/lepidocrocite weight ratio was observed from a NaCl concentration of 0.05 wt % (Figure 30).

Akaganeite was reduced to an amorphous intermediate oxide during the wet stage of the cycle and reproduced in the dry stage, giving rise to the proposal of the following rusting model of iron in wet and dry corrosion in the presence of NaCl (Figure 31).

Later, in the year 2000, continuing with the in-situ XRD technique but here in combination with alternating current impedance, Nishimura et al. observed the transition of akaganeite from GR1 in the dry process; the amount of GR1 also depended on the $Cl^-$ ion concentration [10]. After dripping a $Cl^-$ solution (3% $Cl^-$) on the steel surface, the dry process progressed with the formation of akaganeite at a high corrosion rate. When a low Cl⁻ concentration was used (0%–0.3% Cl⁻), lepidocrocite was formed from $Fe(OH)_2$ instead of akaganeite and the corrosion rate was low.

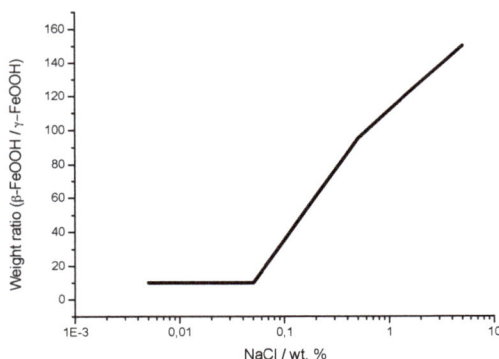

**Figure 30.** Weight ratio of β-FeOOH to γ-FeOOH as a function of NaCl concentration [39].

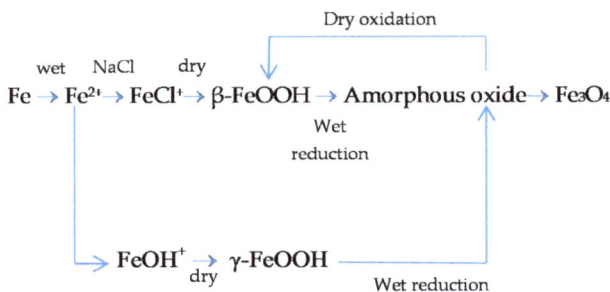

**Figure 31.** Rusting model of iron in wet and dry corrosion condition containing NaCl [39].

The integral intensity of akaganeite decreased after dripping a Cl⁻ solution, which implied that akaganeite was consumed in the wet process. After 60 min in the dry process of the cycle the presence of GR1 was detected. GR1 could still be detected after 180 min (Figure 32), but disappeared after 12 h of testing, indicating that the transformation to akaganeite was complete.

Quantitative analysis of the identified phases was carried out using an XRD standard method. Akaganeite was the most abundant crystalline phase in the iron rust, and its proportion grew considerably as the Cl⁻ concentration increased, as was previously seen [39]. The other large phase was goethite. In contrast, the amounts of lepidocrocite and magnetite were low. These two phases were not affected by the Cl⁻ concentration. Nishimura et al. concluded that in a Cl⁻-rich environment the corrosion process was dominated by the formation of akaganeite, rather than lepidocrocite, which acts as the oxidation agent that accelerates corrosion in Cl⁻-free environments [38].

The authors carried out XPS and TEM observations on those portions of iron rust that could not be detected by XRD. It was determined that they contained large amounts of spinel oxide (magnetite structure) with bivalent/trivalent iron. This spinel oxide may have been formed by reduction of akaganeite during the wet process of the cycle.

These findings mark a turning point in the knowledge of the MAC mechanisms of steel, determining that steel corrosion progresses by the formation of akaganeite from GR1 (dry stage or "drying-out stage" following the terminology of Stratmann [38]) and its reduction (wet stage).

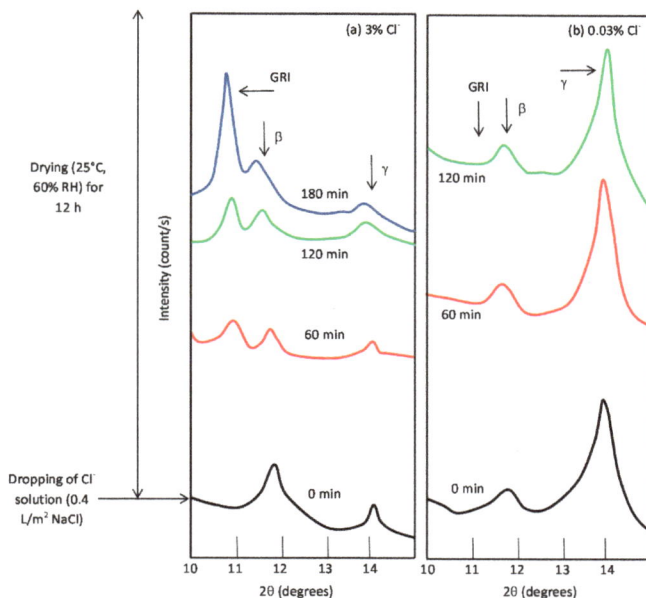

**Figure 32.** In-situ X-ray diffraction (XRD) results in wet and dry corrosion test using (**a**) 3% Cl⁻ and (**b**) 0.03% Cl⁻ solution [10].

## 7.3. Initial Stages of MAC

In marine atmospheres, corrosion is generally driven by the deposition of hygroscopic sea salt aerosols that absorb moisture from the environment and form salt droplets. These aerosols range from a few angstroms to several hundred microns in diameter. Lan et al. [175] point out that coarse sea salt particles are those that contribute most to CS corrosion in marine atmospheres. In coastal marine locations (<2 km from the shoreline) the most common aerosols deposited are in the "coarse mode" size range: 1–100 μm in diameter [57].

Li and Hihara [55] studied salt particle deposition and the initial stage of MAC at severe marine test sites. They found both: (a) small sea-salt particles (D < 5 μm) and sea-salt clusters (D < 10 μm) formed by dehydration on the steel substrate that did not corrode under relatively small seawater droplets (D < 30 μm); and (b) sea-salt clusters integrated with iron corrosion products formed on the steel substrate that did corrode from larger seawater droplets (D > 30 μm). The corrosion that occurred under larger seawater droplets (D > 30 μm) showed the typical characteristics of droplet corrosion, with Cl⁻ and Na⁺ ions migrated to the central anode and peripheral cathode, respectively. The corrosion products were identified as lepidocrocite.

These same researchers [176,177], in a laboratory study in which they manually deposited NaCl droplets of different diameters on CS steel, used RS to analyse the very initial stage of NaCl particle-induced corrosion. They found that corrosion did not initiate under NaCl droplets with diameters of less than 45 μm after 6 h (at 80% RH). At larger NaCl droplet diameters, corrosion initiated quickly under the droplets in the form of pitting. In-situ and ex-situ Raman spectra show the formation of GR in regions close to the anodic sites and the precipitation of lepidocrocite clusters over cathodic sites surrounding the GR region. Magnetite was detected mostly in the rust clusters formed in the transitional region from GR to lepidocrocite. Upon exposure to ambient air, GR transformed to the more stable lepidocrocite due to oxidation. Li and Hihara underline the need for more research effort in droplet electrochemistry [178].

Risteen et al. [179] recently used a new methodology to study corrosion under NaCl solution droplets ranging in diameter from 20 to 1000 μm. They also observed the dependence of the occurrence of corrosion on drop size, noting that this behaviour appears to be strongly dependent on the microstructure and surface finish: corrosion initiation on 1010 steel was dominated by manganese sulfide inclusions when a mirror surface finish was maintained. In contrast, for high purity iron, initiation was dominated by surface roughness.

Ohtsuka and Tanaka have recently used RS to carry out a study of changes in rust composition on CS over six days of cyclic exposure: 4 h wet (90% RH)/4 h dry (10% RH) in the presence of NaCl droplets (0.93 and 0.11 mg/cm$^2$) [180]. The NaCl solution was first dripped and immediately dried in a vacuum desiccator. A Raman spectrum was recorded every 15 min.

The Raman spectra of the rust surface in the presence of 0.93 mg/cm$^2$ NaCl deposits corresponded to lepidocrocite and magnetite in the initial 12 h of exposure. After 12 h of exposure, akaganeite started to form, and its molar ratio on the rust surface increased to 90% at 30 h of exposure (Figure 33). They assume that in order for akaganeite to form, a lepidocrocite + magnetite rust layer of some thickness is required. The Raman spectra further changed after 30 h of exposure, when lepidocrocite again emerged. The reappearance of lepidocrocite is assumed to be caused by the capture of Cl$^-$ ions in the akaganeite, resulting in a decrease in the free Cl$^-$ ions in the aqueous adlayer.

When the amount of NaCl deposit is decreased to 0.11 mg/cm$^2$, the steel surface does not reach high enough concentrations to form akaganeite. Only after repeated wet/dry cycles may a spot with a high concentration of NaCl emerge on the surface and akaganeite form on that spot.

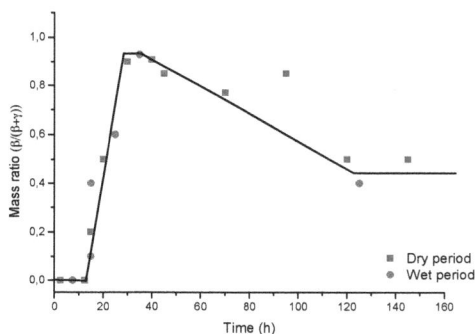

**Figure 33.** Change of mass ratio of (β-FeOOH /(β-FeOOH + γ-FeOOH)) of rust on weathering steel during the exposure of dry and wet periods in the presence of NaCl deposition at 0.93 mg/cm$^2$ [180].

### 7.4. Formation and Growth of the Corrosion Layer

Once the corrosion process has started on the steel surface it will be necessary to consider the possible mechanisms that take place in the formation and growth of the rust layer, where, as is known, the corrosion products transform from one compound to a more stable form and may involve any of a number of processes including hydrolysis, nucleation, crystallisation, precipitation, dehydration, thermal transformation, dehydroxylation, etc. [91]. Temperature, time and pH are the main factors governing such transformations [97].

In 1965, Evans [181] formulated the first electrochemical method for atmospheric rusting, in which the oxidation of iron (wet periods)

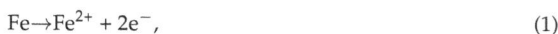

$$Fe \rightarrow Fe^{2+} + 2e^-, \tag{1}$$

Is balanced by the reduction of ferric rust to magnetite

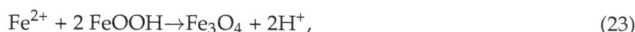

$$Fe^{2+} + 2\,FeOOH \rightarrow Fe_3O_4 + 2H^+, \tag{23}$$

Later, after partial drying of the pore structure of the rust (dry period), magnetite is reoxidised by oxygen that now has free access through the pores due to gas diffusion

$$Fe_3O_4 + 3/2\ O_2 + H_2O \rightarrow 3\ \gamma\text{-FeOOH}, \tag{24}$$

The autocatalytic cycle responsible for the fact that rust promotes further rusting involves alternate reduction and reoxidation of the preexisting rust.

Subsequently, Stratmann et al. [182], in an electrochemical study of phase transitions in rust layers, experimentally showed that the oxidation of magnetite to lepidocrocite, as proposed by Evans, was not possible. Stratmann et al. used a combination of magnetic and volumetric measurements to show that when a prerusted iron sample is wetted, iron dissolution is not immediately balanced by a reaction with oxygen, but rather by reduction of the preexisting rust

$$\gamma\text{-FeOOH (lepidocrocite)} + H^+ + e^- \rightarrow \gamma\text{-Fe.OH.OH (reduced lepidocrocite)}, \tag{25}$$

With later reoxidation of the reduced species

$$2\ \gamma\text{-Fe.OH.OH} + 1/2\ O_2 \rightarrow 2\ \gamma\text{-FeOOH} + H_2O, \tag{26}$$

Thus, Stratmann [38] proposed dividing the AC mechanism of pure iron into the following three stages: wetting of the dry surface, wet surface, and drying-out of the surface (see Figure 2).

Misawa [170] notes the following mechanism for the rusting process (Figure 34):

(a) In the first stage of rusting the aerial oxidation of ferrous ions, dissolved from the steel into a slightly acidic thin water layer formed by rain on the steel surface, leads to the precipitation of lepidocrocite. Fine weather accelerates the precipitation and crystallisation of lepidocrocite by drying.

(b) The lepidocrocite is formed on the steel surface and transformed to amorphous ferric oxyhydroxide and goethite during the atmospheric rusting process. The amorphous ferric oxyhydroxide transforms to goethite by deprotonation using hydroxyl ions provided by the rainwater.

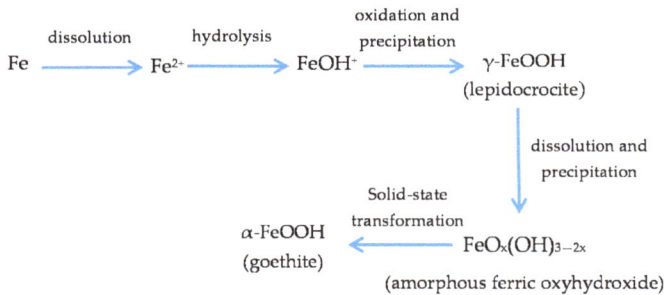

**Figure 34.** Mechanism for the rusting process according to Misawa [170].

Another important aspect to consider is how the rust layer grows. Horton [183] in 1964 observed that rust layers grow by several mechanisms: (i) by iron ions diffusing outward through the rust to form fresh rust at the air-rust interface; (ii) at the steel-rust surface; and (iii) within the rust layer to fill pores and cracks. It was the first time that this observation was reported in scientific literature. Years later, Burger et al. [184] using an ingenious technique known as the "gold marker method", addressed the following two aspects:

(a) the location at which precipitation of corrosion products occurs within the corrosion system (steel/rust/atmosphere), and

(b) the structural evolution of the corrosion product layer during wet/dry cycles.

With regard to (a), they observed a significant contribution of inward diffusion of oxidant through the corrosion product layer. With regard to (b) they note that the continuous decrease in the reactivity of the corrosion product layer seems to be related with a two-step process in the corrosion mechanisms: the preliminary formation of ferrihydrite (a highly reactive hydrated iron oxide) close to the metal/rust interface, followed by its progressive transformation into goethite, a more stable oxyhydroxide. This progressive transformation may be the consequence of incremented cyclic reduction/reoxidation reactions which are not completely reversible. As these cyclic electrochemical reactions require electrical contact between the reactive phase and the metallic substrate, and given the complex morphology of the corrosion patterns, an important outlook is to take into account the connectivity and conductivity of the different phases constituting the corrosion product layer and their influence on its structural evolution. Due to the expansive nature of the corrosion products, mechanical stresses may develop in these materials, thus inducing two opposing effects: pore blocking and formation of cracks/spalling in the rust layer.

In the last decade great advances have been made in the understanding of AC mechanisms. As has been mentioned above, many of these advances have been due to the French research groups of Professors Legrand and Dillmann [40,41,44,47,93,131,185,186]. Both groups have made important advances in: (a) the electrochemical reactivity of the ferric phases that constitute rust; (b) the localisation of oxygen reduction sites; (c) the decoupling of anodic and cathodic reactions; (d) in-situ characterisation of reduction and reoxidation processes, etc.

With a view to the development of a model of the AC process that can predict the long term AC behaviour of iron, they note the need to consider several important parameters in order to describe rust layer: average lepidocrocite fraction, thickness, average porosity, tortuosity and specific area, connectivity of the phases inside the rust layer, etc. [41,47]. As these researchers note, there is still a long way to go before long-term AC mechanisms are fully clarified.

As has been mentioned several times in this paper, considerable advances have been made in the knowledge of AC mechanisms in atmospheres polluted with $SO_2$ (e.g., urban and industrial atmospheres) while less progress has been made on corrosion mechanisms in marine atmospheres. In addition to the proposals of Nishimura et al. [10,39] referred to in 7.2., other authors have made contributions relating to the subject of MAC which will be enumerated below, and more are sure to appear in the forthcoming years, considering the growing interest of researchers in this field of knowledge.

In 1988 Nomura et al. [187] applied conversion electron MS to study the formation of akaganeite on iron in a NaCl (3 wt %) solution. On the basis of the study on early stages of $Fe(OH)_2$ formation, the reaction that takes place on the iron surface in a $Cl^-$ solution can be expressed as follows:

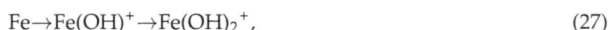

$$O_2 + 2H_2O + 4e^- \rightarrow 4OH^-, \tag{2}$$

$$Fe \rightarrow Fe(OH)^+ \rightarrow Fe(OH)_2{}^+, \tag{27}$$

In conditions of high dissolved oxygen (initial stages)

$$Fe(OH)_2{}^+ + 2OH^- \rightarrow Fe^{3+}(OH)_4{}^-, \tag{28}$$

$$2Fe(OH)_4{}^- \rightarrow 2FeOOH + 2H_2O + 2OH^-, \tag{29}$$

Poorly crystalline FeOOH is considered to deposit on the iron surface by the initial corrosion reaction listed above.

However, in conditions of low dissolved oxygen, when a first rust layer is formed, the supply of $OH^-$ (Equation (28)) is suppressed, and $Cl^-$ ions begin to have a relatively strong affinity to the iron(III) ion, thus iron(III) oxyhydroxide complexes containing $Cl^-$ may be formed.

$$Fe(OH)_2{}^+ + (2 - X)OH^- + XCl^- \rightarrow Fe[OH_{4-X}Cl_X]^-, \tag{30}$$

Before the start of the polymerisation process to form rust according to the following equation:

$$2Fe[OH_{4-x}Cl_x]^- \text{ (polymerisation)} \rightarrow 2\beta\text{-FeOOH (Cl}_x^-) + H_2O + 2_{(1-x)}OH^-, \tag{31}$$

In short, when unstable oxyhydroxide is first formed on iron, lepidocrocite is formed on the surface where dissolved oxygen has easy access, and then akaganeite and magnetite start to be produced by the transformation of the $Fe(OH)_2$ complex containing $Cl^-$ at the intermediate surface between the lepidocrocite layer and the iron substrate and by the slow oxidation of iron, respectively, because the supply of dissolved oxygen to the intermediate layers is restricted by the top lepidocrocite layer.

It is relevant to note at this point the important laboratory studies carried out by Refait and Genin [113,115] and subsequently by Remazeilles and Refait [112,114] on the formation conditions of Fe(II) hydroxychlorides, GR1 and akaganeite previously mentioned in Section 5.

More recently, Ma et al. [188,189], using XRD and IRS, detected the formation of akaganeite in the inner rust layer accompanied by an acceleration of the corrosion rate on steel exposed to very severe marine atmospheres. After six months of exposure the akaganeite content and the corrosion rate decrease and akaganeite is gradually transformed into maghemite until it completely disappears. The authors speculate on the need to exceed a critical chloride threshold in order for akaganeite to form.

In atmospheres with less $Cl^-$ pollution akaganeite was not formed, though the $Cl^-$ content facilitated the transformation of lepidocrocite into goethite (Figure 35). The wet/dry cycle accelerates these transformation processes, and especially in the dry cycle HCl is released into the environment.

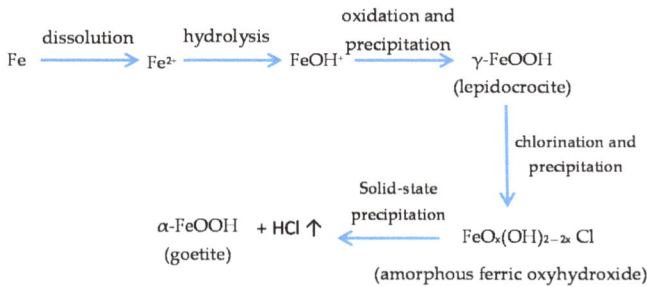

**Figure 35.** In marine atmospheres with low chloride deposition rates, $Cl^-$ content facilitates the transformation of lepidocrocite into goethite [188,189].

## 7.5. Proposal of an Overall Mechanism for the MAC Process of Steel

The composition of the rust layer depends on the conditions in the aqueous adlayer and thus varies according to the type of atmosphere. It is unanimously accepted that lepidocrocite is the primary crystalline corrosion product formed in the atmosphere. As the exposure time increases and the rust layer becomes thicker, the active lepidocrocite is partially transformed into goethite and magnetite.

In mildly acidic solutions lepidocrocite is transformed into goethite. Schwertmann and Taylor established that the transformation occurs in solution through different steps: dissolution of lepidocrocite, formation of goethite nuclei, and nuclei growth [190].

Magnetite may be formed by oxidation of $Fe(OH)_2$ or intermediate ferrous-ferric species such as GR [119], but also by lepidocrocite reduction in the presence of a limited oxygen supply [119,120]:

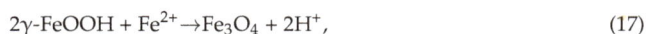

$$2\gamma\text{-FeOOH} + Fe^{2+} \rightarrow Fe_3O_4 + 2H^+, \tag{17}$$

Thus it is not surprising that magnetite is usually detected in the inner part of rust adhering to the steel surface, where oxygen depletion may occur.

With a broader view, Ishikawa et al. [121] and Tanaka et al. [122] found that the formation of magnetite particles was caused by the reaction of dissolved ferric species of oxyhydroxides with ferrous species in the solution. The formation of magnetite rust can be represented by the following cathodic reaction:

$$Fe^{2+} + 8FeOOH + 2e^- \rightarrow 3Fe_3O_4 + 4H_2O, \tag{18}$$

In marine atmospheres, where the surface electrolyte contains chlorides, akaganeite is also formed. How does akaganeite form? The high $Cl^-$ concentration in the aqueous adlayer on the steel surface gives rise to the formation of $FeCl_2$, which hydrolyses the water [25]:

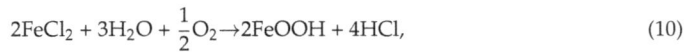

$$2FeCl_2 + 3H_2O + \frac{1}{2}O_2 \rightarrow 2FeOOH + 4HCl, \tag{10}$$

Notably raising the acidity of the electrolyte. At the steel/corrosion products interface, where $Cl^-$ ions can accumulate, large $Cl^-$ concentrations and acidic conditions give rise to akaganeite formation after the precipitation of ferrous hydroxychloride ($\beta$-$Fe_2(OH)_3Cl$), a very slow process requiring the transformation of metastable precursors [112,113].

As Remazeilles and Refait point out, large amounts of dissolved Fe(II) species and high $Cl^-$ concentrations are both necessary for akaganeite formation [114]. The oxidation process of ferrous hydroxychloride which leads to akaganeite formation passes through different steps via the formation of GR1 intermediate compounds. The whole oxidation process can be summarised as follows [99,112–115]:

$$FeCl_2 \rightarrow \beta\text{-}Fe_2(OH)_3Cl \rightarrow GR1 \rightarrow \beta\text{-}FeOOH, \tag{16}$$

Thus requiring a relatively long time. This time will depend on the environmental conditions: temperature, $Fe^{2+}$, $Cl^-$ and $OH^-$ conditions, $O_2$ flow, etc. The acid environment at the steel/rust interface also leads to an acceleration of corrosion of the underlying steel and pitting [23,191], where the dominant cathodic reaction is hydrogen evolution.

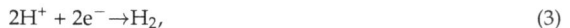

$$2H^+ + 2e^- \rightarrow H_2, \tag{3}$$

There has been much speculation about the need to exceed a critical atmospheric salinity threshold for akaganeite formation to take place. Morcillo et al. provide a general indication of the environmental conditions in the atmosphere which lead to akaganeite formation: annual average RH around 80% or higher and simultaneously an annual average $Cl^-$ deposition rate of around 60 mg/m$^2$·d or higher [192] (Figure 36). This confirms the laboratory experiments carried out by Remazeilles and Refait [114], which indicate that a high $Cl^-$ concentration is not the only condition for akaganeite formation. The medium must also be characterised by large dissolved $Fe^{2+}$ concentrations, as occur during the high TOW of the metallic surface in high RH atmospheres.

With regard to the corrosion products that form on CS in this type of atmospheres, Table 4 has been prepared using field data obtained by the authors in different studies carried out at various sites in Spain [64,150,193] and allows the following facts to be deduced: (a) in marine atmospheres with extremely low $Cl^-$ deposition rates (Ponte do Porto) only lepidocrocite and goethite phases form, the latter in a practically insignificant proportion; (b) above a certain atmospheric salinity, akaganeite and spinel (magnetite/maghemite) phases appear; (c) the akaganeite and spinel contents increase notably as the atmospheric salinity rises (Cabo Vilano-2 and Cabo Vilano-3); and (d) it can clearly be seen how an increase in the $Cl^-$ deposition rate is accompanied by a drop in the lepidocrocite phase content of the rust and a rise in the goethite, akaganeite and spinel contents. This fact is well seen in Figure 37, which shows rusts formed after 3 months exposure of CS in marine atmospheres with different salinities [95].

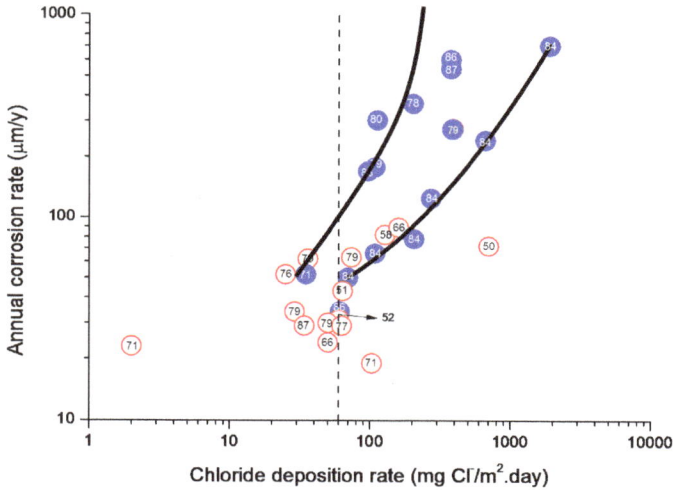

**Figure 36.** Corrosion rate of mild steel during the first year of atmospheric exposure as a function of the annual average chloride deposition rate at the exposure site. The points of the graph, represented by circles, include an indication of the annual average RH at exposure site. Blue circle represent test site where akaganeite had been identified, and white circle represent test sites where it had not been possible to identify it by XRD [192].

**Table 4.** Variation of rust phases content on mild steel exposed during one year in test sites with different chloride deposition rate [64,150,193].

| Test Site | Annual Average Chloride Deposition Rate, mg/m²·d | wt % | | | |
|---|---|---|---|---|---|
| | | Lepidocrocite | Goethite | Akaganeite | Spinel |
| Ponte do Porto | 4 | 100 | * | 0 | 0 |
| Cabo Vilano-1 | 30 | 80.0 | 16.0 | 0 | 4.0 |
| Cabo Vilano-2 | 70 | 59.6 | 20.0 | 17.6 | 2.7 |
| Cabo Vilano-3 | 665 | 35.8 | 27.0 | 12.5 | 24.7 |

* The authors not ruling out a small contribution of this phase.

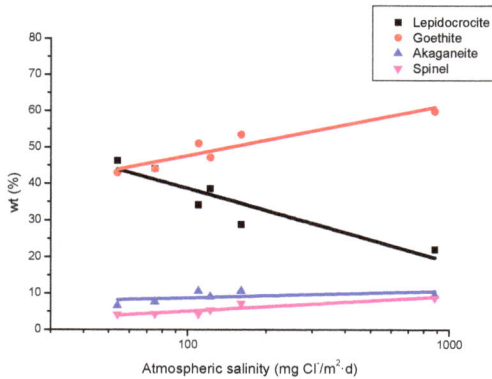

**Figure 37.** Variation of rust phases content in rusts formed on mild steel exposed during 3 months in marine atmospheres with different levels of salinity [95].

Akaganeite could be reduced electrochemically in the corrosion process, being consumed in the wetting of the metallic surface [10]. Lair et al. [40] experimentally saw the high reducing capacity of akaganeite in comparison with other oxyhydroxides, reporting the following order: akaganeite > lepidocrocite >> goethite.

This explains the high magnetite contents found in the rust formed on steel exposed to atmospheres with heavy $Cl^-$ ion pollution. As Hiller noted some time ago, the rust formed in atmospheres with high $Cl^-$ deposition rates contains more magnetite than that formed in $Cl^-$ free atmospheres [123].

Rust layers present considerable porosity and cracking. Ishikawa et al., examining the textures of rusts, note that the rusts formed at coastal sites are agglomerates of large particles and have larger pores than rusts formed at rural and urban sites. NaCl promotes rust particle growth, resulting in the formation of larger pores and voids between larger particles in the rust layer and facilitating further corrosion [145,163]. The SSA of rusts decreases as salinity increases, enlarging the diameter of the pores and forming less and less compact rust layers with low protective properties.

Thus the compactness of the corrosion product layers formed is dependent on the salinity of the atmosphere at the exposure site. At low $Cl^-$ deposition rates, even when the time of wetness (TOW) of the metallic surface is high, relatively consistent (less porous) layers, whose thickness does not usually exceed 100 μm, are formed. However, high $Cl^-$ deposition rates lead to the formation of very porous rust layers showing cracks and even flaking and exfoliation [129].

In Figure 38 it is possible to see the variation in the structure of the rust layer as the atmospheric salinity rises. While at relatively low atmospheric salinities (44 and 110 mg $Cl^-$/m²·d) the rust layers are fairly compact (though they can show the presence of longitudinal cracks), at higher atmospheric salinities (173 and 245 mg $Cl^-$/m²·d) the rust layers present abundant cracking which facilitates their subsequent detachment (exfoliation), as can clearly be seen at the highest atmospheric salinities (889 and 1136 mg $Cl^-$/m²·d) [64].

Figure 38 also indicates how the content of the different phases in the rust layers varies as the atmospheric salinity rises. The lepidocrocite phase decreases while the goethite and akaganeite phase contents increase [64].

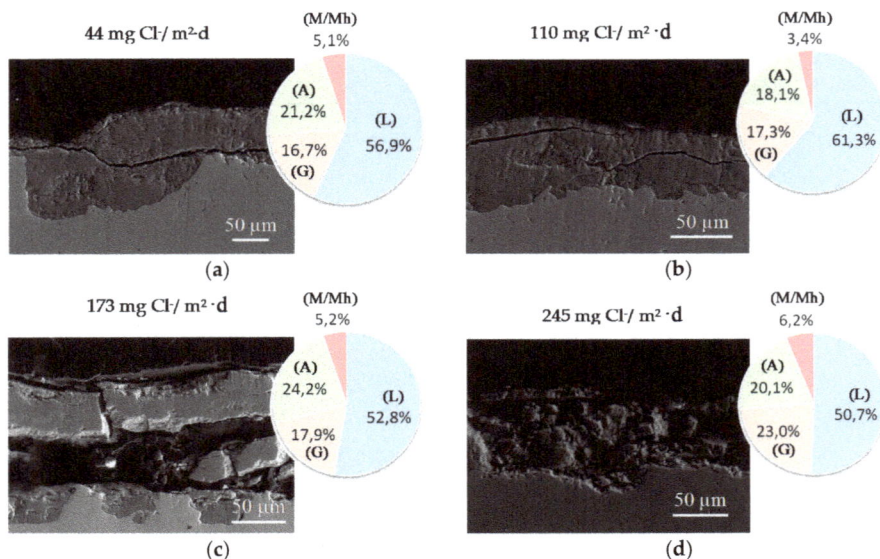

Figure 38. *Cont.*

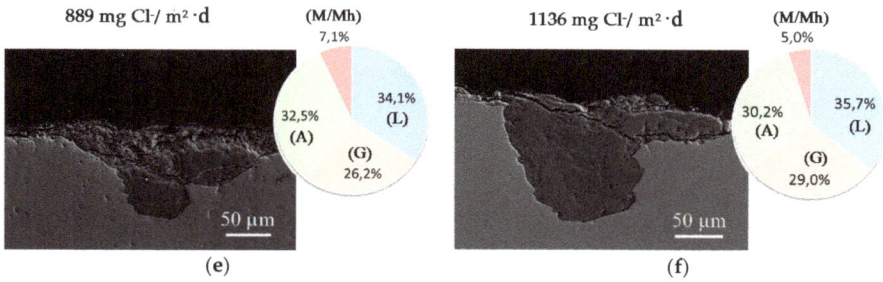

**Figure 38.** Variation in rust layer structure with atmospheric salinity: 44 mg Cl$^-$/m$^2$·d (**a**); 110 mg Cl$^-$/m$^2$·d (**b**); 173 mg Cl$^-$/m$^2$·d (**c**); 245 mg Cl$^-$/m$^2$·d (**d**); 889 mg Cl$^-$/m$^2$·d (**e**); 1136 mg Cl$^-$/m$^2$·d (**f**). Carbon steel specimens were exposed for three months in different marine atmospheres. The circles indicate the content of different phases in the rust, information obtained by XRD (RIR) of powdered rust [64]. L: lepidocrocite; G: goethite; A: akaganeite; M: magnetite and Mh: maghemite.

The base steel shows the formation of pits when exposed to marine atmospheres. As the atmospheric salinity rises, pitting becomes more significant and the Cl signal obtained by EDS inside the pits also rises (Figure 39) [64]. As has been seen in Section 6 (Figure 22), there is a strong presence of akaganeite in the interior of the pits formed on the base steel in severe marine environments.

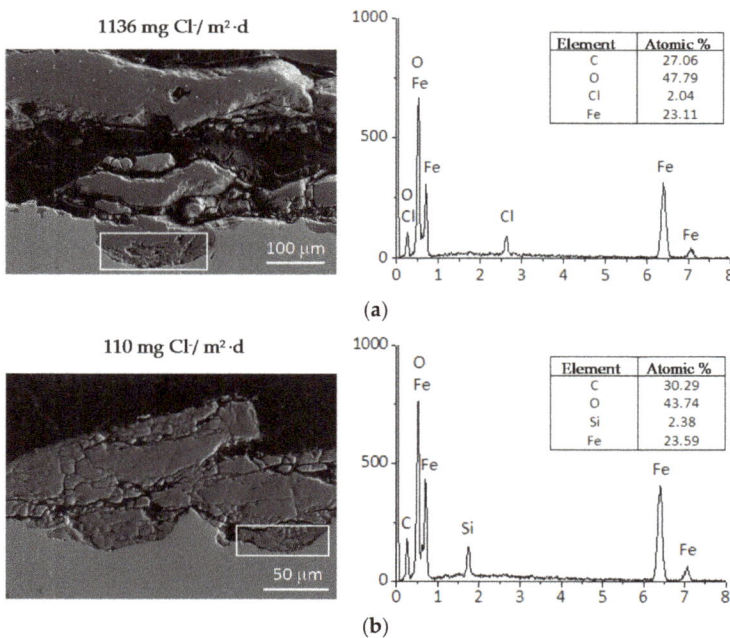

**Figure 39.** Formation of pits in steel substrate exposed to two atmospheres of different salinities. The EDS signal for Cl is more intense in the atmosphere of higher salinity [64].

Thus there seem to be two notably different situations with regard to the mechanisms involved in the MAC of CS: (a) establishment of a consistent (consolidated), adherent and continuous rust layer (at low $Cl^-$ deposition rates); and (b) formation of a thick rust layer that is easily detached (exfoliated) from the base steel, leaving large areas uncovered (at high $Cl^-$ deposition rates) [193].

When a consolidated layer of corrosion products remains on the steel surface, the conditions are right for a diffusion-controlled corrosion mechanism to act, in which the aggressive species from the atmosphere ($O_2$, $H_2O$ and $Cl^-$) pass through the rust layer to interact with the underlying steel (Figure 38a). This situation seems to occur in relatively low $Cl^-$-containing atmospheres. The steel corrosion process consists of the following reactions:

Steel starts to corrode according to the anodic reaction:

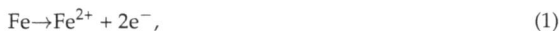

$$Fe \rightarrow Fe^{2+} + 2e^-, \tag{1}$$

where the cathodic process consists of reduction of oxygen dissolved in the aqueous adlayer:

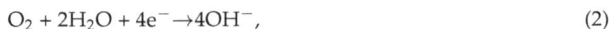

$$O_2 + 2H_2O + 4e^- \rightarrow 4OH^-, \tag{2}$$

The $OH^-$ ions formed migrate towards the anodic zones forming $Fe(OH)_2$ as the initial rust product:

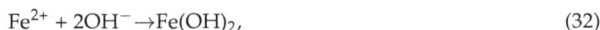

$$Fe^{2+} + 2OH^- \rightarrow Fe(OH)_2, \tag{32}$$

Under this basic mechanism, the steel corrosion rate will be highly influenced by the concentration of ionisable substances in the aqueous adlayer, as in the case of chlorides present in marine atmospheres. This explains the notable increase in the steel corrosion rate at station Cabo Vilano-2 compared to the Ponte Do Porto background station (Table 4), as the $Cl^-$ deposition rate rises from 3.6 mg $Cl^-/m^2{\cdot}d$ at Ponte Do Porto to 70 mg $Cl^-/m^2{\cdot}d$ at Cabo Vilano-2.

In contrast, the exposure of CS to severe marine atmospheres can lead in certain circumstances to the formation of thick rust layers. High times of wetness of the metallic surface and an atmosphere with a high $Cl^-$ deposition rate lead to the formation of this type of rust. These thick rust layers tend to become detached from the steel substrate, leaving it uncovered and without protection and thus accelerating the metallic corrosion process. The formation of anomalous thick rust layers and the accompanying exfoliation phenomenon has also been observed in studies carried out by the authors and other researchers [42,129,153,161].

In studies by Chico et al. on CS in $Cl^-$-rich atmospheres, the average $Cl^-$ deposition rate needed to exceed a critical threshold of close to 300 mg $Cl^-/m^2{\cdot}d$ for exfoliation to take place. The annual steel corrosion at that atmospheric salinity was higher than 100 μm [129].

Exfoliated rust layers are composed of multiple rust strata, as can clearly be seen in the cross-section of Figure 23. The characteristics of the different rust sublayers within the rust multilayer are described in Figure 24: the outermost rust layer (OR) (rich in lepidocrocite and goethite), and a succession of alternating strata of fragile compact rust (CR) and loose interlayer rust (LIR) layers [12,129,153,162].

The CRs present high goethite and maghemite contents, low lepidocrocite contents and the practical absence of akaganeite. The mechanism that is proposed for the formation of CRs consists of two stages: (i) the formation of magnetite by electrochemical reduction of lepidocrocite and akaganeite phases (wet stage); and (ii) the solid-state transformation of magnetite into maghemite (dry stage). The LIR presents high goethite and akaganeite contents along with low lepidocrocite and spinel contents. It is proposed that the akaganeite and lepidocrocite phases will be electrochemically reduced to magnetite (maghemite at a later stage) and the formation of the CR layer takes place by consumption of the akaganeite and lepidocrocite phases leading to the complete disappearance of the interlayer rust stratum. An extremely dry period may cause the corrosion process to end without fully exhausting the interlayer rust stratum.

Subsequently, once the extremely dry period has come to an end and a new wet period starts, the formation of a second CR layer would begin, and so on, giving rise to the formation of a sandwich-type structure constituted by alternate CR and LIR layers. A scheme of a feasible multilayered rust formation and rust exfoliation mechanism for CS exposed to severe marine atmospheres is shown in Figure 40 [129,153].

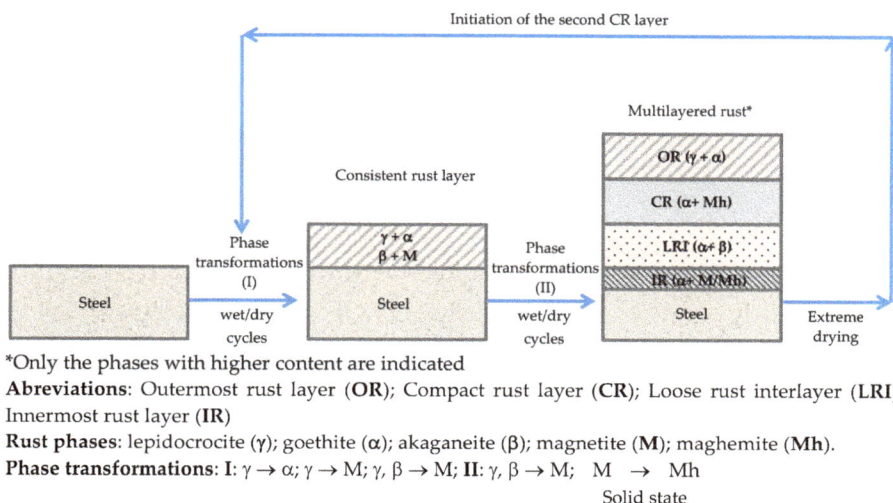

*Only the phases with higher content are indicated
**Abreviations**: Outermost rust layer (**OR**); Compact rust layer (**CR**); Loose rust interlayer (**LRI**), Innermost rust layer (**IR**)
**Rust phases**: lepidocrocite ($\gamma$); goethite ($\alpha$); akaganeite ($\beta$); magnetite (**M**); maghemite (**Mh**).
**Phase transformations**: **I**: $\gamma \rightarrow \alpha$; $\gamma \rightarrow M$; $\gamma$, $\beta \rightarrow M$; **II**: $\gamma$, $\beta \rightarrow M$; $M \rightarrow Mh$
Solid state

**Figure 40.** Scheme of a feasible multilayered rust formation mechanism of carbon steel exposed to severe marine atmospheres [129,153].

The detachment (exfoliation) of multilayered rust from the steel substrate takes place after complete drying of the whole rust layer, creating expansion stresses that exceed the adhesion forces which keep the multilayered rust joined to the steel substrate.

The difference in density between the sublayers involved (CR and LIR) suggests that compactness combined with mechanical properties may play an important role in the triggering of rust exfoliation. A closer look at the molar volume of the rust phases, i.e., the ratio between the molar mass and the density of each phase expressed in $cm^3/mol$, indeed shows great variations [162]. Table 5 displays a factor of 5 when going from the most compact rust phases involved (goethite, lepidocrocite; molar volume around 20 $cm^3/mol$), over medium compact phases (magnetite, maghemite; around 40 $cm^3/mol$), to the least compact phase (akaganeite; around 100 $cm^3/mol$) [194]. The much lower compactness of akaganeite is due to the presence of tunnels of the akaganeite lattice into which the $Cl^-$ ions can enter and become integrated, resulting in a much less dense structure than the other rust phases [162].

**Table 5.** Molar volume of different rust phases. Data from Crystallographic and Crystallochemical Database for Minerals and their Structural Analogues, WWW-MINCRYST. Institute of Experimental Mineralogy, Russian Academy of Sciences [194].

| Phase | Molar Volume ($cm^3/mol$) |
|---|---|
| Goethite ($\alpha$-FeOOH) | 20.84 |
| Lepidocrocite ($\gamma$-FeOOH) | 22.40 |
| Maghemite ($\gamma$-$Fe_2O_3$) | 43.71 |
| Magnetite ($Fe_3O_4$) | 44.56 |
| Akaganeite ($\beta$-FeOOH) | 101.62 |

Taking this difference in molar volume into consideration, it is possible to anticipate a great volume contraction and consequent void formation when the least compact phase akaganeite is structurally transformed into the much more compact spinel phase during primarily wet periods. Similarly, great volume expansion and stress introduction is induced when lepidocrocite is transformed into spinel. Hence, considering the molar volume data, it is not surprising that the compact rust sublayer contains the rust phases with the lowest molar volume (goethite and spinel) while the loose rust interlayer is dominated by akaganeite with a higher molar volume than the main phases in the solid rust sublayer [162].

Thus it is suggested that rust exfoliation is the result of frequent phase transformations, together with great variations in compactness between the rust phases involved. At some critical point the changes in compactness, compressive stresses and void formation become too large and the whole rust sublayer collapses mechanically and results in a fracture along the innermost rust layer [162].

## 8. Coastal-Industrial Atmospheres

The atmosphere of many coastal cities in developing countries is polluted with $SO_2$ due to the growth of industry, and in many cases formerly pure marine atmospheres can now be categorised as marine-industrial. The effect of $SO_2$ on the corrosion behaviour of steel in atmospheres containing $Cl^-$ has not been widely studied. The first information on this subject was published by Copson [195] in 1945, who reported that a combined influence of $Cl^-$ deposited on the surface and $SO_2$ in the atmosphere was considered to cause extensive corrosion. The first laboratory research was carried out by Ericsson [196], who observed a synergic effect of the combined influence of $SO_2$ (1 mg $SO_2$/cm$^2$·h) and NaCl (8 mg NaCl/cm$^2$) at 90% RH which was not seen at 70% RH.

The small amount of research that has been performed on this matter has been carried out in field tests by Corvo [80], Allam [197], Almeida et al. [198], Feliu and Morcillo [5], Liang et al. [199] and Wang et al. [200].

Corvo [80], after 6 months of atmospheric exposure of steel at marine testing stations in Cuba, found the following damage function:

$$C(g/m^2) = 64.9 + 6.9\,[Cl^-] + 0.15\,[SO_2]^2 - 0.17\,[Cl^-]\,[SO_2]^2, \tag{33}$$

From which a very significant influence of $Cl^-$ is deduced, according to the coefficient obtained. $SO_2$ also influences weight loss with a lower coefficient, but in a quadratic form. However, the combined influence of $SO_2$ and $Cl^-$ has a negative sign, indicating a decrease in corrosion. The results corresponding to multilinear stepwise regression and correlation for annual data are however different, where coefficients affecting $SO_2$ alone or combined with $Cl^-$ are not significant.

$$C(g/m^2) = 243.7 + 6.7\,[Cl^-], \tag{34}$$

Allam [197], on the basis of results obtained in a study carried out at the shoreline on the western coast of the Arabian Gulf, with the presence of $SO_2$ (10 ppb) and $H_2S$ (70 ppb) in the atmosphere, formulated the following mechanism for tests from 10 h to 12 months in duration, in which advanced surface analysis was used to characterise the corrosion products.

During the initial stage, the formation of iron sulfate ($FeSO_4$) takes place concurrently with the formation of iron chlorides. The negatively charged sulfate ions may compete with $Cl^-$ ions for the ferrous ions ($Fe^{2+}$) produced by anodic reaction. This competitive effect has been reported in different papers [70,201,202].

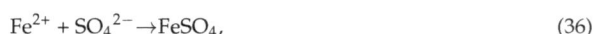

$$Fe^{2+} + 2Cl^- \rightarrow FeCl_2, \tag{35}$$

$$Fe^{2+} + SO_4^{2-} \rightarrow FeSO_4, \tag{36}$$

Iron chlorides, as a major constituent in the initially formed blister covers, indicate that Cl$^-$ ions are more aggressive than sulfate ions during the initial stages of AC. During the formation of FeSO$_4$ at anodic sites, sodium ions (Na$^+$) in the electrolyte are expected to migrate to the cathodic sites at the periphery of the anodic sites (blister site) to form sodium sulfate (Na$_2$SO$_4$). As blisters grow to form a thick continuous corrosion product layer, the formation of iron chlorides will eventually decrease compared to that seen during the initial stages. In contrast, further formation of iron sulfates takes place at the metal/rust interface during prolonged exposure times.

Feliu and Morcillo [5] in mixed atmospheres with SO$_2$ and NaCl contents ranging from 0.2–1.6 mg SO$_2$/dm$^2$·d and 0.5–3.8 mg NaCl/dm$^2$·d, respectively, observe an additivity of effects as both pollutants act together.

Almeida et al. [198] carried out a study involving a large number of atmospheres in the Ibero-American region in which both pollutants were found: Cl$^-$ ranging from 4.4 to 203.0 mg Cl$^-$/m$^2$·d and SO$_2$ ranging from 16.7 to 65.2 mg SO$_2$/m$^2$·d. Figure 41 shows the evolution of CS corrosion with the atmospheric concentration of both pollutants, in which it is possible to see the significantly more corrosive action of Cl$^-$ compared to SO$_2$. At low Cl$^-$ contents the presence of SO$_2$ shows a beneficial effect on the corrosion of the base steel, as SO$_2$ favours the transformation of lepidocrocite into goethite. As the Cl$^-$ concentration rises, the attack is intensifies with the SO$_2$ content (perhaps a synergic effect) and among the corrosion products it is possible to see the presence of akaganeite together with the lepidocrocite and goethite phases. After passing a certain threshold in the concentration of both pollutants, the attack of the steel seems to decrease, but this observation will need to be confirmed in a greater number of atmospheres with very high concentrations of both pollutants.

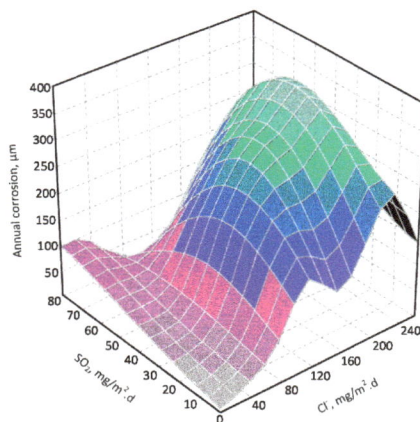

**Figure 41.** 3D representation of annual steel corrosion rate as a function of SO$_2$ and Cl$^-$ contents in the atmosphere [198].

Liang et al. [199] carried out a sixteen-year AC exposure study of steels and found that SO$_2$ only had an obvious deteriorating effect in the initial stages of atmospheric exposure. Wang et al. [200] also confirm this effect, indicating that Cl$^-$ is the ion that plays the main role in later stages, contrarily to the findings of Allam [197].

Some laboratory studies have also been carried out on the joint effect of both pollutants acting in combination. Attention is drawn to the contributions of Knotková et al. [203], Bastidas [204] and Chen et al. [205].

Knotková et al. [203] studied the joint action of both pollutants by alternate exposure of steel in cabinets containing SO$_2$ and sprayed NaCl solution (100–200 mg NaCl/m$^2$.d or equivalent Cl$^-$ contents in artificial seawater), respectively. They also studied the effect of SO$_2$ using Na$_2$SO$_4$ solutions.

They saw that (a) the corrosion products formed exhibited sulfate nests and akaganeite, the latter in smaller quantities than when the $Cl^-$ pollutant acted individually. From the morphological point of view, the influence of $Cl^-$ was predominant; and (b) a synergic effect on steel corrosion which would disappear by the end of the test, and their effect would only be additive.

These researchers conclude that the steel corrosion process when the two pollutants act together is a complex process involving two fundamental factors: the activity of $H^+$ in the aqueous adlayer formed on the metal

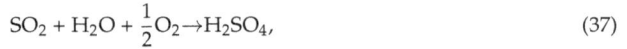

$$SO_2 + H_2O + \frac{1}{2}O_2 \rightarrow H_2SO_4, \tag{37}$$

Which acquires an acid pH, and the ion capturing capacity of the corrosion products that are formed. In this respect they note that when $SO_4^{2-}$ is introduced in the form of sodium sulfate ($Na_2SO_4$) solution there is an increase in the pH value, an opposing effect to the situation noted above, as a consequence of the formation of a very stable NaOH solution:

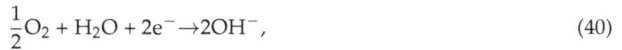

$$Na_2SO_4 \rightarrow 2Na^+ + SO_4^{2-}, \tag{38}$$

$$Fe + SO_4^{2-} \rightarrow FeSO_4 + 2e^-, \tag{39}$$

$$\frac{1}{2}O_2 + H_2O + 2e^- \rightarrow 2OH^-, \tag{40}$$

Bastidas [204], in alternate exposure to NaCl (20, 50 and 100 mg/m$^2$·d) and iron sulfate at the same concentrations, observe that $Cl^-$ is more harmful than an identical concentration of $SO_2$, and that the presence of one pollutant increases the attack caused by the other; the higher the concentration, the greater the combined effect (Figure 42). They find an additivity of effects when both pollutants act together.

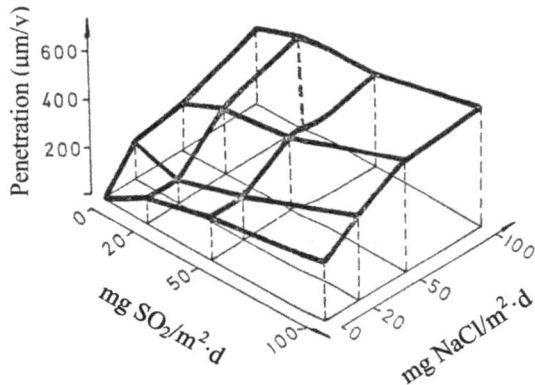

**Figure 42.** Effect of chlorides, $SO_2$ and the combination of both pollutants on iron corrosion kinetics. Electrochemical testing in conditions of 100% RH and 30 °C temperature [204].

Finally, Chen et al. [205] have recently carried out a wet/dry cyclic corrosion test using electrolytes prepared by adding different amounts of $Na_2SO_3$ (from 25.6 to 256 mg/m$^2$·d) into a NaCl solution (710 mg/m$^2$·d). They note that in the co-presence of $Cl^-$ and $SO_2$, the steel corrosion mass gain increases as the $SO_2$ content rises up to a certain level, and beyond this level the steel shows a slower mass gain. From the kinetic point of view this finding is not in accordance with the commonly held idea [30] that a higher $SO_2$ content in the atmosphere should lead to a higher steel corrosion rate. $Cl^-$ dominates the corrosion process in the initial stage and the effect of $SO_2$ comes later, accelerating the corrosion process as Allam reported [197].

Thus, this is a topic of great practical importance where, as has been seen, there is still considerable controversy on the effect of $SO_2$ in relation with the effect of $Cl^-$ in the initial stages of the AC process and the total magnitude of the corrosive attack with exposure time. Accordingly, greater research efforts are needed on both aspects.

## 9. Long-Term Behaviour of Carbon Steel Exposed to Marine Atmospheres

For socio-economically advanced societies with heavy infrastructure investments in coastal regions, steel corrosion may be a considerable problem. Thus it is fundamental for engineers and political policy-makers to be able to predict AC well into the future (25, 50, 100 years). It must be considered that in some highly developed countries efforts are now being made to design civil structures such as bridges and other load-bearing structures for 50–100 years of service without any maintenance. Data mining and modelling tools can help to improve AC forecasts and anti-corrosive designs, but despite great progress in the development of damage functions (dose-response) in wide-scale international cooperative research programmes there is still a way to go for such long-term modelling of AC processes.

### 9.1. Nature of Corrosion Products

The nature of the rust constituents is barely affected by the exposure time; in fact, the same species are usually detected at a given site however long the exposure. The time factor only alters the proportions of the constituents, or at most determines the appearance or disappearance of intermediate or minor compounds [97]. Thus, lepidocrocite, goethite, akaganeite and spinel phase (magnetite/maghemite) are usually the main corrosion products found on steel after long-term marine atmospheric exposure. Akaganeite is a typical component of rust developed in marine atmospheres. On contacting the steel surface, akaganeite is gradually transformed into magnetite [121,122], in such a way that in severe marine atmospheres this substance can become the main component of the corrosion layer [124].

### 9.2. First Year Steel Corrosion

There are numerous published damage functions on CS corrosion and environmental parameters, both meteorological (air temperature, RH, rainfall, TOW, etc.) and atmospheric-pollution-related (mainly $SO_2$ and airborne salinity). In this respect, attention is drawn to the efforts made by ISO with regard to atmospheric corrosivity classification (ISO 9223 [30]) and the international cooperative programmes on AC: ISOCORRAG [29], ICP Materials [206], and MICAT [207].

In Section 4 it was seen that the annual corrosion of steel accelerates as the saline content in the atmosphere rises (Figure 7); the magnitude of the attack in marine atmospheres normally exceeds that found in other types of atmospheres.

### 9.3. Long-Term Steel Corrosion

For long-term AC, most of the experimental data has been found to adhere to the following kinetic relationship:

$$C = At^n, \tag{41}$$

where C is the corrosion after time $t$, and A and $n$ are constants.

Thus, corrosion penetration data is usually fitted to a power function involving logarithmic transformation of the exposure time and corrosion penetration:

$$\log C = \log A + n \log t, \tag{42}$$

This power function (also called the bilograrithmic law) is widely used to predict the AC behaviour of metallic materials even after long exposure times, and its accuracy and reliability have been demonstrated by a great number of authors. Figure 43, obtained from CS corrosion data after different exposure times in marine atmospheres at different test sites [87], confirms the verification of the power function (Equation (41)).

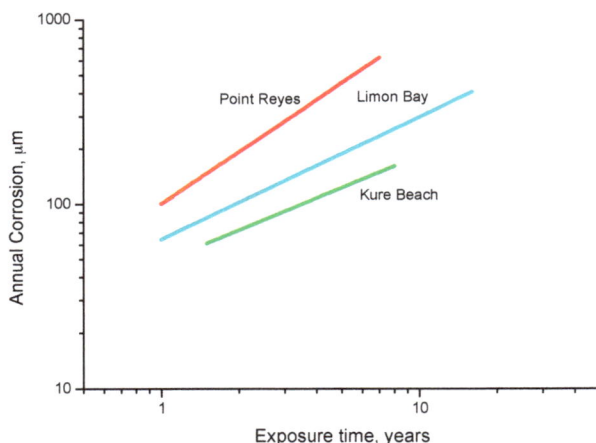

**Figure 43.** Typical log-log plots for carbon steel corrosion versus exposure time at different marine sites: Kure Beach (USA), Limon Bay (Panamá) and Point Reyes (USA). Data obtained from the reference [87].

According to Benarie and Lipfert [208], Equation (41) is a mass-balance equation, showing that the diffusion process is rate-determining, and this rate depends on the diffusive properties of the layer separating the reactants. The exponential law with $n$ close to 0.5, can result from an ideal diffusion-controlled mechanism when all the corrosion products remain on the metal surface. This situation seems to occur in slightly polluted inland atmospheres. On the other hand, $n$ values of more than 0.5 arise due to acceleration of the diffusion process (e.g., as a result of rust detachment by erosion, dissolution, flaking, cracking, etc.). This situation is typical of marine atmospheres, even those with low $Cl^-$ contents. Conversely, $n$ values of less than 0.5 result from a decrease in the diffusion coefficient with time through recrystallisation, agglomeration, compaction, etc. of the rust layer. Therefore, the exponent $n$ value can be used as an indicator of the physico-chemical behaviour of the corrosion product layer and thus of its interaction with the local atmosphere, exposure conditions, nature of wetting/drying cycles, etc.

According to Benarie and Lipfert [208], as a rule $n < 1$ and there is no physical sense for $n > 1$, as $n = 1$ is the limit for unimpeded diffusion (high permeable corrosion products or no layer at all). Thus, values of $n > 1$, very frequent in severe marine atmospheres, have been dismissed in many MAC studies as being due to outliers or errors in mass loss determinations.

However, many of these may be real values and not be due to error in mass loss determinations. The reason for this behaviour lies in the fact that in highly severe marine atmospheres, Equation (41), based on diffusion mechanisms, can sometimes not be applicable. When applied, it is common to find exponent $n$ values of close to 1 due to the existence of highly permeable (and barely protective) corrosion layers or the absence of corrosion layers because of their detachment by delamination (exfoliation), or even values of $n > 1$ due to acceleration of the corrosive attack as a result of an "autocatalytic" mechanism [23], in contrast to the diffusion mechanism upon which Equation (41) is based. Figure 44 shows the evolution of steel corrosion with exposure time in marine atmospheres with high deposited $Cl^-$ ion contents. The acceleration of the attack as exposure time advances is evident [12].

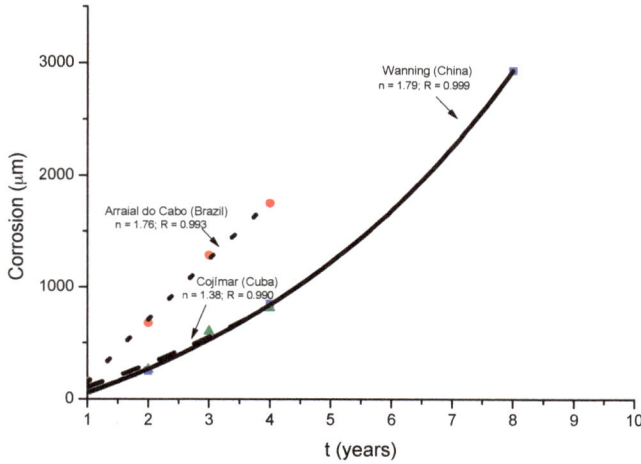

**Figure 44.** Evolution of mild steel corrosion with exposure time in severe marine atmospheres: Wanning, China, Arraial do Cabo, Brazil and Cojímar, Cuba. Values of exponent n and correlation coefficient (R) have been obtained from log C vs. log t plots for each site [12].

In an attempt to relate the exponent *n* value with atmospheric salinity, Figure 45, prepared using data obtained in the ISOCORRAG [29] and MICAT [207] programmes, for marine atmospheres with low $SO_2$ levels (<35 mg $SO_2/m^2 \cdot d$), shows the tendency for the exponent *n* value to increase with atmospheric salinity and how *n* can acquire values of more than unity. It would be important to have a greater volume of data for *n* > 1 in order to perfect Figure 45 [12].

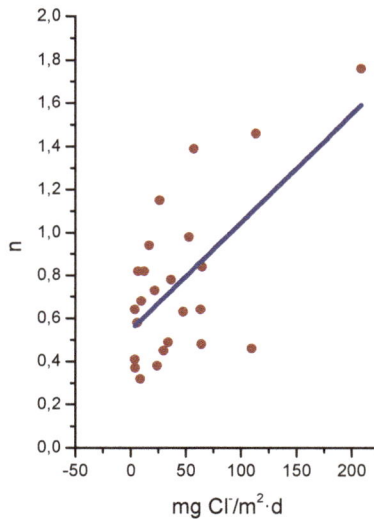

**Figure 45.** Variation in exponent n with atmospheric salinity at marine test sites in the MICAT [207] and ISOCORRAG [29] programmes.

Highly permeable corrosion layers or the absence of thick corrosion layers at all because of their detachment by delamination (exfoliation) (see Figure 38) must be constituted by macropores ($2 < D < 10$ nm) or perhaps even mesopores (D ~10–30 nm), instead of micropores (D < 2 nm), as occurs in consolidated rust layers formed in marine atmospheres of less aggressivity. According to Ishikawa et al. [146], macropores are considered to be too large to govern diffusion of molecules and ions through the rust layers.

*9.4. Modelling of the Long-Term Atmospheric Corrosion Process*

Data on the corrosion resistance of metals over long periods of time is important for determining the service life of metal structures and for developing the methods and means for their protection and preservation. Reliable estimates of corrosion resistance can be provided by corrosion tests under natural conditions. Such tests are time-consuming and expensive. In view of this, researchers pay great attention to the development of models that allow long-term forecasts without requiring testing under natural conditions.

It has been seen that the power function (Equation (41)) is widely used in long-term forecasts of the AC of metals. Table 6 sets out average values of exponent *n* for plain CS in different types of atmospheres, and Figure 46 shows the corresponding box-whisker plots of *n* values. It is possible to see a clear tendency towards higher *n* values in marine atmospheres.

**Table 6.** Average values of exponent n in bi-logarithmic plots of the power function ($C = At^n$) for plain carbon steel in non-marine (rural-urban-industrial) and marine atmospheres [87].

| Non-Marine (Rural-Urban-Industrial) Atmospheres | | Marine Atmospheres | |
|---|---|---|---|
| *Av. n* | Range of *n* in Equation (41) | *Av. n* | Range of *n* in Equation (41) |
| 0.49 | 0.26–0.76 | 0.73 | 0.37–0.98 |

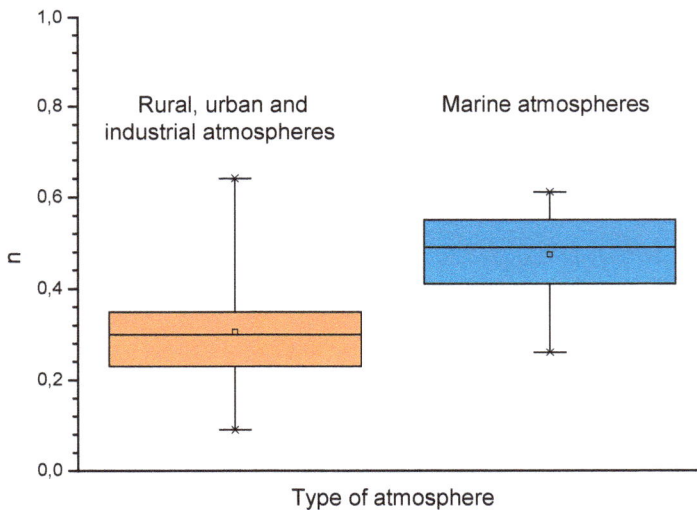

**Figure 46.** Box-whisker plots of n values in power function ($C = At^n$) for carbon steel in different types of atmospheres [87].

Panchenko et al. [209] propose a modification of the model for the long-term forecasting of corrosion losses of metals in any type of atmosphere after the establishment of the steady state. They find a stochastic relationship between exponent *n* of the power function (Equation (41)) and corrosion losses over the first year, and make a forecast of corrosion losses based on a power function using the *n* values calculated from the identified stochastic relationships.

McCuen and Albrecht [210] proposed improving the power model by replacing it with two different approaches: numerical model and power-linear model, the latter consisting of a power function at the initial stage (Equation (41)) and a linear function:

$$C_t = C_0 + \alpha_t, \tag{43}$$

at the steady-state stage.

As to whether this law provides a better prediction of the AC of WS for exposure times of at least 20 years, McCuen et al. compared both models (the power model and the power-linear model) using AC data reported for WS in the United States and concluded that the experimental data fitted the power-linear model better than the power model and thus provided more accurate predictions of long-term AC [210].

Panchenko et al. [211] propose methods for the calculation of *n* and *α* (Equation (43)) and make a comparative estimate of long-term predictions using the power-linear function and ISOCORRAG standard, obtaining comparable results in atmospheric corrosivity categories C1–C3 [30].

Albrecht and Hall [212], by refinement of the power-linear model, have proposed a new bi-linear model based on ISO 9224 [213], called modified ISO 9224, as well as an adjustment of this new bi-linear model that accounts for a modified corrosion rate during the first year of exposure and a steady state in subsequent years.

Finally, Melchers [214,215] suggests a bi-modal model for long-term forecasts of the corrosion loss of WS and grey cast iron in marine atmospheres. The model consists of a number of sequential corrosion phases, each representing the corrosion process that is dominant at that time and which controls the instantaneous corrosion rate. The phases are summarised in Figure 47. The important difference from conventional models is that the bi-modal model has longer-term corrosion governed by microbial activity. Melchers has successfully applied this model to different sets of data points for long-term exposures at different test sites.

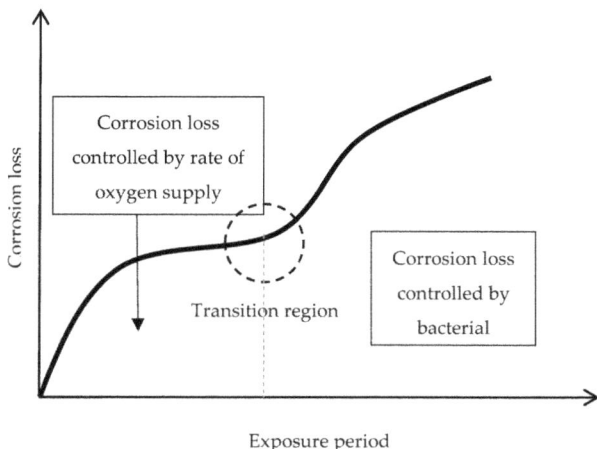

**Figure 47.** Bi-modal model for corrosion loss showing the changing behavior of corrosion process, according to Melchers [214,215].

*9.5. Towards an Estimation of the Marine Atmospheric Corrosion of Steel Based on Existing Environmental Parameters*

Following the philosophy of ancient ISO 9223 [216], an estimation of the atmospheric corrosivity of a particular coastal region could be made from the knowledge of three fundamental parameters: TOW of the metallic surface, $SO_2$ concentration in the atmosphere, and $Cl^-$ deposition rate due to airborne sea-salt particles.

Nowadays most countries have ample meteorological databases covering their entire territories which would allow estimations of TOW. Furthermore, information on atmospheric $SO_2$ concentrations is increasingly available. However, there tends to be very little information available on atmospheric salinity in different coastal areas, and reliable databases on this subject do not generally exist. It would be desirable to include this data in the numerous published damage functions between steel corrosion and environmental factors. This would make it possible to estimate AC simply from environmental data, without having to carry out natural corrosion tests at a specific site, which involve long waiting times and considerable expense.

9.5.1. Time of Wetness

TOW can generally be defined as the amount of time a metal surface remains wet during atmospheric exposure. The most commonly used practical definition of TOW is that given by ISO 9223 [30], which defines it as the total time when the RH of the ambient environment is equal to or greater than 80% at temperatures above 0 °C.

A number of limitations of the ISO definition have been pointed out in the literature and even in the ISO 9223 standard itself: (a) One major issue is that by definition precipitation and dewing events are excluded in ISO 9223. With regard to the latter, temperature differences between the ambient air and the surface can cause quite large deviations between the RH of the surface and the surrounding air; and (b) many atmospheric pollutants deliquesce at humidity levels far below 80%. A study by Cole et al. [13] clearly shows wetness measured on resistance-type sensors occurring well below the 80% threshold. Wetting phenomena associated with AC along with TOW definitions and determination methods were overviewed by Schindelholz and Kelly [14–16].

According to Cole et al. [13], in the case of salt deposits on the metallic surface, as occurs in marine atmospheres, it is necessary to take into account the simple principle that a metal surface is wetted when the surface RH exceeds the deliquescent RH (DRH) of any salts on the surface, as the ambient RH increasingly exceeds the DRH. They note the need for a more flexible method for predicting not only TOW but also the cycles of moisture accumulation and depletion. Given the established role of hygroscopic salts in promoting wetting in particulates, whether airborne or on surfaces, a model of surface wetting should address the role of deposited salts. The basic principle of the model is that the TOW of metal surfaces fully exposed to the environment can be approximated by the time of condensation (TCD) plus rain periods, i.e.,

$$TOW = TCD + \text{rain period}, \tag{44}$$

Thus, these researchers propose a method for estimating the wetting of a surface based on a comparison of surface RH and the deliquescence of salts that may pollute the surface, deriving relatively simple rules for wetting based on the DRH model and ISO classifications. These rules predict the total TOW to a high degree of accuracy.

9.5.2. Chloride Deposition Rate

Studies have shown that the two main sources of salt aerosol carried by the wind are ocean waves and breaking surf [60,66,217–219]. The magnitude of atmospheric salinity ($Cl^-$ deposition rate) at any given place depends on numerous factors. To mention just a few (see Section 4): wave height at high sea and at the coast; presence or absence of a surf zone on the coast; direction, speed and persistence of marine winds; height above sea level; distance from the shore; topographical effects; etc. Thus it has been seen in Section 4 that for instance wave height values of 1.5–2.0 m are sufficient to produce high monthly average salinity values, or that the wind blowing for a relatively short time at speeds above 3 m/s in a direction with great influence on the entrainment of marine aerosol is sufficient for atmospheric salinity to acquire important values.

There is abundant literature on this topic. Perhaps a less studied aspect has been the degree of shielding or sheltering on wind speed. In relation with this matter, Nakajima [220] has developed a mapping method based on grids with topographic factors to assess the influence of several factors on average wind speed at locations near a sea coast. The concept integrates the geographical texture of the environment in all directions with its effect on airflow. The most significant factor affecting sea wind is the degree of shielding.

Similarly, Klassen and Roberge [221] have measured and modelled by CLIMAT units [222] the influence of wind effects on local atmospheric corrosivity considering various degrees of wind sheltering. They found a 34-fold difference between the average mass loss of the most wind-protected and the least wind-protected points.

As noted by Roberge et al. [223], the dispersion of airborne salinity is highly dependent upon the local geography and wind patterns and is therefore rather difficult to model with simplistic functions. In this respect, Cole et al. [218] developed an aerosol penetration model applying the principles of aerosol transport using fluid dynamics principles. Salt aerosol sinks, such as gravity, rain and trees (obstacles), were accounted for. The study also noted that a high surface RH and a lower cloud height led to decreased penetration inland, and further reductions occurred in areas of high rainfall. It was also noted that structures in the airflow, such as forests and urban environments, reduced the $Cl^-$ concentration. This work illustrated that complex variables such as airspeed, ground roughness (vegetation), surface air RH, cloud height and rainfall could be incorporated into a model. A good correlation was reported to exist between the model and empirical results from a limited data set.

Despite the high number of factors involved, it is hoped that the growing flow of knowledge on this subject will soon lead to the desirable goal of being able to make a rough estimate of atmospheric salinity in a given geographic area without the need for measurements involving marine aerosol capturing techniques.

The prediction of atmospheric corrosivity at a given site is even more complex, as it is necessary to take into account an even greater number of variables. Thus, Roberge et al. paint a pessimistic view when they note that a single transferable and comprehensive environmental corrosivity prediction model still has to be published and may not be possible due to the complexity of the issues [223].

## 10. Issues Pending

AC has been extensively researched over the last one hundred years, and as a result the effects of meteorological and pollution variables on AC are now well known. Even so, our knowledge on this issue still holds many gaps, such as how to accurately estimate the total TOW of metallic structures, and the effects of climate change and acid rain, etc.

The issue of steel corrosion in coastal regions is particularly relevant in view of the latter's great importance to human society, considering that about half of the world's population lives in coastal regions. Thus it is surprising that marine atmospheric corrosion (MAC) has until recently received relatively little attention by corrosion scientists. This is therefore a relatively young scientific field, where there continue to be great gaps in knowledge.

In this review, we have noted a number of aspects in relation with which greater research efforts would seem to be necessary. To mention just a few:

1. Experimentation in this field is carried out at atmospheric corrosion testing stations and in the laboratory by means of wet/dry cyclic tests. A great amount of research is under way using both methods, but there are two issues of enormous importance that are not being paid sufficient attention: (a) experimentation in marine atmospheres with high $Cl^-$ ion deposition

rates (>500 mg/m²·d), where very little information is available; and (b) the standardisation of wet/dry laboratory cyclic tests by means of more specific codes in order to make research results more comparable.

It would also be necessary to undertake more research in the case of marine-industrial atmospheres, where there are great discrepancies among researchers. This is a particularly important issue in developing countries where factories are often located in coastal regions.

2. One currently unresolved question is concerned with the presence of the amorphous phase in the rust layer and the evolution of its amount with exposure time. With regard to the role played in corrosion mechanisms by the less crystalline phases of rust (ferrihydrite, feroxyhyte, etc.), despite the enormous research effort that has been carried out in recent years by a number of research teams using highly sophisticated analytical techniques, there are still numerous gaps in knowledge.

    Another matter that generates a great deal of uncertainty is the differentiation of magnetite and maghemite phases, both of which are very similar in many of their characteristics, but which can play a different role in the MAC process.

3. In the rust layers formed, aspects such as decoupling of the anodic and cathodic corrosion reactions, localisation, connectivity and reactivity of the different rust phases inside the corrosion layers, as well as characteristics such as porosity, tortuosity, etc., are also of primary importance in the determination of corrosion mechanisms.

    At high $Cl^-$ ion deposition rates, rust layers can exfoliate and become detached from the steel substrate. Although great advances have recently been made in this field, there are still a number of basic aspects that remain to be clarified in order for a complete comprehension of rust exfoliation phenomena.

4. Finally, a matter of enormous technical importance for engineers and political policy-makers is to be able to predict steel corrosion rate well into the future (20, 50, 100 years). Data mining and modelling tools can help to improve forecasts and anti-corrosive designs, but despite great progress in the development of damage functions (dose-response) in wide-scale international cooperative research programmes, there is still a long way to go for such long-term modelling of atmospheric corrosion processes.

    In this sense, better scientific knowledge is needed towards the desirable goal of being able to estimate atmospheric salinity in a given geographic area without the need for measurements involving marine aerosol capturing techniques.

**Conflicts of Interest:** The authors declare no conflict of interest.

## References

1. Evans, U.R. *The Corrosion and Oxidation of Metals: Scientific Principles and Practical Applications*; Edward Arnold Ltd.: London, UK, 1960.
2. Tomashov, N.D. *Theory of Corrosion and Protection of Metals. The Science of Corrosion*; The MacMillan Co.: New York, NY, USA, 1966.
3. Rozenfeld, I.L. *Atmospheric Corrosion of Metals*; NACE: Houston, TX, USA, 1972.
4. Barton, K. *Protection against Atmospheric Corrosion*; John Wiley and Sons: New York, NY, USA, 1976.
5. Feliu, S.; Morcillo, M. *Corrosión y Protección de los Metales en la Atmósfera*; Bellaterra: Barcelona, Spain, 1982.
6. Graedel, T.E.; Mc Gill, R. Degradation of materials in the atmosphere. *Environ. Sci. Technol.* **1986**, *20*, 1093–1100. [CrossRef]
7. Kucera, V.; Mattsson, E. Atmospheric corrosion. In *Corrosion Mechanisms*; Mansfeld, F., Ed.; Marcel Dekker: New York, NY, USA, 1987; pp. 211–284.
8. Leygraf, C.; Odnevall Wallinder, I.; Tidblad, J.; Graedel, T. *Atmospheric Corrosion*, 2nd ed.; The Electrochemical Society Series; John Wiley and Sons: Hoboken, NJ, USA, 2016.
9. Ambler, H.R.; Bain, A.A.J. Corrosion of metals on the tropics. *J. Appl. Chem.* **1955**, *5*, 437–527. [CrossRef]

10.  Nishimura, T.; Katayama, H.; Noda, K.; Kodama, T. Electrochemical behavior of rust formed on carbon steel in a wet/dry environment containing chloride ions. *Corrosion* **2000**, *56*, 935–941. [CrossRef]
11.  Morcillo, M.; Díaz, I.; Chico, B.; Cano, H.; de la Fuente, D. Wethearing steels: From empirical development to scientific desing. A review. *Corros. Sci.* **2014**, *83*, 6–31. [CrossRef]
12.  Morcillo, M.; Chico, B.; Alcántara, J.; Díaz, I.; Simancas, J.; de la Fuente, D. Atmospheric corrosion of mild steel in chloride—Rich environments. Questions to be answered. *Mater. Corros.* **2015**, *66*, 882–892.
13.  Cole, I.S.; Ganther, W.D.; Sinclair, J.D.; Lau, D.; Paterson, D.A. A study of the wetting of metal surfaces in order to understand the process controlling atmospheric corrosion. *J. Electrochem. Soc.* **2004**, *151*, B627–B635. [CrossRef]
14.  Schindelholz, E.; Risteen, B.E.; Kelly, R.G. Effect of relative humidity on corrosion of steel under sea salt aerosol proxies: I. NaCl. *J. Electrochem. Soc.* **2014**, *161*, C450–C459. [CrossRef]
15.  Schindelholz, E.; Risteen, B.E.; Kelly, R.G. Effect of relative humidity on corrosion of steel under sea salt aerosol proxies: II. MgCl$_2$, artificial seawater. *J. Electrochem. Soc.* **2014**, *161*, C460–C470. [CrossRef]
16.  Schindelholz, E.; Kelly, R.G. Wetting phenomena and time of wetness in atmospheric corrosion: A review. *Corros. Rev.* **2012**, *30*, 135–170. [CrossRef]
17.  Costa, J.M.; Morcillo, M.; Feliu, S. Effect of environmental parameters on atmospheric corrosion of metals. In *Encyclopedia of Environmental Control Technology: Air Pollution Control*; Cheremisinoff, P.N., Ed.; Gulf Publishing Company: Houston, TX, USA, 1989; Volume 2, pp. 197–238.
18.  Rozenfeld, I.L. Atmospheric corrosion of metals. Some questions of theory. In Proceedings of the 1st International Congress on Metallic Corrosion, London, UK, 10–15 April 1961; pp. 243–253.
19.  Tidblad, J.; Kucera, V.; Ferm, M.; Kreislova, K.; Brüggerhoff, S.; Doytchinov, S.; Screpanti, A.; Grøntoft, T.; Yates, T.; de la Fuente, D.; et al. Effects of air pollution on materials and cultural heritage: ICP materials celebrates 25 years of research. *Int. J. Corros.* **2012**, *2012*. [CrossRef]
20.  Haagenrud, S.; Ottar, B. Long range transport of air pollutants and corrosion effects. In Proceedings of the 7th Scandinavian Corrosion Congress, Trondheim, Norway, May 1975; p. 102.
21.  Kucera, V. Effects of sulphur dioxide and acid precipitation on metals and anti-rust painted steel. *Ambio* **1976**, *5*, 243–248.
22.  Feliu, S.; Morcillo, M. Corrosión atmosférica. In *Corrosión y Protección metáLicas*; Andrade, M.C., Feliu, S., Eds.; CSIC (Colección Nuevas Tendencias): Madrid, Spain, 1991; Volume II.
23.  Burstein, G.T. Passivity and localised corrosion. In *Corrosion. Metal/Environment Reactions*, 3rd ed.; Shreir, L.L., Jarman, R.A., Burstein, G.T., Eds.; Butterworth-Heinemann: Oxford, UK, 1994; Volume 1, p. 1:146.
24.  Feitknecht, W. The breakdown of oxide films on metal surfaces in acidic vapors and the mechanism of atmospheric corrosion. *Chimia* **1952**, *6*, 3–13.
25.  Askey, A.; Lyon, S.B.; Thompson, G.E.; Johnson, J.B.; Wood, G.C.; Cooke, M.; Sage, P. The corrosion of iron and zinc by atmospheric hydrogen chloride. *Corros. Sci.* **1993**, *34*, 233–247. [CrossRef]
26.  Schikorr, G. On the mechanism of atmospheric corrosion of iron. *Werkst. Korros.* **1963**, *14*, 63–80.
27.  International Organization for Standardization. *ISO 8565, Metals and Alloys—Atmospheric Corrosion Testing—General Requirements*; International Organization for Standardization: Geneva, Switzerland, 2011.
28.  American Society for Testing and Materials. *ASTM G50, Conducting Atmospheric Corrosion Tests on Metals*; American Society for Testing and Materials: Philadelphia, PA, USA, 1991.
29.  Knotkova, D.; Kreislova, K.; Dean, S.W.J. *ISOCORRAG. International Atmospheric Exposure Program: Summary of Results*; ASTM: West Conshohocken, PA, USA, 2010.
30.  European Committee for Standardization. *EN ISO 9223, Corrosion of Metals and Alloys—Corrosivity of Atmospheres—Classification, Determination and Estimation*; European Committee for Standardization: Brussels, Belgium, 2012.
31.  Compton, K.G.; Mendizza, A.; Bradley, W.W. Atmospheric galvanic couple corrosion. *Corrosion* **1955**, *11*, 35–44. [CrossRef]
32.  Doyle, D.P.; Godard, H.P. Rapid determination of corrosivity of an atmosphere to aluminium. In *Proceedings of the 3rd International Congress on Metallic Corrosion*; MIR Publishers: Moscow, Russia, 1969; Volume IV, pp. 429–437.
33.  Doyle, D.P.; Godard, H.P. A rapid method for determining the corrosivity of the atmosphere at any location. *Nature* **1963**, *200*, 1167–1168. [CrossRef]

34. Haynes, G. Cabinet. In *Corrosion Tests and Standards. Application and Interpretation*; Baboian, R., Ed.; American Society for Testing and Materials: Philadelphia, PA, USA, 1995; pp. 91–97.

35. American Society for Testing and Materials. *ASTM B117, Test Method of Salt Spray (Fog) Testing*; American Society for Testing and Materials: Philadelphia, PA, USA, 2011.

36. International Organization for Standardization. *ISO 11130, Corrosion of Metals and Alloys—Alternate Immersion Test in Salt Solution*; International Organization for Standardization: Geneve, Switzerland, 1999.

37. González, J.A. *Control de la Corrosión. Estudio y Medida por téCnicas Electroquímicas*; CSIC: Madrid, Spain, 1989.

38. Stratmann, M. The atmospheric corrosion of iron steel. *Metal Odlew* **1990**, *16*, 46–52.

39. Nishimura, T.; Tanaka, K.; Shimizu, Y. Effect of NaCl on rusting of steel in wet and dry corrosion cycle. *J. Iron Steel Inst. Jpn.* **1995**, *81*, 1079–1084.

40. Lair, V.; Antony, H.; Legrand, L.; Chaussé, A. Electrochemical reduction of ferric corrosion products and evaluation of galvanic coupling with iron. *Corros. Sci.* **2006**, *48*, 2050–2063. [CrossRef]

41. Hœrlé, S.; Mazaudier, F.; Dillmann, P.; Santarini, G. Advances in understanding atmospheric corrosion of iron. II. Mechanistic modelling of wet–dry cycles. *Corros. Sci.* **2004**, *46*, 1431–1465.

42. Cook, D.C. Spectroscopic identification of protective and non-protective corrosion coatings on steel structures in marine environments. *Corros. Sci.* **2005**, *47*, 2550–2570. [CrossRef]

43. De Faria, D.L.A.; Venâncio Silva, S.; de Oliveira, M.T. Raman microspectroscopy of some iron oxides and oxyhydroxides. *J. Raman Spectrosc.* **1997**, *28*, 873–878. [CrossRef]

44. Monnier, J.; Réguer, S.; Foy, E.; Testemale, D.; Mirambet, F.; Saheb, M.; Dillmann, P.; Guillot, I. XAS and XRD in situ characterisation of reduction and reoxidation processes of iron corrosion products involved in atmospheric corrosion. *Corros. Sci.* **2014**, *78*, 293–303. [CrossRef]

45. Kimura, M.; Mizoguchi, T.; Kihira, H.; Kaneko, M. Various scale analyses to create functioning corrosion products. In *Characterization of Corrosion Products on Steel Surfaces*; Waseda, Y., Suzuki, S., Eds.; Advances in Materials Research; Springer: Heildelberg, Germany, 2006; pp. 245–272.

46. Konishi, H.; Yamashita, M.; Uchida, H.; Mizuki, J. Characterization of rust layer formed on Fe, Fe-Ni and Fe-Cr alloys exposed to Cl-rich environment by Cl and Fe k-edge XANES measurements. *Mater. Trans.* **2005**, *46*, 329–336. [CrossRef]

47. Dillmann, P.; Mazaudier, F.; Hoerlé, S. Advances in understanding atmospheric corrosion of iron. I. Rust characterization of ancient ferrous artefacts exposed to indoor atmospheric corrosion. *Corros. Sci.* **2004**, *46*, 1401–1429.

48. Ishikawa, T. Asessment of rust layers formed on WS in saline environment by gas adsorption. *Mater. Corros.* **2015**, *66*, 1460–1466. [CrossRef]

49. Cole, I.S.; Lau, D.; Paterson, D.A. Holistic model for atmospheric corrosion part 6—From wet aerosol to salt deposit. *Corros. Eng. Sci. Technol.* **2004**, *39*, 209–218. [CrossRef]

50. Johnson, E.; Stanners, J.F. *The Characterisation of Corrosion Test Sites in the Community*; Commission of the European Communities: Luxembourg, 1981.

51. Blanchard, D.C.; Woodcock, A.H. The production, concentration and vertical distribution of the sea-salt aerosol. *Ann. N. Y. Acad. Sci.* **1980**, *338*, 330–347. [CrossRef]

52. Meira, G.R.; Pinto, W.T.A.; Lima, E.E.P.; Andrade, C. Vertical distribution of marine aerosol salinity in a brazilian coastal area—The influence of wind speed and the impact on chloride accumulation into concrete. *Constr. Build. Mater.* **2017**, *135*, 287–296. [CrossRef]

53. Zezza, F.; Macrì, F. Marine aerosol and stone decay. *Sci. Total Environ.* **1995**, *167*, 123–143. [CrossRef]

54. Whitby, K.T. The physical characteristics of sulfate aerosols. *Atmos. Environ.* **1978**, *12*, 135–159. [CrossRef]

55. Li, S.; Hihara, L.H. Aerosol salt particle deposition of metals exposed to marine environments: A study related to marine atmospheric corrosion. *J. Electrochem. Soc* **2014**, *161*, C268–C275. [CrossRef]

56. Gustafsson, M.E.R.; Franzén, L.G. Dry deposition and concentration of marine aerosols in a coastal area. *Atmos. Environ.* **1996**, *30*, 977–989. [CrossRef]

57. Cole, I.S.; Azmat, N.S.; Kanta, A.; Venkatraman, M. What really controls the atmospheric corrosion of zinc? Effect of marine aerosols on atmospheric corrosion of zinc. *Int. Mater. Rev.* **2009**, *54*, 117–133.

58. Fitzgerald, J.W. Marine aerosol: A review. *Atmos. Environ.* **1991**, *25A*, 533–545. [CrossRef]

59. Wu, J. Evidence of sea spray produced by bursting bubbles. *Science* **1981**, *212*, 324–326. [CrossRef] [PubMed]

60. Feliu, S.; Morcillo, M.; Chico, B. Effect of distance from sea on atmospheric corrosion rate. *Corrosion* **1999**, *55*, 883–891. [CrossRef]

61. Ohba, R.; Okabayashi, K.; Yamamoto, M.; Tsuru, T. A method for predicting the content of sea salt particles in the atmosphere. *Atmos. Environ.* **1990**, *24A*, 925–935. [CrossRef]

62. Ten Harkel, M.J. The effects of particle-size distribution and chloride depletion of sea-salt aerosols on estimating atmospheric deposition at a coastal site. *Atmos. Environ.* **1997**, *31*, 417–427. [CrossRef]

63. McKay, W.A.; Garland, J.A.; Livesley, D.; Halliwell, C.M.; Walker, M.I. The characteristics of the shore-line sea spray aerosol and the landward transfer of radionuclides discharged to coastal sea water. *Atmos. Environ.* **1994**, *28*, 3299–3309. [CrossRef]

64. Alcántara, J.; Chico, B.; Díaz, I.; de la Fuente, D.; Morcillo, M. Airborne chloride deposit and its effect on marine atmospheric corrosion of mild steel. *Corros. Sci.* **2015**, *97*, 74–88. [CrossRef]

65. Strekalov, P.V.; Panchenko, Y.M. The role of marine aerosols in atmospheric corrosion of metals. *Prot. Met.* **1994**, *30*, 254–263.

66. Morcillo, M.; Chico, B.; Mariaca, L.; Otero, E. Salinity in marine atmospheric corrosion: Its dependence on the wind regime existing in the site. *Corros. Sci.* **2000**, *42*, 91–104. [CrossRef]

67. Meira, G.R.; Andrade, C.; Alonso, C.; Padaratz, I.J.; Borba, J.C., Jr. Salinity of marine aerosols in a brazilian coastal area—Influence of wind regime. *Atmos. Environ.* **2007**, *41*, 8431–8441. [CrossRef]

68. Strekalov, P.V. Wind regimes, chloride aerosol particle sedimentation and atmospheric corrosion of steel and copper. *Prot. Met.* **1988**, *24*, 630–641.

69. Preston, R.S.J.; Sanyal, B. Atmospheric corrosion by nuclei. *J. Appl. Chem.* **1956**, *6*, 26–44. [CrossRef]

70. Evans, U.R.; Taylor, C.A.J. Mechanism of atmospheric rusting. *Corros. Sci.* **1972**, *12*, 227–246. [CrossRef]

71. Morcillo, M.; Chico, B.; Otero, E.; Mariaca, L. Effect of marine aerosol on atmospheric corrosion. *Mater. Perform.* **1999**, *38*, 72–77.

72. Pascual Marqui, R.D. Influencia de la concentración de ion cloruro sobre la corrosión atmosférica de un acero al carbono bajo capa de fase de humedad. *Rev. Corros. Prot.* **1980**, *XI*, 37–40.

73. Espada, L.; González, A.M.; Sánchez, A.; Merino, P. Estudio de la velocidad de corrosión de aceros de bajo contenido de carbono en nieblas salinas de distinta concentración. *Rev. Iberoam. Corros. Prot.* **1988**, *19*, 227–229.

74. Hache, A. Contribution à l'étude de la corrosion de l'acier en solutions salines. *Rev. Metall.* **1956**, *53*, 76–80. [CrossRef]

75. Strekalov, P.V. Estimation of atmospheric salinity from analysis of dry and wet chloride precipitates. *Prot. Met.* **1994**, *30*, 59–64.

76. State System for Standardization of Russian Federation. *GOST 9.039–74, Unified System of Corrosion and Ageing Protection. Corrosive Aggressiveness of atMosphere*; State System for Standardization of Russian Federation: Moscow, Russia, 1974.

77. Beruksitis, G.K.; Klark, G.B. *Corrosion Fatigue of Metals and Metals Coatings under Atmospheric Conditions*; Nauka: Moscow, Russia, 1971.

78. European Committee for Standardization. *EN ISO 9225, Corrosion of Metals and Alloys—Corrosivity of Atmospheres—Measurement of Environmental Parameters Affecting Corrosivity of Atmospheres*; European Committee for Standardization: Brussels, Belgium, 2012.

79. Foran, M.R.; Gibbons, E.V.; Wellington, J.R. The measurement of atmospheric sulfur dioxide and chlorides. *Chem. Can.* **1958**, *10*, 33–41.

80. Corvo, F. Atmospheric corrosion of steel in humid tropical climate. Influence of pollution, humidity, temperature, rainfall and sun radiation. *Corrosion* **1984**, *40*, 170–175.

81. Dean, S.W. Atmospheric. In *Corrosion Tests and Standards. Application and Interpretation*; Baboian, R., Ed.; ASTM: Philadelphia, PA, USA, 1995; pp. 116–125.

82. Industrial Galvanizers (INGALV). Corrosion Mapping System. Available online: http://www.valmontcoatings.com/locations/asia-pacific (accessed on 12 April 2017).

83. Japan Industrial Standard. *JIS-Z-2381, Recommended Practice for Weathering Test*; Japan Industrial Standard: Tokyo, Japan, 1987.

84. Hujioara, M.; Danakka, O. Weathering due to the chloride of ferro-concrete structure. *Concr. Eng.* **1987**, *25*, 44–47.

85. Li, Q.X.; Wang, Z.Y.; Han, W.; Han, E.H. Characterization of the rust formed on weathering steel exposed to Qinghai salt lake atmosphere. *Corros. Sci.* **2008**, *50*, 365–371. [CrossRef]

86. Wang, J.; Wang, Z.Y.; Ke, W. A study of the evolution of rust on weathering steel submitted to the Qinghai salt lake atmospheric corrosion. *Mater. Chem. Phys.* **2013**, *139*, 225–232. [CrossRef]

87. Morcillo, M.; Chico, B.; Díaz, I.; Cano, H.; de la Fuente, D. Atmospheric corrosion data of weathering steels. A review. *Corros. Sci.* **2013**, *77*, 6–24. [CrossRef]

88. Takebe, M.; Ohya, M.; Ajiki, S.; Furukawa, T.; Adachi, R.; Gan-ei, R.; Kitagawa, N.; Ota, J.; Matsuzaki, Y.; Aso, T. Estimation of quantity of Cl⁻ from deicing salts on weathering steel used for bridges. *Int. J. Steel Struct.* **2008**, *8*, 73–81.

89. Takebe, M.; Matsuzaki, Y.; Ohya, M.; Ajiki, S.; Furukawa, T.; Aso, T. Study of corrosion level and composition of accumulating salt on weathering bridges. *J. JSCE* **2007**, *63*, 172–180. [CrossRef]

90. Hara, S.; Miura, M.; Uchiumi, Y.; Fujiwara, T.; Yamamoto, M. Suppression of deicing salt corrosion of weathering steel bridges by washing. *Corros. Sci.* **2005**, *47*, 2419–2430. [CrossRef]

91. Cornell, R.M.; Schwertmann, U. *The Iron Oxides: Structure, Properties, Reactions, Occurrences and Uses*, 2nd ed.; Wiley-VCH Verlag GmbH: Weinheim, Germany, 2003.

92. Gilberg, M.R.; Seeley, N.J. The identity of compounds containing chloride ions in marine iron corrosion products: A critical review. *Stud. Conserv.* **1981**, *26*, 50–56. [CrossRef]

93. Monnier, J.; Neff, D.; Réguer, S.; Dillmann, P.; Bellot-Gurlet, L.; Leroy, E.; Foy, E.; Legrand, L. A corrosion study of the ferrous medieval reinforcement of the amiens cathedral. Phase characterisation and localisation by various microprobes techniques. *Corros. Sci.* **2010**, *52*, 695–710. [CrossRef]

94. Neff, D.; Reguer, S.; Bellot-Gurlet, L.; Dillmann, P. Structural characterization of corrosion products on archaeological iron: An integrated analytical approach to establish corrosion forms. *J. Raman Spectrosc.* **2004**, *35*, 739–745. [CrossRef]

95. Pino, E.; Alcántara, J.; Chico, B.; Díaz, I.; Simancas, J.; de la Fuente, D.; Morcillo, M. Atmospheric corrosion of mild steel in marine atmospheres. *Corros. Prot. Mater.* **2015**, *34*, 35–41.

96. Bernal, J.D.; Dasgupta, D.R.; Mackay, A.L. The oxides and hydroxides of iron and their structural inter-relationships. *Clay Miner. Bull.* **1959**, *4*, 15–30. [CrossRef]

97. Arroyave, C.; Morcillo, M. Atmospheric corrosion products in iron and steels. *Trends Corros.* **1997**, *2*, 1–16.

98. Butler, G.; Beynon, J.G. The corrosion of mild steel in boiling salt solutions. *Corros. Sci.* **1967**, *7*, 385–404. [CrossRef]

99. Refait, P.; Abdelmoula, M.; Génin, J.M.R. Mechanisms of formation and structure of green rust one in aqueous corrosion of iron in the presence of chloride ions. *Corros. Sci.* **1998**, *40*, 1547–1560. [CrossRef]

100. Stampfl, P.P. Ein basisches eisen-II-III-karbonat in rost. *Corros. Sci.* **1969**, *9*, 185–187. [CrossRef]

101. Ståhl, K.; Nielsen, K.; Jiang, J.; Lebech, B.; Hanson, J.C.; Norby, P.; van Lanschot, J. On the akaganéite crystal structure, phase transformations and possible role in post-excavational corrosion of iron artifacts. *Corros. Sci.* **2003**, *45*, 2563–2575. [CrossRef]

102. Watson, J.H.L.; Cardell, R.R.; Heller, W. The internal structure of colloidal crystals of β-Feooh and remarks on their assemblies in schiller layers. *J. Phys. Chem.* **1962**, *66*, 1757–1763. [CrossRef]

103. Keller, P. Occurrence, formation and phase transformation of β-Feooh in rust. *Werkst. Korros.* **1969**, *20*, 102–108. [CrossRef]

104. Rezel, D.; Genin, J.M.R. The substitution of chloride ions to OH⁻-ions in the akaganeite beta ferric oxyhydroxide studied by Mössbauer effect. *Hyperfine Interact.* **1990**, *57*, 2067–2075. [CrossRef]

105. Gallagher, K.J. The atomic structure of tubular subcrystals of β-iron (III) oxide hydroxide. *Nature* **1970**, *226*, 1225–1228. [CrossRef] [PubMed]

106. Post, J.E.; Buchwald, V.F. Crystal-structure refinement of akaganeite. *Am. Mineral.* **1991**, *76*, 272–277.

107. Mackay, A.L. β-ferric oxyhydroxide. *Mineral. Mag.* **1960**, *32*, 545–557. [CrossRef]

108. Mackay, A.L. β-ferric oxyhydroxide—Akaganeite. *Mineral. Mag.* **1962**, *33*, 270–280. [CrossRef]

109. Shiotani, K.; Tanimoto, W.; Maeda, C.; Kawabata, F.; Amano, K. Structural analysis of the rust layer on a bare weathering steel bridge exposed in a coastal industrial zone for 27 years. *Corros. Eng.* **2000**, *49*, 99–109. [CrossRef]

110. Morcillo, M.; Chico, B.; Alcántara, J.; Díaz, I.; Wolthuis, R.; de la Fuente, D. SEM/micro-Raman characterization of the morphologies of marine atmospheric corrosion products formed on mild steel. *J. Electrochem. Soc.* **2016**, *163*, C426–C439. [CrossRef]

111. De la Fuente, D.; Alcántara, J.; Chico, B.; Díaz, I.; Jiménez, J.A.; Morcillo, M. Characterisation of rust surfaces formed on mild steel exposed to marine atmospheres using XRD and SEM/micro-Raman techniques. *Corros. Sci.* **2016**, *110*, 253–264. [CrossRef]

112. Rémazeilles, C.; Refait, P. Formation, fast oxidation and thermodynamic data of Fe(II) hydroxychlorides. *Corros. Sci.* **2008**, *50*, 856–864. [CrossRef]

113. Refait, P.; Genin, J.M.R. The mechanism of oxidation of ferrous hydroxychloride $\beta$-Fe$_2$(OH)$_3$Cl in aqueous solution: The formation of akaganeite vs goethite. *Corros. Sci.* **1997**, *39*, 539–553. [CrossRef]

114. Rémazeilles, C.; Refait, P. On the formation of $\beta$-FeOOH (akaganéite) in chloride-containing environments. *Corros. Sci.* **2007**, *49*, 844–857. [CrossRef]

115. Refait, P.; Génin, J.M.R. The oxidation of ferrous hydroxide in chloride-containing aqueous media and Pourbaix diagrams of green rust one. *Corros. Sci.* **1993**, *34*, 797–819. [CrossRef]

116. Wells, A.F. *Structural Inorganic Chemistry*, 4th ed.; Oxford University Press: London, UK, 1975.

117. Fasiska, E.J. Structural aspects of the oxides and oxyhydrates of iron. *Corros. Sci.* **1967**, *7*, 833–839. [CrossRef]

118. Antunes, R.A.; Costa, I.; de Faria, D.L.A. Characterization of corrosion products formed on steels in the first months of atmospheric exposure. *Mater. Res.* **2003**, *6*, 403–408. [CrossRef]

119. Schwarz, H. Über die Wirkung des magnetits beim atmosphärischen Rosten und beim Unterrosten von Anstrichen. *Werkst. Korros.* **1972**, *23*, 648–663. [CrossRef]

120. Singh, A.K. Mössbauer and X-ray diffraction phase analysis of rusts from atmospheric test sites with different environments in Sweden. *Corros. Sci.* **1985**, *25*, 931–945. [CrossRef]

121. Ishikawa, T.; Kondo, Y.; Yasukawa, A.; Kandori, K. Formation of magnetite in the presence of ferric oxyhydroxides. *Corros. Sci.* **1998**, *40*, 1239–1251. [CrossRef]

122. Tanaka, H.; Mishima, R.; Hatanaka, N.; Ishikawa, T.; Nakayama, T. Formation of magnetite rust particles by reacting iron powder with artificial $\alpha$-, $\beta$- and $\gamma$-FeOOH in aqueous media. *Corros. Sci.* **2014**, *78*, 384–387. [CrossRef]

123. Hiller, J.E. Phasenumwandlungen im rost. *Werkst. Korros.* **1966**, *17*, 943–951. [CrossRef]

124. Jeffrey, R.J.; Melchers, R.E. The changing composition of the corrosion products of mild steel in severe marine atmospheres. In Proceedings of the Annual Conference of the Australasian Corrosion Association, Melbourne, Australia, 11–14 November 2012.

125. Haces, C.; Corvo, F.; Furet, N.R. Mecanismo de la corrosión atmosférica del acero en una zona de alta salinidad. In Proceedings of the 3rd Iberoamerican Congress of Corrosion and Protection, ABRACO, Río de Janeiro, Brazil, 26–30 June 1989; Volume 1, pp. 415–419.

126. Asami, K. Characterization of rust layers on a plain-carbon steel and weathering steels exposed to industrial and coastal atmosphere for years. In *Characterization of Corrosion Products on Steel Surfaces*; Waseda, Y., Suzuki, S., Eds.; Advances in Materials Research; Springer: Heidelberg, Germany, 2006; pp. 159–197.

127. Graham, M.J.; Cohen, M. Analysis of iron corrosion products using Mössbauer spectroscopy. *Corrosion* **1976**, *32*, 432–438. [CrossRef]

128. Leidheiser, H., Jr.; Musíc, S. The atmospheric corrosion of iron as studied by Mössbauer spectroscopy. *Corros. Sci.* **1982**, *22*, 1089–1096. [CrossRef]

129. Chico, B.; Alcántara, J.; Pino, E.; Díaz, I.; Simancas, J.; Torres-Pardo, A.; de la Fuente, D.; Jiménez, J.A.; Marco, J.F.; González-Calbet, J.M.; et al. Rust exfoliation on carbon steels in chloride-rich atmospheres. *Corros. Rev.* **2015**, *33*, 263–282. [CrossRef]

130. Oh, S.J.; Cook, D.C.; Townsend, H.E. Atmospheric corrosion of different steels in marine, rural and industrial environments. *Corros. Sci.* **1999**, *41*, 1687–1702. [CrossRef]

131. Antony, H.; Perrin, S.; Dillmann, P.; Legrand, L.; Chaussé, A. Electrochemical study of indoor atmospheric corrosion layers formed on ancient iron artefacts. *Electrochim. Acta* **2007**, *52*, 7754–7759. [CrossRef]

132. Matsubara, E.; Suzuki, S.; Waseda, Y. Corrosion mechanism of iron from an X-ray structural viewpoint. In *Characterization of Corrosion Products on Steel Surfaces*; Waseda, Y., Suzuki, S., Eds.; Advances in Materials Research; Springer: Heidelberg, Germany, 2006; pp. 105–129.

133. Environmental Geochemistry of Ferric Polymers in Aqueous Solutions and Precipitates. Available online: http://geoweb.princeton.edu/research/geochemistry/research/aqueous-polymers.html (accessed on 11 April 2017).

134. Waseda, Y.; Suzuki, S.; Saito, M. Structural characterization for a complex system by obtaining middle-range ordering. In *Characterization of Corrosion Products on Steel Surfaces*; Waseda, Y., Suzuki, S., Eds.; Advances in Materials Research; Springer: Heidelberg, Germany, 2006; pp. 77–104.

135. Raman, A.; Nasrazadani, S.; Sharma, L. Morphology of rust phases formed on weathering steels in various laboratory corrosion test. *Metallography* **1989**, *22*, 79–96. [CrossRef]

136. Morcillo, M.; Wolthuis, R.; Alcántara, J.; Chico, B.; Díaz, I.; de la Fuente, D. Scanning Electron Microscopy/microRaman: A very useful technique for characterizing the morphologies of rust phases formed on carbon steel in atmospheric exposures. *Corrosion* **2016**, *72*, 1044–1054.

137. Alcántara, J.; Chico, B.; Simancas, J.; Díaz, I.; de la Fuente, D.; Morcillo, M. An attempt to classify the morphologies presented by different rust phases formed during the exposure of carbon steel to marine atmospheres. *Mater. Charact.* **2016**, *118*, 65–78. [CrossRef]

138. Kihira, H. Colloidal aspects of rusting of weathering steel. In *Electrical Phenomena at Interfaces, Fundamentals, Measurements and Applications*, 2nd ed.; Ohshima, H., Furusawa, K., Eds.; Marcel Dekker, Inc.: New York, NY, USA, 1998.

139. Licheri, G.; Pinna, G. EXAFS and X-ray diffraction in solutions. In *EXAFS and Near Edge Structure*; Springer: Frascati, Italy, 13–17 September 1982; pp. 240–247.

140. Jolviet, J.P. *Metal Oxide Chemistry and Synthesis*; John Wiley & Sons Ltd.: West Sussex, UK, 2000.

141. Ishikawa, T.; Yoshida, T.; Kandori, K.; Nakayama, T.; Hara, S. Assessment of protective function of steel rust layers by $N_2$ adsorption. *Corros. Sci.* **2007**, *49*, 1468–1477. [CrossRef]

142. Hara, S.; Kamimura, T.; Miyuki, H.; Yamashita, M. Taxonomy for protective ability of rust layer using its composition formed on weathering steel bridge. *Corros. Sci.* **2007**, *49*, 1131–1142. [CrossRef]

143. Calero, J.; Alcántara, J.; Chico, B.; Díaz, I.; Simancas, J.; de la Fuente, D.; Morcillo, M. Wet/dry accelerated laboratory test to simulate the formation of multilayered rust on carbon steel in marine atmospheres. *Corros. Eng. Sci. Technol.* **2017**. [CrossRef]

144. Ishikawa, T.; Maeda, A.; Kandori, K.; Tahara, A. Characterization of rust on Fe-Cr, Fe-Ni, and Fe-Cu binary alloys by Fourier Transform Infrared and $N_2$ adsorption. *Corrosion* **2006**, *62*, 559–567. [CrossRef]

145. Ishikawa, T.; Katoh, R.; Yasukawa, A.; Kandori, K.; Nakayama, T.; Yuse, F. Influences of metal ions on the formation of β-FeOOH particles. *Corros. Sci.* **2001**, *43*, 1727–1738. [CrossRef]

146. Ishikawa, T.; Kumagai, M.; Yasukawa, A.; Kandori, K. Characterization of rust on weathering steel by gas adsorption. *Corrosion* **2001**, *57*, 346–352. [CrossRef]

147. LaQue, F.L. Corrosion testing. *Proc. ASTM* **1951**, *51*, 495–582.

148. Díaz, I. Corrosión Atmosférica de Aceros Patinables de Nueva Generación. Ph.D. Thesis, Complutense University, Madrid, Spain, October 2012.

149. Cano, H. Aceros Patinables (Cu, Cr, Ni): Resistencia a la Corrosión Atmosférica y Soldabilidad. Ph.D. Thesis, Complutense University, Madrid, Spain, December 2013.

150. Díaz, I.; Cano, H.; de la Fuente, D.; Chico, B.; Vega, J.M.; Morcillo, M. Atmospheric corrosion of Ni-advanced weathering steels in marine atmospheres of moderate salinity. *Corros. Sci.* **2013**, *76*, 348–360. [CrossRef]

151. Suzuki, I.; Hisamatsu, Y.; Masuko, N. Nature of atmospheric rust on iron. *J. Electrochem. Soc.* **1980**, *127*, 2210–2215. [CrossRef]

152. De la Fuente, D.; Diaz, I.; Simancas, J.; Chico, B.; Morcillo, M. Long-term atmospheric corrosion of mild steel. *Corros. Sci.* **2011**, *53*, 604–617. [CrossRef]

153. Morcillo, M.; Alcántara, J.; Díaz, I.; Chico, B.; Simancas, J.; de la Fuente, D. Marine atmospheric corrosion of carbon steels. *Rev. Metal. Madrid* **2015**, *51*, e045. [CrossRef]

154. Honzák, J. *Schutz von Stahlkonstruktion Gegen Atmospharische Korrosion (Protection of Steel Structures againts Atmospheric Corrosion)*; 57 Veranstaltung EFK: Prague, Czech Republic, 1971.

155. Copson, H.R. Atmospheric corrosion of low alloy steels. *Proc. ASTM* **1952**, *52*, 1005–1026.

156. García, K.E.; Morales, A.L.; Barrero, C.A.; Greneche, J.M. On the rusts products formed on weathering and carbon steels exposed to chloride in dry/wet cyclical processes. *Hyperfine Interact.* **2005**, *161*, 127–137. [CrossRef]

157. García, K.E.; Barrero, C.A.; Morales, A.L.; Greneche, J.M. Lost iron and iron converted into rust in steels submitted to dry–wet corrosion process. *Corros. Sci.* **2008**, *50*, 763–772. [CrossRef]

158. Barrero, C.A.; García, K.E.; Morales, A.L.; Greneche, J.M. A proposal to evaluate the amount of corroded iron converted into adherent rust in steels exposed to corrosion. *Corros. Sci.* **2011**, *53*, 769–775. [CrossRef]

159. Cano, H.; Neff, D.; Morcillo, M.; Dillmann, P.; Díaz, I.; de la Fuente, D. Characterization of corrosion products formed on Ni 2.4 wt%–Cu 0.5 wt%–Cr 0.5 wt% weathering steel exposed in marine atmospheres. *Corros. Sci.* **2014**, *87*, 438–451. [CrossRef]

160. Asami, K.; Kikuchi, M. In-depth distribution of rusts on a plain carbon steel and weathering steels exposed to coastal–industrial atmosphere for 17 years. *Corros. Sci.* **2003**, *45*, 2671–2688. [CrossRef]

161. Hara, S. A X-ray diffraction analysis on constituent distribution of heavy rust layer formed on weathering steel using synchrotron radiation. *Corros. Eng. Jpn.* **2008**, *57*, 70–75. [CrossRef]

162. Morcillo, M.; Chico, B.; de la Fuente, D.; Alcántara, J.; Odnevall Wallinder, I.; Leygraf, C. On the mechanism of rust exfoliation in marine environments. *J. Electrochem. Soc.* **2017**, *164*, C8–C16. [CrossRef]

163. Ishikawa, T.; Isa, R.; Kandori, K.; Nakayama, T.; Tsubota, T. Influences of metal chlorides and sulfates on the formation of β-FeOOH particles by aerial oxidation. *J. Electrochem. Soc* **2004**, *151*, B586–B594. [CrossRef]

164. Brunauer, S.; Emmett, P.H.; Teller, E. Adsorption of gases in multimolecular layers. *J. Am. Chem. Soc.* **1938**, *60*, 309–319. [CrossRef]

165. Rouguerol, F.; Rouguerol, J.; Sing, K. *Adsorption by Powders and Porous Solids*; Academic Press: London, UK, 1999.

166. Yamashita, M.; Misawa, T. Recent progress in the study of protective rust-layer formation on weathering steel. In Proceedings of the Corrosion' 98, San Diego, CA, USA, 22–27 March 1998; Technical Publication; NACE International: Houston, TX, USA, 31 December 1998; p. 357.

167. Kamimura, T.; Hara, S.; Miyuki, H.; Yamashita, M.; Uchida, H. Composition and protective ability of rust layer formed on weathering steel exposed to various environments. *Corros. Sci.* **2006**, *48*, 2799–2812. [CrossRef]

168. Henriksen, J.F. Distribution of NaCl on Fe during atmospheric corrosion. *Corros. Sci.* **1969**, *9*, 573–577. [CrossRef]

169. Misawa, T.; Kyuno, Y.; Suëtaka, W.; Shimodaira, S. The mechanism of atmospheric rusting and the effect of Cu and P on the rust formation of low-alloy steels. *Corros. Sci.* **1971**, *11*, 35–48. [CrossRef]

170. Misawa, T.; Hashimoto, K.; Shimodaira, S. The mechanism of formation of iron oxide and oxyhydroxides in aqueous solutions at room temperature. *Corros. Sci.* **1974**, *14*, 131–149. [CrossRef]

171. Misawa, T.; Asami, K.; Hashimoto, K.; Shimodaira, S. Mechanism of atmospheric rusting and protective amorphous rust on low-alloy steel. *Corros. Sci.* **1974**, *14*, 279–289. [CrossRef]

172. Mikhailovskii, N. *Atmospheric Corrosion of Metals and Protection Methods*; Metallurgiia: Moscow, Russia, 1989.

173. Worch, H.; Forker, W.; Rahner, D. Rust formation on iron. A model. *Werkst. Korros.* **1983**, *34*, 402–410. [CrossRef]

174. Schikorr, G. Korrosionsverhalten von zink und zinküberzügen und der atmosphäre. *Werkst. Korros.* **1964**, *15*, 537–543. [CrossRef]

175. Lau, T.T.N.; Thoa, N.T.P.; Nishimura, R.; Tsujino, Y.; Yokoi, M.; Maeda, Y. Atmospheric corrosion of carbon steel under field exposure in the southern part of Vietnam. *Corros. Sci.* **2006**, *48*, 179–192.

176. Li, S.; Hihara, L.H. In situ Raman spectroscopy study of NaCl particle-induced marine atmospheric corrosion of carbon steel. *J. Electrochem. Soc.* **2012**, *159*, C147–C154. [CrossRef]

177. Li, S.; Hihara, L.H. Atmospheric corrosion initiation on steel from predeposited NaCl salt particles in high humidity atmospheres. *Corros. Eng. Sci. Technol.* **2010**, *45*, 49–56. [CrossRef]

178. Li, S.; Hihara, L.H. Atmospheric corrosion electrochemistry of NaCl droplets on carbon steel. *J. Electrochem. Soc.* **2012**, *159*, C461–C468. [CrossRef]

179. Risteen, B.E.; Schindelholz, E.; Kelly, R.G. Marine aerosol drop size effects on the corrosion behavior of low carbon steel and high purity iron. *J. Electrochem. Soc.* **2014**, *161*, C580–C586. [CrossRef]

180. Ohtsuka, T.; Tanaka, S. Monitoring the development of rust layers on weathering steel using in situ Raman spectroscopy under wet-and-dry cyclic conditions. *J. Solid State Electrochem.* **2015**, *19*, 3559–3566. [CrossRef]

181. Evans, U.R. Electrochemical mechanism of atmospheric rusting. *Nature* **1965**, *206*, 980–982. [CrossRef]

182. Stratmann, M.; Bohnenkamp, K.; Engell, H.J. An electrochemical study of phase-transitions in rust layers. *Corros. Sci.* **1983**, *23*, 969–985. [CrossRef]

183. Horton, J.B. The Composition, Structure and Growth of the Atmospheric Rust on Various Steels. Ph.D. Thesis, Lehigh University, Bethlehem, PA, USA, May 1964.

184. Burger, E.; Fenart, M.; Perrin, S.; Neff, D.; Dillmann, P. Use of the gold markers method to predict the mechanisms of iron atmospheric corrosion. *Corros. Sci.* **2011**, *53*, 2122–2130. [CrossRef]

185. Antony, H.; Legrand, L.; Maréchal, L.; Perrin, S.; Dillmann, P.; Chaussé, A. Study of lepidocrocite γ-FeOOH electrochemical reduction in neutral and slightly alkaline solutions at 25 °C. *Electrochim. Acta* **2005**, *51*, 745–753. [CrossRef]

186. Monnier, J.; Burger, E.; Berger, P.; Neff, D.; Guillot, I.; Dillmann, P. Localisation of oxygen reduction sites in the case of iron long term atmospheric corrosion. *Corros. Sci.* **2011**, *53*, 2468–2473. [CrossRef]

187. Nomura, K.; Tasaka, M.; Ujihira, Y. Conversion electron Mössbauer spectrometric study of corrosion products of iron immersed in sodium chloride solution. *Corrosion* **1988**, *44*, 131–135. [CrossRef]

188. Ma, Y.; Li, Y.; Wang, F. The effect of β-FeOOH on the corrosion behavior of low carbon steel exposed in tropic marine environment. *Mater. Chem. Phys.* **2008**, *112*, 844–852. [CrossRef]

189. Ma, Y.; Li, Y.; Wang, F. Corrosion of low carbon steel in atmospheric environments of different chloride content. *Corros. Sci.* **2009**, *51*, 997–1006. [CrossRef]

190. Schwertmann, V.; Taylor, R.M. The transformation of lepidocrocite to goethite. *Clays Clay Miner.* **1972**, *20*, 151–158. [CrossRef]

191. Saha, J.K. *Corrosion of Constructional Steels in Marine and Industrial Environment*; Springer: New Delhi, India, 2013.

192. Morcillo, M.; González-Calbet, J.M.; Jiménez, J.A.; Díaz, I.; Alcántara, J.; Chico, B.; Mazarío-Fernández, A.; Gómez-Herrero, A.; Llorente, I.; de la Fuente, D. Environmental conditions for akaganeite formation in marine atmosphere mild steel corrosion products and its characterisation. *Corrosion* **2015**, *71*, 872–886. [CrossRef]

193. De la Fuente, D.; Díaz, I.; Alcántara, J.; Chico, B.; Simancas, J.; Llorente, I.; García-Delgado, A.; Jiménez, J.A.; Adeva, P.; Morcillo, M. Corrosion mechanisms of mild steel in chloride-rich atmospheres. *Mater. Corros.* **2015**, *67*, 227–238. [CrossRef]

194. Institute of Experimental Mineralogy. *Crystallographic and Crystallochemical Database for Minerals and Their Structural Analogues*; www.Mincryst; Institute of Experimental Mineralogy, Russian Academy of Sciences: Chernogolovka, Russia, 2016; Available online: http://database.iem.ac.ru/mincryst/index.php (accessed on 11 April 2017).

195. Copson, H.R. A theory of the mechanism of rusting of low-alloy steels in the atmosphere. *Proc. ASTM* **1945**, *45*, 554–580.

196. Ericsson, R. Influence of sodium-chloride on atmospheric corrosion of steel. *Werkst. Korros.* **1978**, *29*, 400–403. [CrossRef]

197. Allam, I.M.; Arlow, J.S.; Saricimen, H. Initial stages of atmospheric corrosion of steel in the arabian gulf. *Corros. Sci.* **1991**, *32*, 417–432. [CrossRef]

198. Almeida, E.; Morcillo, M.; Rosales, B. Atmospheric corrosion of mild steel part II—Marine atmospheres. *Mater. Corros.* **2000**, *51*, 865–874. [CrossRef]

199. Liang, C.; Hou, W. Sixteen year atmospheric corrosion exposure study of steels. *J. Chin. Soc. Corros. Prot.* **2005**, *25*, 1–6.

200. Wang, Z. Study of the corrosion behaviour of weathering steels in atmospheric environments. *Corros. Sci.* **2013**, *67*, 1–10. [CrossRef]

201. Scully, J.C. *The Fundamentals of Corrosion*, 3rd ed.; Pergamon Press: Oxford, UK, 1990; p. 106.

202. Brown, P.W.; Masters, L.W. Factors affecting the corrosion of metals in the atmosphere. In *Atmospheric Corrosion*; Ailor, W.H., Ed.; Wiley: New York, NY, USA, 1982; pp. 34–49.

203. Knotkova, D.; Barton, K.; van Tu, B. Atmospheric corrosion in maritime industrial atmospheres: Laboratory research. In *Degradation of Metals in the Atmosphere*; ASTM STP 965; Dean, S.W., Lee, T.S., Eds.; American Society for Testing and Materials: Philadelphia, PA, USA, 1987; pp. 290–305.

204. Bastidas, J.M. Corrosión del Al, Cu, Fe y Zn en atmóSferas Controladas. Ph.D. Thesis, Complutense University, Madrid, Spain, 1981.

205. Chen, W.; Hao, L.; Dong, J.; Ke, W. Effect of sulphur dioxide on the corrosion of a low alloy steel in simulated coastal industrial atmosphere. *Corros. Sci.* **2014**, *83*, 155–163. [CrossRef]

206. *UNECE International Cooperative Programme on Effects on Materials Including Historic and Cultural Monuments, Report No. 01: Technical Manual*; Swedish Corrosion Institute: Stockholm, Sweden, 1988.

207. Morcillo, M.; Almeida, E.; Rosales, B.; Uruchurtu, J.; Marrocos, M. *Corrosion y Protección de Metales en las Atmósferas de Iberoamérica. Parte I—Mapas de Iberoamérica de Corrosividad Atmosférica (Proyecto MICAT, XV.1/CYTED)*; CYTED: Madrid, Spain, 1998.

208. Benarie, M.; Lipfert, F.L. A general corrosion function in terms of atmospheric pollutant concentrations and rain pH. *Atmos. Environ.* **1986**, *20*, 1947–1958. [CrossRef]

209. Panchenko, Y.M.; Marshakov, A.I.; Igonin, T.N.; Kovtanyuk, V.V.; Nikolaeva, L.A. Long-term forecast of corrosion mass losses of technically important metals in various world regions using a power function. *Corros. Sci.* **2014**, *88*, 306–316. [CrossRef]

210. McCuen, R.H.; Albrecht, P.; Cheng, J.G. A new approach to power-model regression of corrosion penetration data. In *Corrosion Forms and Control for Infrastructure*; Chaker, V., Ed.; ASTM STP 1137; American Society for Testing and Materials: Philadelphia, PA, USA, 1992; Volume 1137, pp. 46–76.

211. Panchenko, Y.M.; Marshakov, A.I. Long-term prediction of metal corrosion losses in atmosphere using a power-linear function. *Corros. Sci.* **2016**, *109*, 217–229. [CrossRef]

212. Albrecht, P.; Hall, T.T. Atmospheric corrosion resistance of structural steels. *J. Mater. Civ. Eng.* **2003**, *15*, 2–24. [CrossRef]

213. European Committee for Standardization. *EN ISO 9224, Corrosion of Metals and Alloys—Corrosivity of Atmospheres—Guiding Values for the Corrosivity Categories*; European Committee for Standardization: Brussels, Belgium, 2012.

214. Melchers, R.E. A new interpretation of the corrosion loss processes for weathering steels in marine atmospheres. *Corros. Sci.* **2008**, *50*, 3446–3454. [CrossRef]

215. Melchers, R.E. Long-term corrosion of cast irons and steel in marine and atmospheric environments. *Corros. Sci.* **2013**, *68*, 186–194. [CrossRef]

216. International Organization for Standardization. *ISO 9223 1st Edition, Corrosion of Metals and alloys—Corrosivity of Atmospheres—Classification*; International Organization for Standardization: Geneve, Switzerland, 1992.

217. Cole, I.S.; Neufeld, A.K.; Kao, P.; Ganther, W.D.; Chotimongkol, L.; Bharmornsut, C.; Hue, N.V.; Bernardo, S.; Purwadaria, S. Development of performance verification methods for the durability of metallic components in tropical countries. In Proceedings of the 11th Asia-Pacific Corrosion Control Conference, Ho Chi Minh City, Vietnam, 1–5 November 1999; Volume 2, pp. 555–570.

218. Cole, I.S.; Furman, S.A.; Neufeld, A.K.; Ganther, W.D.; King, G.A. A holistic approach to modelling in atmospheric corrosion. In Procededings of the 14th International Corrosion Congress, Cape Town, South Africa, 26 September–1 October 1999.

219. Cole, I.S.; King, G.A.; Trinidad, G.S.; Chan, W.Y.; Paterson, A. An Australia-wide map of corrosivity: A gis approach. In Proceedings of the 8th International Conference on Durability of Building Materials and Components, Vancouver, BC, Canada, 30 May–3 June 1999.

220. Nakajima, M. Mapping method for salt attack hazzard using topographic effects analysis. In Proceedings of the 1st Asia/Pacific Conference on Harmonisation of Durability Standards and Performance Tests for Components in the Building Industry, Bangkok, Thailand, 8–10 September 1999.

221. Klassen, R.D.; Roberge, P.R. The effects of wind on local atmospheric corrosivity. In *Corrosion 2001*; NACE International: Houston, TX, USA, 2001.

222. Doyle, D.P.; Wright, T.E. Rapid method for determining atmospheric corrosivity and corrosion resistance. In *Atmospheric Corrosion*; Ailor, W.H., Ed.; Wiley: New York, NY, USA, 1982; pp. 227–243.

223. Roberge, P.R.; Klassen, R.D.; Haberecht, P.W. Atmospheric corrosivity modeling—A review. *Mater. Des.* **2002**, *23*, 321–330. [CrossRef]

*materials* MDPI

*Article*

# Rust Formation Mechanism on Low Alloy Steels after Exposure Test in High Cl$^-$ and High SO$_x$ Environment

**Toshiyasu Nishimura**

Corrosion Resistant Steel Group, National Institute for Materials Science (NIMS), Tsukuba, Ibaraki 305-0047, Japan; NISHIMURA.Toshiyasu@nims.go.jp; Tel.: +81-29-859-2127

Academic Editor: Manuel Morcillo
Received: 14 December 2016; Accepted: 15 February 2017; Published: 17 February 2017

**Abstract:** Exposure tests were performed on low alloy steels in high Cl$^-$ and high SO$_x$ environment, and the structure of the rust were analyzed by TEM (Transmission Electron Microscopy) and Raman Spectroscopy. In the exposure test site, the concentrations of Cl$^-$ and SO$_x$ were found to be high, which caused the corrosion of the steels. The conventional weathering steel (SMA: 0.6% Cr-0.4% Cu-Fe) showed higher corrosion resistance as compared to the carbon steel (SM), and Ni bearing steel exhibited the highest one. Raman spectroscopy showed that the inner rust of Ni bearing steel was mainly composed of α-FeOOH and spinel oxides. On the other hand, SMA contained β- and γ-FeOOH in inner rust, which increased the corrosion. TEM showed that nano-scale complex iron oxides containing Ni or Cr were formed in the rust on the low alloy steels, which suppressed the corrosion of steels in high Cl$^-$ and high SO$_x$ environment.

**Keywords:** atmospheric corrosion; low alloy steel; rust; Cl$^-$; SO$_x$; nickel; chromium; transmission electron microscopy; Raman spectroscopy

## 1. Introduction

As the economy of East Asia grows rapidly, the corrosion of infrastructure is becoming a serious problem. The corrosion by airborne salt particles in coastal areas is reported as severer, which is caused by Cl$^-$ ion from the sea. However, although the corrosion in high SO$_x$ environment is thought to be heavy, there is little information on this case. Besides, there is no information on the corrosion behavior of steels under high Cl$^-$ and high SO$_x$ environment. As there are many cities located in high Cl$^-$ and high SO$_x$ environment in Asia, it is important to investigate the corrosion resistance of steels under this environment. The corrosivity and corrosion map was already identified in ISO 9223, where the exposure test using steel samples is conducted [1]. Besides, the adjustment of the classification system has been presented in progress based on ISO 9223 [2]. However, the characterization of the rust on steels at each site has not been conducted sufficiently.

Weathering steels are advantageous for reducing the maintenance cost of bridges and other infrastructure structures [3]. In addition, with the conventional weathering steel (SMA: 0.6% Cr-0.4% Cu-Fe), Ni bearing weathering steels have been proposed for applications in coastal environments [4]. Thus, there is the possibility for Ni bearing steel to also show high corrosion resistance in high SO$_x$ condition. However, there is few data concerning the corrosion performance of SMA and Ni bearing steel in high Cl$^-$ and high SO$_x$ environment.

While there has been extensive analysis of rust on steels in mild environments [5–12], numerous questions remain regarding the basic mechanism of rust formation [4] and the effects of alloying elements [13–15] in severe environments. Indeed, several symposia have been held on the atmospheric corrosion of low alloy steels [15–21]. However, there have been no reports yet on the detail structure of the rust on low alloy steels [22–24] in high Cl$^-$ and high SO$_x$ environment.

In this study, the rust formation on low alloy steels was investigated through the use of an actual exposure test in high Cl$^-$ and high SO$_x$ environment. In particular, the nano structure of the rust on low alloy steel was examined by Raman spectroscopy and TEM (Transmission Electron Microscopy). Finally, the relationship between the formation of the rust and corrosion behavior was examined for low alloy steels exposed in high Cl$^-$ and high SO$_x$ environment.

## 2. Materials and Methods

### 2.1. Test Samples and Exposure Corrosion Test

The low alloy steel was rough rolled at 1553 K, and then rolled at 1327 K to produce a 5 mm thick plate. The chemical composition (mass %) of low alloy steels was shown in Table 1. Here, KA1 and KA2 are 1% and 3% Ni steel, respectively; SMA is a conventional weathering steel (SMA: 0.6% Cr-0.4% Cu-Fe); and Carbon steel (SM) is for comparison.

**Table 1.** Chemical composition of the low alloy steels (mass %).

| Number | Samples | C | Si | Ni | Cr | Cu |
|--------|---------|-----|-----|-----|-----|-----|
| KA1 | 1% Ni Steel | 0.1 | 0.2 | 1 | - | - |
| KA2 | 3% Ni Steel | 0.1 | 0.2 | 3 | - | - |
| SMA | 0.6% Cr-0.4% Cu | 0.1 | 0.2 | 0.1 | 0.6 | 0.4 |
| SM | Carbon Steel | 0.1 | 0.2 | - | - | - |

The exposure test was conducted for three years at the exposure test site on Hainan Island in China. The corrosivity of Hainan area is C5 by ISO standards, which shows very high corrosivity. The test samples were exposed with a slope of 45 degree against the horizontal line.

The annual climate data (temperature, RH (relative humidity), amount of rain) for the test site are shown in Figure 1. The average temperature is 19–29 °C, RH is almost 80%, and the amount of rain is high in summer, which is identified as a high humidity climate in the subtropical zone.

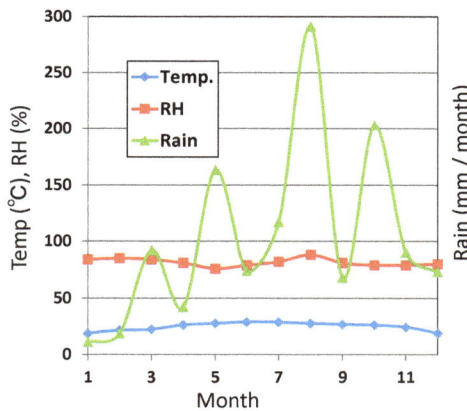

**Figure 1.** Climate dates of temperature, relative humidity (RH) and amount of rain in a year at the exposure site.

Figure 2 shows corrosion factors of: (a) airborne particles; and (b) rain water in a year at the exposure site. The airborne particles are estimated as a weight in 100 cm$^2$ in a day, showing that Cl$^-$ is very high and SO$_x$ is to some extent high. The concentrations of SO$_4{}^{2-}$ and Cl$^-$ in rainwater are examined in the unit of mg/m$^3$, showing that concentration of SO$_4{}^{2-}$ and Cl$^-$ are very high in winter. Accordingly, pH in rainwater is low in winter. Thus, the environment at the test site is defined as high Cl$^-$ and high SO$_x$ condition in the subtropical climate zone.

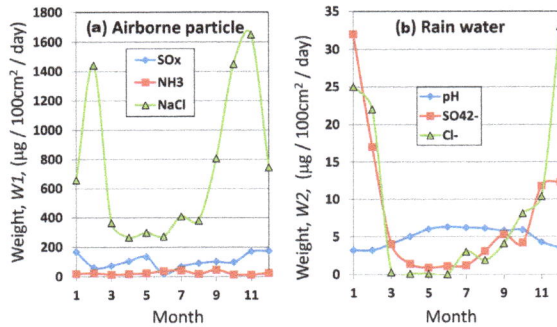

**Figure 2.** Corrosion factors of: (**a**) airborne particles; and (**b**) rainwater in a year at the exposure site.

The airborne particles are measured by collection using $10 \times 10$ cm$^2$ gauze every month. Thus, these values are different from the contents in the rain as airborne particles come from the sea with the wind, which is different from the condition in the rain.

After the exposure test, the extent of corrosion was determined by the reduction in thickness of the steel plate after removing the rust. The rust was first taken from the steel using a steel stick. Then, the steels were exposed in a diammonium hydrogen citrate solution (300 g/L, 60 °C) up to the surface of steels.

## 2.2. Physical Analysis of Rust

Surface analysis of the rust was conducted after the exposure test. The cross section of the rust was measured by SEM (Scanning Electron Microscopy). The EDS (Electron Dispersing Spectroscopy) was applied to investigate the concentration of various elements in the rust. As for the rust of Ni steel, Fe, Ni and Si were measured. In the case of SMA, Fe, Cr and Cu were measured. In addition, micro Raman spectroscopy was carried out with a 532 nm laser beam and a slit width of 25 μm. The frequency region was 4000–200 cm$^{-1}$ to detect Fe oxides. The inner and outer rust were examined and compared. The measured peak positions were identified by using those of standard chemicals of iron oxides.

Nanostructure observation of the rust was performed by TEM analysis. The rust was cut by FIB (focused ion beam) from the inner rust. EELS (Electron Energy Loss Spectroscopy) analysis was conducted in order to identify the chemical state of elements and the nano structure in inner rust. In the case of Ni steel, the chemical shift of Ni was examined by Ni-L peak using standard chemicals of Ni and NiO. In the case of SMA, the chemical shift of Cr was examined by Cr-L peak. Additionally, the chemical shift of Oxygen and Fe were measured by O-K and Fe-L peaks. Finally, the nano structures of the rust of low alloy steels were discussed.

## 3. Results and Discussion

### 3.1. Corrosion of Low Alloy Steels in High SO$_x$ Environment

The corrosion resistance of the steels was estimated after the exposure test for three years at the test site. Figure 3 shows the exposure test results for KA1 (1% Ni), KA2 (3% Ni), SMA (0.6% Cr-0.4% Cu) and SM (carbon steel). The amount of corrosion of SM increases greatly with exposure time, and that of SMA is a little low compared to SM. On the other hand, the amount of corrosion of Ni bearing steel is less. Especially, KA2 (3% Ni) shows much less corrosion than other steels. Thus, Ni bearing steel exhibits excellent corrosion resistance in the exposure test as compared to SM. In other words, Ni bearing steel is recognized to be resistant to corrosion in high $Cl^-$ and high $SO_x$ environment.

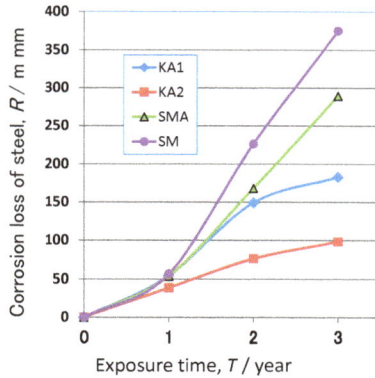

**Figure 3.** Exposure test results for low alloy steels at the test site for three years.

*3.2. Surface Analysis of the Rust Formed on Low Alloy Steels*

To identify the corrosion preventing mechanism of rust on low alloy steels, the rust was estimated by surface analysis. Figure 4 shows SEM-EDS mapping for the rust on KA2 (3% Ni steel) after the exposure test for three years. Figure 4a is a cross section of SEM that indicates the location of the rust in the left side and the steel in the right one. Figure 4b–d is EDS mappings of the rust, showing the presence of Fe, Si and Ni. In Figure 4d, Ni is contained in the rust, and slightly enriched in inner rust. The experimental spots for Raman spectroscopy and area for FIB in TEM are shown in the figure. In Figure 4c, Si is contained in the rust; however, there is little enrichment. From above results, it is found that the rust on KA2 (3% Ni steel) contains Fe, Ni and Si.

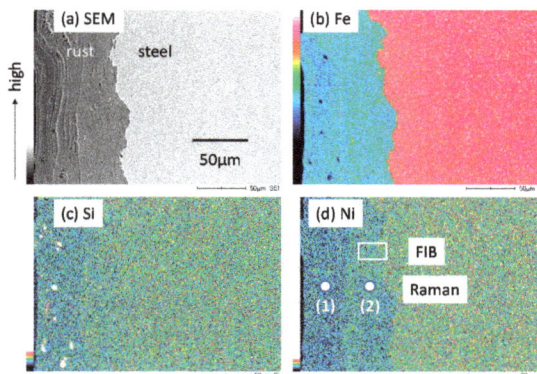

**Figure 4.** EDS (Electron Dispersing Spectroscopy) mapping of the rust formed on KA2 (3% Ni steel) after the exposure test for three years. (**a**) SEM (Scanning Electron Microscopy); (**b**) Fe; (**c**) Si; (**d**) Ni. FIB: focused ion beam.

Figure 5 shows SEM-EDS mapping for the rust on SMA (0.6% Cr-0.4% Cu-Fe) after the exposure test for three years. Figure 5a is a cross section of SEM, and Figure 5b–d is EDS mappings of the rust, showing the presence of Fe, Cr and Cu. In Figure 6c, Cr is enriched in inner rust. The corrosion resistance of SMA is thought to relate the enrichment of Cr in inner rust. The experimental spots for Raman spectroscopy and area for FIB are shown in the figure. In Figure 5d, Cu is slightly enriched in inner rust. Thus, the rust on SMA contains Fe, Cr and Cu.

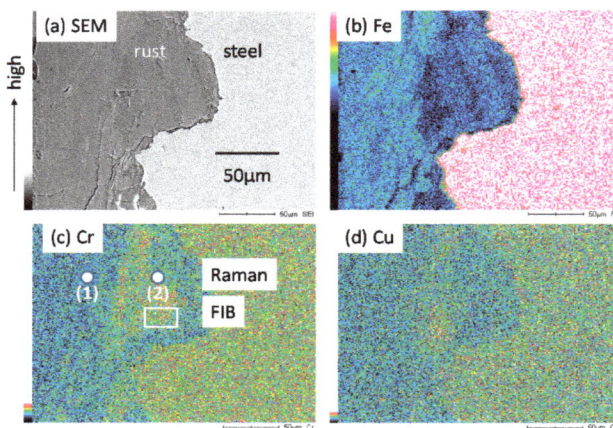

**Figure 5.** EDS mapping of the rust formed on SMA (0.6% Cr-0.4% Cu) after the exposure test for three years. (**a**) SEM; (**b**) Fe; (**c**) Cr; (**d**) Cu.

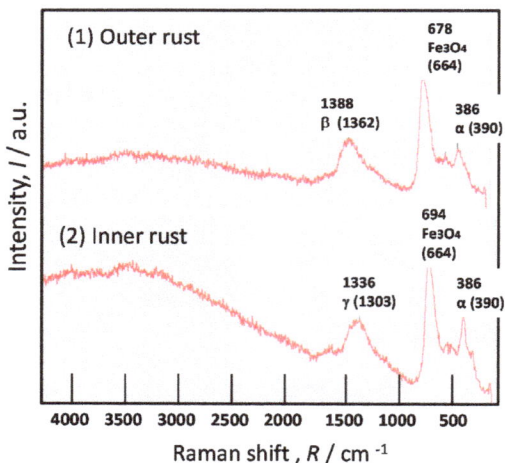

**Figure 6.** Raman results of the rust formed on KA2 (3% Ni steel) after the exposure test for three years. (**1**) Outer rust; (**2**) inner rust.

To identify the Fe oxides in inner and outer rust separately, the micro Raman spectra were measured for KA2 and SMA. Figure 6 shows Raman spectra at: (1) outer; and (2) inner rust on KA2 (3% Ni steel) after the exposure test for three years. At spot (2) in inner rust, the $\alpha$-FeOOH ($\alpha$), $\gamma$-FeOOH ($\gamma$) and $Fe_3O_4$ are detected. As the intensities of $\alpha$-FeOOH and $Fe_3O_4$ are strong, the amounts of these oxides are thought high in inner rust. Besides, this $Fe_3O_4$ is thought to be the spinel Fe oxide, which is very fine particle. In general, $\alpha$-FeOOH and $Fe_3O_4$ are thought to make the protective rust in inner rust. Thus, the inner rust of Ni bearing steel is thought to be composed of nano-size $\alpha$-FeOOH and the spinel oxide, which increases the corrosion resistance of the rust. Similarly, at Spot (1) in outer rust, the $\alpha$-FeOOH ($\alpha$), $\beta$-FeOOH ($\beta$) and $Fe_3O_4$ are observed. $\beta$-FeOOH is thought made by the Chloride particles from the sea [25–27]. Although $\beta$-FeOOH decreases the corrosion resistance of steel, the protective inner rust can protect the steel. Thus, Ni bearing steel thought to have the corrosion resistant rust which is mainly composed of nano-size $\alpha$-FeOOH and the spinel oxide in inner rust.

Figure 7 shows Raman spectra at spot (1) and (2) for the rust on SMA (0.6% Cr-0.4% Cu) after the exposure test three years. In the Raman results, the measured peak positions and those of standard chemicals are shown using values of Raman shift. The Raman peak positions for the standard chemicals of iron oxides have been measured in my laboratory. The peak positions between results and chemicals are well fit. The peaks of $\beta$- and $\gamma$-FeOOH around 1300–1360 are very close. However, they can be distinguished by the peak values $\beta$-FeOOH has the peak at 1362, and $\gamma$-FeOOH has one at 1303 cm$^{-1}$.

At spot (2) in inner rust, $\alpha$-FeOOH ($\alpha$) is detected. Thus, the inner rust of SMA is thought to be composed of nano-size $\alpha$-FeOOH, which increases the corrosion resistance of steel. However, $\beta$-FeOOH ($\beta$) and $\gamma$-FeOOH ($\gamma$) are recognized in inner rust, and there is no $Fe_3O_4$. Thus, the protection of the inner rust is considered less than Ni steel. At Spot (1) in outer rust, the strong peak of $\gamma$-FeOOH ($\gamma$) and $\beta$-FeOOH ($\beta$) are detected with peaks of $\alpha$-FeOOH ($\alpha$). As $\gamma$-FeOOH and $\beta$-FeOOH accelerates the corrosion of steels, the corrosion resistance of SMA is found to be less than Ni steel. Therefore, the corrosion resistance of the rust can be assumed using the estimation of Fe oxides measured by Raman spectra.

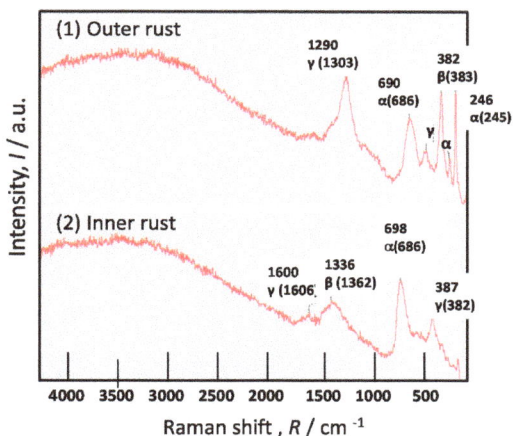

**Figure 7.** Raman results of the rust formed on SMA after the exposure test for three years. (1) Outer rust; (2) inner rust.

FIB-TEM analysis was conducted on the rust of low alloy steels after exposure test for three years. The sample of KA2 (3% Ni steel) was cut from the rust very near to the steel using a FIB, as shown in Figure 4. The experimental positions (1) and (2) for TEM-EELS (Electron Energy Loss Spectroscopy) are indicated in Figure 8. Position 1 shows white and Position 2 is dark, which reflects the chemical state of element at each position. In the following, EELS measurement was conducted at Positions 1 and 2.

**Figure 8.** FIB-TEM (Transmission Electron Microscopy) observation taken from the position in Figure 4d for the rust of KA2 (3% Ni steel).

Figure 9a shows the TEM-EELS spectra of Ni-L in inner rust of KA2 (3% Ni steel) as shown at Positions 1 and 2 in Figure 8. Besides, Figure 9b shows Ni-L spectra of the standard chemicals of NiO and Ni used for comparison. Ni-L spectrum at Position 1 has a sharper peak than that at Position 2. Thus, Ni content is higher at Position 1 than Position 2. The spectra have a strong peak of *Ni-L$_3$* at 855 eV, which is just the same as those of NiO and Ni. Though the peak of *Ni-L$_2$* of KA2 at 874 eV is the same as those of NiO and Ni, the shape of the spectrum is similar to that of NiO. Thus, Ni is thought to exist as Ni(II) oxide state in inner rust of Ni bearing steel.

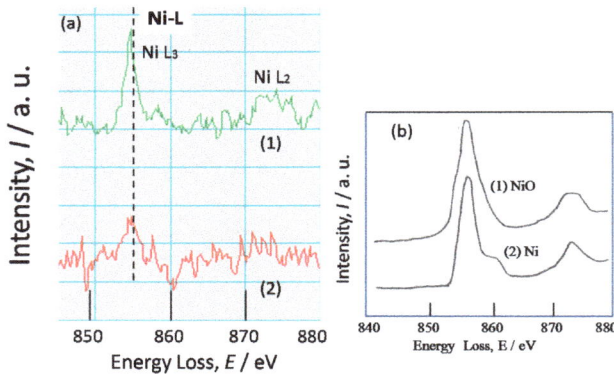

**Figure 9.** TEM-EELS (Electron Energy Loss Spectroscopy) spectra of Ni-L for the rust on: (**a**) KA2 (3% Ni steel); and (**b**) chemicals ((**1**) NiO and (**2**) Ni).

FIB-TEM result for the rust of SMA (0.6% Cr-0.4% Cu-Fe) near the steel is indicated in Figure 10 corresponding to the position in Figure 5c. EELS measuring positions are shown in the figure at 1–5. As shown in the figure, a layer-by-layer structure is formed. Position 1 is in the narrow layer in white. Position 2 is in the wide layer. Position 3 is in the narrow layer in black. Position 4 is in the wide layer. Position 5 is in the narrow layer in white. They are thought to have each style and color reflecting the chemical state of element in the rust.

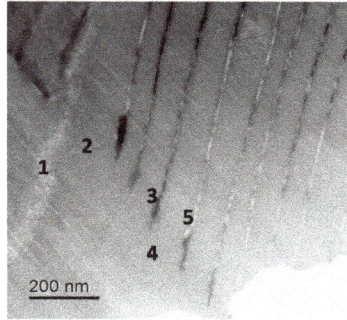

**Figure 10.** FIB-TEM observation taken from the position in Figure 5c for the rust of SMA.

Figure 11 shows the EELS spectra of Cr-L at positions corresponding to those in Figure 10. The EELS spectra of Cr-L have peaks of Cr-L$_3$ at 578 and Cr-L$_2$ at 587 eV. In more detail, the peak of spectrum 3 is shifted, showing that the valence of Cr is different. In general, Cr-L$_3$ and Cr-L$_2$ of Cr(II)O are sifted to lower energy region as compared to those of Cr(III)O. From the previous paper [28], Cr(III)O has the peak of Cr-L$_3$ at 580 and Cr-L$_2$ at 589 eV, which are higher than the test results. Thus, Cr in Figure 11 is thought to contain Cr(III) and Cr(II) oxide state in the rust. There is no peak of Cr-L at Position 1, showing that the content of Cr is very less at this position. Therefore, the chemical statement of Cr in inner rust can be estimated by EELS measurement.

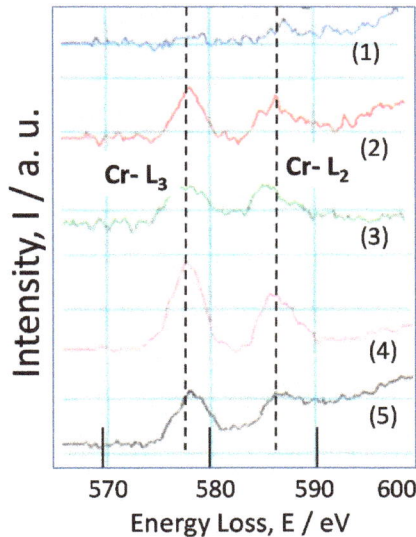

**Figure 11.** TEM-EELS spectra of Cr-L for the rust of SMA.

In order to investigate the chemical state of Fe oxide in the rust, EELS spectra of O-K and Fe-L were observed as followings. TEM-EELS spectra of O-K (Figure 12a) and Fe-L (Figure 12b) for the rust of KA2 (3% Ni steel) at Positions 1 and 2 in Figure 8 are indicated in Figure 12. Both spectra of O-K have peaks at 532 and 542 eV, showing the oxidized state. In more detail, the shape of each spectrum differs a little from each other, demonstrating the concentration of Ni at each position is different.

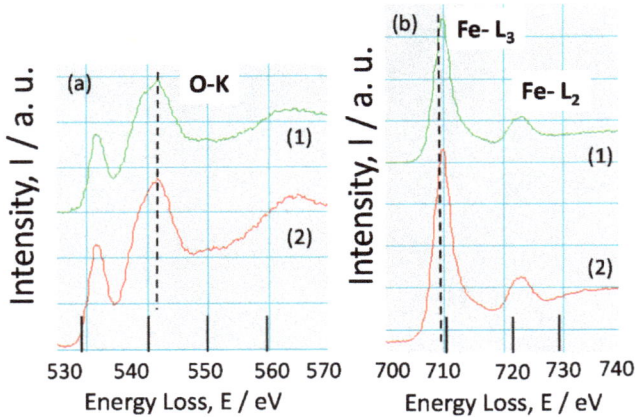

**Figure 12.** TEM-EELS spectra of: (**a**) O-K; and (**b**) Fe-L for the rust of KA2 (3% Ni steel).

Figure 12b shows the EELS spectra of Fe-L at positions corresponding to those in Figure 8. Both spectra have peaks of Fe-L$_3$ and Fe-L$_2$, indicating the presence of Fe in the rust. In more detail, both spectra have peaks of Fe-L$_3$ at 710 and Fe-L$_2$ at 724 eV. The EELS spectra of Fe oxides including Fe(II)O and Fe(III)O has been reported in the previous papers [29]. Besides, the peaks of Fe-L$_3$ and Fe-L$_2$ are shifted to higher energy region in the case of Fe(III)O than Fe(II)O. The shapes of the spectra obtained here show the peak similar to that of Fe(III)O. Thus, Fe is thought to exist in inner rust of low alloy steel mainly as Fe(III) oxide state.

Figure 13 indicates TEM-EELS spectra of: (**a**) O-K; and (**b**) Fe-L, for the rust of SMA at positions 1–5 in Figure 10. Except spectrum 3, most spectra of O-K (Figure 13a) have peaks at 532 and 542 eV, showing the oxidized state. In more detail, the shape of each spectrum differs a little from each other, reflecting the concentration of elements at each position.

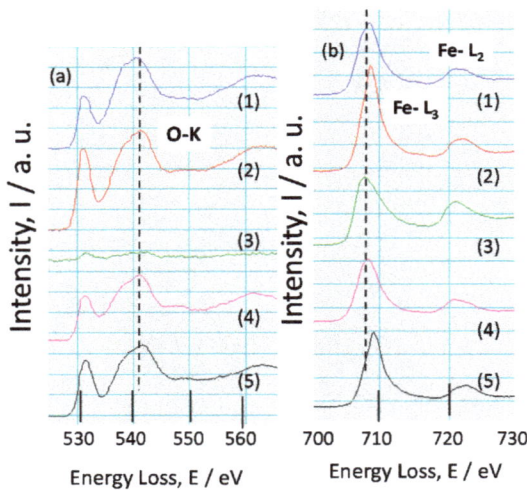

**Figure 13.** TEM-EELS spectra of: (**a**) O-K; and (**b**) Fe-L for the rust of SMA.

EELS spectra of Fe-L are shown in Figure 13b corresponding to those at positions in Figure 10. All spectra have peaks of Fe-$L_3$ and Fe-$L_2$, indicating the presence of Fe in the rust. In more detail, spectra 2, 4, and 5 have peaks of Fe-$L_3$ at 709 and Fe-$L_2$ at 723 eV. However, spectra 1 and 3 have positions of Fe-$L_3$ at 708 and Fe-$L_2$ at 722. Thus, the peak positions for spectra 1 and 3 are shifted to lower energy direction, which indicates that Fe(II)O is contained. The shape of the spectra obtained here are mainly similar to that of Fe(III) O, and a few are that of Fe(II)O in spectra 1 and 3. Thus, Fe is thought to exist in inner rust of low alloy steel as mainly Fe(III) and little Fe(II) oxide state. Raman results detected $\alpha$-FeOOH in inner rust in Figure 7. As $\alpha$-FeOOH contains only Fe(III)O, the higher content of $\alpha$-FeOOH inner rust increases the presence of Fe(III)O. Thus, Fe is thought to exist in inner rust of low alloy steel as mainly Fe(III) and little Fe(II) oxide state.

### 3.3. Corrosion Resistance Mechanism of Low Alloy Steels in High Cl⁻ and High SO$_x$ Environment

The climate of the test site indicates high temperature and high RH (relative humidity), showing that of the sub tropical zone. Although the amount of airborne particles of $NH_3$ is low, that of NaCl is very high. Besides, the amount of airborne particles of SO$_x$ is more than 100 mg/100 cm$^2$/day. In rainwater, Cl⁻ and SO$_4$ $^{2-}$ ions are high in winter season. The rainwater pH is low (4–6) in winter, showing that the corrosion of steels is increased by the acidic rain. Thus, the reason for high corrosion of carbon steel (SM) is thought to be caused by high Cl⁻ and high SO$_x$ in the environment. Moreover, RH at the exposure site is very high throughout one year, which promotes the increase of the corrosion of SM. Thus, the corrosion factors at the exposure test site are thought to be high Cl⁻, high SO$_x$ and high RH, which makes a very severe corrosion condition.

Ni bearing steel showed much higher corrosion resistance than SM in the high Cl⁻ and high SO$_x$ environment. The primary reason for this fact is likely to be the protective rust formed on Ni bearing steel during the exposure test. Thus, the high corrosion resistance of Ni steel is maintained in the Cl⁻ and SO$_x$ rich environment. By Raman spectroscopy, the rust of KA2 (Ni steel) was found to be mainly composed of nano-size $\alpha$-FeOOH and spinel oxides. Even in outer rust of KA2, nano-size $\alpha$-FeOOH and the spinel oxides were the primary components. Thus, Ni bearing steel had protective rust in this severe environment. On the other hand, in inner rust of SMA, $\beta$- and $\gamma$-FeOOH was detected, showing that corrosion resistance was not so high as compared to KA2. Probably, as the structure of inner rust of SMA is a little porous structure, chloride ions penetrates into the rust, and makes $\beta$- and $\gamma$-FeOOH. However, as SMA has much higher corrosion resistance than SM in the exposure test result in Figure 3, the protect ability of inner rust of SMA is thought effective using nano-size $\alpha$-FeOOH.

TEM-EELS measurements were conducted to identify the chemical state of elements in inner rust of low alloy steels. The EELS spectra of Ni show that Ni exists as Ni(II) oxide state in inner rust of Ni bearing steel. In addition, SEM EDS indicated that Ni exists throughout the rust, and there is no localized enrichment. Ni likely exists as Ni(II) oxide state in all of the rust.

The EELS spectra of Cr show that Cr exists as mainly Cr(III) and a little Cr(II) oxide state in inner rust of SMA. Cr likely exists as Cr(III) oxide state in the $\alpha$-FeOOH. The EELS spectra of Fe indicated the presence of Fe(III)O, suggesting that Fe exists as mainly Fe(III) and a little Fe(II) oxide state in inner rust. From Raman results, $\alpha$-FeOOH is detected in inner rust which implies that Fe(III) oxide exists mainly in $\alpha$-FeOOH in inner rust.

Finally, it was demonstrated that the rust layer was composed of nano-size complex iron oxides containing Ni or Cr, which could prevent the corrosion of steel from Cl⁻ and SO$_x$. Thus, these low alloy steels could form protective rusts against the corrosion in high Cl⁻ and high SO$_x$ environments.

## 4. Conclusions

Exposure tests were performed on low alloy steels in high $Cl^-$ and high $SO_x$ environment, and the structure of the rust was analyzed by TEM and Raman spectroscopy.

1.  In the exposure tests, $Cl^-$ and $SO_x$ were dominant factors in the corrosion of steels, and a high relative humidity also had an effect. Besides, the corrosivity at the test site was C5 by ISO standards, which corresponds to the results of this paper.
2.  The conventional weathering steel (SMA) showed lower corrosion weight loss as compared to the carbon steel (SM), and Ni bearing steel exhibited the lowest one.
3.  Raman spectroscopy showed that the inner rusts on Ni bearing steel and SMA had $\alpha$-FeOOH. Besides, Ni bearing steel had the spinel oxide in inner rust, which suppressed the corrosion.
4.  TEM showed that the rust layer was composed of nano-size complex iron oxides containing Ni or Cr, which indicated that these low alloy steels formed protective rusts against the corrosion in high $Cl^-$ and high $SO_x$ environments.

**Acknowledgments:** The author thanks Nippon steel and Sumikin Technology corp. for the exposure test help.

**Conflicts of Interest:** The author declares no conflicts of interest.

## References

1.  *Corrosion of Metals and Alloys, Corrosivity of Atmospheres, Classification;* ISO 9223:1992; International Organization for Standardization: Geneva, Switzerland, 1992.
2.  Knotkova, D. Atmospheric corrosion—Research, testing, and standardization. *Corrosion* **2005**, *61*, 723–738. [CrossRef]
3.  Kihira, H.; Kimura, M. Advancements of weathering steel technologies in Japan. *Corrosion* **2011**, *67*, 095002. [CrossRef]
4.  Nishimura, T.; Katayama, H.; Noda, K.; Kodama, T. Electrochemical behavior of rust formed on carbon steel in wet/dry environment containing chloride ions. *Corrosion* **2000**, *56*, 935–941. [CrossRef]
5.  Evans, U.R. Mechanism of rusting. *Corros. Sci.* **1969**, *9*, 813–821. [CrossRef]
6.  Misawa, T.; Asami, K.; Hashimoto, K.; Shimodaira, A. The mechanism of atmospheric rusting and the protective amorphous rust on low alloy steel. *Corros. Sci.* **1974**, *14*, 279–289. [CrossRef]
7.  Suzuki, I.; Hisamatsu, Y.; Masuko, N. Nature of Atmospheric Rust on Iron. *J. Electrochem. Soc.* **1980**, *127*, 2210–2215. [CrossRef]
8.  Stratmann, M.; Bohnenkamp, K.; Ramchandran, T. The influence of copper upon the atmospheric corrosion of iron. *Corros. Sci.* **1987**, *27*, 905–926. [CrossRef]
9.  Dunnwald, J.; Otto, A. An investigation of phase transitions in rust layers using Raman spectroscopy. *Corros. Sci.* **1989**, *29*, 1167–1176. [CrossRef]
10. Townsend, H.E. Effects of alloying elements on the corrosion of steel in industrial atmospheres. *Corrosion* **2001**, *57*, 497–501. [CrossRef]
11. Yamashita, M.; Miyuki, H.; Mastuda, Y.; Nagano, H.; Misawa, T. The long term growth of the protective rust layer formed on weathering steel by atmospheric corrosion during a quarter of a century. *Corros. Sci.* **1994**, *36*, 283–299. [CrossRef]
12. Konishi, H.; Yamashita, M.; Uchida, H.; Mizuki, J. Characterization of rust layer formed on Fe, Fe-Ni and Fe-Cr alloys exposed to Cl-rich environment by Cl and Fe K-Edge XANES measurements. *Mater. Trans.* **2005**, *46*, 329–336. [CrossRef]
13. Nishimura, T.; Noda, K.; Kodama, T. Corrosion behavior of W-bearing steel in a wet/dry environment containing chloride ions. *Corrosion* **2001**, *57*, 753–758. [CrossRef]
14. Nishimura, T.; Tahara, A.; Kodama, T. Effect of Al on the corrosion behavior of low alloy steel in wet/dry environment. *Mater. Trans.* **2001**, *42*, 478–483. [CrossRef]
15. Nishimura, T. Corrosion behavior of Silicon-bearig steel in a wet/dry environment containing chloride ions. *Mater. Trans.* **2007**, *48*, 1438–1443. [CrossRef]

16. Yamashita, M.; Konishi, H.; Kozakura, T.; Mizuki, J.; Uchida, H. In situ observation of initial rust formation process on carbon steel under $Na_2SO_4$ and NaCl solution films with wet/dry cycles using synchrotron radiation X-rays. *Corros. Sci.* **2005**, *47*, 2492–2498. [CrossRef]

17. Kimura, M.; Kihira, H.; Ohta, N.; Hashimoto, M.; Senuma, T. Control of $Fe(O,OH)_6$ nano-network structures of rust for high atmospheric-corrosion resistance. *Corros. Sci.* **2005**, *47*, 2499–2509. [CrossRef]

18. Ishikawa, T.; Miyamoto, S.; Kandori, K.; Miyamoto, S. Influence of anions on the formation of β-FeOOH rusts. *Corros. Sci.* **2005**, *47*, 2510–2520. [CrossRef]

19. Nakayama, T.; Ishikawa, T.; Konno, T. Structure of titanium-doped goethite rust. *Corros. Sci.* **2005**, *47*, 2521–2530. [CrossRef]

20. Ohtsuka, T.; Komatsu, T. Enhancement of electric conductivity of the rust layer by adsorption of water. *Corros. Sci.* **2005**, *47*, 2571–2577. [CrossRef]

21. Nishikata, A.; Suzuki, F.; Tsuru, T. Corrosion monitoring of nickel-containing steels in marine atmospheric environment. *Corros. Sci.* **2005**, *47*, 2578–2588. [CrossRef]

22. Kim, K.Y.; Hwang, Y.H.; Yoo, J.Y. Effect of silicon content on the corrosion properties of calcium-modified weathering steel in a chloride environment. *Corrosion* **2002**, *58*, 570–583. [CrossRef]

23. Asami, K.; Kikuchi, M. Characterization of rust layers on weathering-steels air-exposed for a long period. *J. Jpn. Inst. Met.* **2002**, *66*, 649–656. [CrossRef]

24. Nishimura, T. Rust formation and corrosion performance of Si and Al-bearing Ultrafine Grained Weathering Steel. *Corros. Sci.* **2008**, *50*, 1306–1312. [CrossRef]

25. Nishimura, T. Electrochemical behavior and structure of rust formed on Si- and Al-bearing steel after atmospheric exposure. *Corros. Sci.* **2010**, *52*, 3609–3614. [CrossRef]

26. Génin, J.-M.R.; Refait, P.H.; Abdelmoula, M. *Green Rusts and Their Relationship to Iron Corrosion; a Key Role in Microbially Influenced Corrosion*; Industrial Applications of the Mössbauer Effect; Springer: Dordrecht, The Netherlands, 2002; pp. 119–131.

27. Drissi, S.H.; Refait, Ph.; Abdelmoula, M.; Génin, J.M.R. The preparation and thermodynamic properties of Fe(II)Fe(III) hydroxide-carbonate (green rust 1); Pourbaix diagram of iron in carbonate-containing aqueous media. *Corrosion Sci.* **1995**, *37*, 2025–2041. [CrossRef]

28. Daulton, T.L.; Little, B.J. Determination of chromium valence over the range Cr(0)-Cr(VI) by electron energy loss spectroscopy. *Ultramicroscopy* **2006**, *106*, 561–573. [CrossRef]

29. Tan, H.; Verbeeck, J.; Abakumov, A.; Tendeloo, G.V. Oxidation state and chemical shift investigation in transition metal oxides by EELS. *Ultramicroscopy* **2012**, *116*, 24–33. [CrossRef]

*materials*

MDPI

*Article*

# Corrosion Prediction with Parallel Finite Element Modeling for Coupled Hygro-Chemo Transport into Concrete under Chloride-Rich Environment

Okpin Na [1,*], Xiao-Chuan Cai [2] and Yunping Xi [3]

[1]  R&D Division, Hyundai E&C, Yongin-si, Gyeonggi-do 16891, Korea
[2]  Computer Science, University of Colorado Boulder, Boulder, CO 80309, USA; xiao-chuan.cai@colorado.edu
[3]  Civil, Environmental, and Architectural Engineering, University of Colorado Boulder, Boulder,
    CO 80309, USA; yunping.xi@colorado.edu
*   Correspondence: nao@colorado.edu; Tel.: +82-10-5269-2375

Academic Editor: Manuel Morcillo
Received: 26 February 2017; Accepted: 23 March 2017; Published: 28 March 2017

**Abstract:** The prediction of the chloride-induced corrosion is very important because of the durable life of concrete structure. To simulate more realistic durability performance of concrete structures, complex scientific methods and more accurate material models are needed. In order to predict the robust results of corrosion initiation time and to describe the thin layer from concrete surface to reinforcement, a large number of fine meshes are also used. The purpose of this study is to suggest more realistic physical model regarding coupled hygro-chemo transport and to implement the model with parallel finite element algorithm. Furthermore, microclimate model with environmental humidity and seasonal temperature is adopted. As a result, the prediction model of chloride diffusion under unsaturated condition was developed with parallel algorithms and was applied to the existing bridge to validate the model with multi-boundary condition. As the number of processors increased, the computational time decreased until the number of processors became optimized. Then, the computational time increased because the communication time between the processors increased. The framework of present model can be extended to simulate the multi-species de-icing salts ingress into non-saturated concrete structures in future work.

**Keywords:** parallel finite element method; diffusion; coupled hygro-chemo; concrete degradation

## 1. Introduction

One of the main long-term durability problems of reinforced concrete structure is the corrosion of reinforcing bars (rebar) in concrete. The corrosion can be resulted from several necessary conditions, and one of them is a high concentration of chloride ions near rebars, where the chloride ions come from deicing chemicals used in the winter on roadways and parking structures or seawater for the off-shore structures. Once the chloride content on the surface of steel reinforcement reaches a threshold value and the moisture and oxygen are sufficiently provided, the corrosion of steel bar is initiated [1,2]. The corrosion initiation period is the time during which substances such as water, chloride ions, oxygen and carbon dioxide penetrate through the concrete cover. The length of period depends on the resistance of concrete to the transport processes and the severity of the environmental conditions they are exposed to [3–6]. In practical engineering, the chloride contaminated RC structures have many internal cracks due to rebar corrosion and the prediction of crack width and propagation is very important for safety and serviceability [7,8]. Recently, the severity of corrosion damage was demonstrated through the corrosion-induced modeling with a statistical approach on bridge structures [9–11]. In concrete bridges, there are various concentrations of moisture and chemicals on the top surface of bridge decks due to the scatter of de-icing chemicals when they are applied. Especially, moisture and chloride concentrations have spatial distributions and seasonal variations. The service life of concrete is affected by a combination of material properties and microclimate. The microclimate is a term for the climatic conditions at the concrete surface or very close to the surface. This condition on the surface has a more decisive effect on the conditions inside the concrete than most other parameters [12]. The model of microclimate such as environmental humidity and temperature was proposed by Bazant et al. [13].

To simulate more realistically the durability performance of reinforced concrete structures, sophisticated numerical methods for fully coupled moisture and chloride transport processes and reliable material models for the transport parameters are needed. In addition, in order to predict accurately the moisture and chloride distributions in a large reinforced concrete structure, a large number of fine finite element meshes are needed. For the use of the large meshes in numerical analysis, in this study, a parallel algorithm will be employed to a parallel processing system with up to 2000 processors. To consider more realistic boundary conditions on concrete structures, humidity model applied with random stochastic process will be demonstrated and approximately 1.5 million nodes and three million elements will be used. Furthermore, multi-boundary conditions to describe the actual chloride concentrations and humidity distribution will be applied instead of only constant boundary conditions.

## 2. Basic Diffusion Formulation of Unsaturated Concrete

### 2.1. Governing Equation

The two fully-coupled partial differential equations governing the coupled chloride and moisture diffusion through non-saturated concrete are simply derived by employing the mass balance equations and Fick's law [14,15].

First, the flux of chloride ions ($J$) through a unit area of porous media depends on the gradient of chloride ions as well as the gradient of moisture as. Thus, Fick's law is modified in this study to include the coupling effects between moisture and chloride transfer. The governing equations are shown in Equations (1) and (2).

$$J_{cl} = -(D_{cl}\nabla C_f + \varepsilon D_H \nabla H) \tag{1}$$

$$J_H = -(\delta D_{cl}\nabla C_f + D_H \nabla H) \tag{2}$$

where $D_{cl}$ = chloride diffusion coefficient (cm$^2$/day); $C_f$ = free chloride concentration (in gram of free chloride per gram of concrete, g/g); $D_H$ = humidity diffusion coefficient; $H$ = pore relative humidity; $\varepsilon$ = humidity gradient coefficient, which represents the coupling effect of moisture diffusion on chloride penetration; and $\delta$ = chloride gradient coefficient, which represents the coupling effect of chloride ions on moisture diffusion.

When chloride ions ingress into the concrete, some of them are bound to the internal surface of the cement paste and aggregates, which are called bound chlorides and freely go through the concrete. Steel corrosion is related only to the free chloride content but not to the total chloride content [14].

The mass balance of chloride ions and moisture can be expressed using Fick's second law as Equations (3) and (4),

$$\frac{\partial C_t}{\partial t} = \frac{\partial C_t}{\partial C_f} \cdot \frac{\partial C_f}{\partial t} = -div(J_{Cl}) = div(D_{Cl}\nabla C_f + \varepsilon D_H \nabla H) \tag{3}$$

$$\frac{\partial w}{\partial t} = \frac{\partial w}{\partial H} \cdot \frac{\partial H}{\partial t} = -div(J_H) = div(\delta D_{Cl}\nabla C_f + D_H \nabla H) \tag{4}$$

where

$C_t$: total chloride concentration (in gram of free chloride per gram of concrete, g/g);
$w$: water content;

$\frac{\partial C_t}{\partial C_f}$: chloride binding capacity; and

$\frac{\partial w}{\partial H}$: moisture binding capacity.

Equations (3) and (4) can be rewritten as,

$$\frac{\partial C_t}{\partial C_f} \cdot \frac{\partial C_f}{\partial t} = \nabla \cdot (D_{Cl}\nabla C_f + \varepsilon D_H \nabla H) = \nabla \cdot (D_{Cl}\nabla C_f + D_\varepsilon \nabla H) \tag{5}$$

$$\frac{\partial w}{\partial H} \cdot \frac{\partial H}{\partial t} = div(\delta D_{Cl}\nabla C_f + D_H \nabla H) = \nabla \cdot (D_\delta \nabla C_f + D_H \nabla H) \tag{6}$$

where $D_\varepsilon$ is coupling parameter, $\varepsilon D_H$, and $D_\delta$ is coupling parameter, $\delta D_{Cl}$.

The general boundary conditions are as below,

$$C_f = C_0 \qquad\qquad\qquad\qquad\qquad \text{on } \Gamma_1 \tag{7}$$

$$D_{cl}\frac{\partial C_f}{\partial n} + J_{Cl} + D_\varepsilon \frac{\partial H}{\partial n} + \alpha_{cl}(C_f - C_{fa}) = 0 \quad \text{on } \Gamma_2 \tag{8}$$

$$H = H_0 \qquad\qquad\qquad\qquad\qquad \text{on } \Gamma_3 \tag{9}$$

$$D_H\frac{\partial H}{\partial n} + J_H + D_\delta \frac{\partial C_f}{\partial n} + \alpha_H(H - H_a) = 0 \quad \text{on } \Gamma_4 \tag{10}$$

where $\alpha_{Cl}$ = convective chloride coefficient, $\alpha_H$ = convective relative humidity coefficient, $C_{fa}$ = ambient chloride ions, and $H_a$ = ambient relative humidity.

$\Gamma_1$ and $\Gamma_3$ are the part of boundary with constant chloride ions and relative humidity and $\Gamma_2$ and $\Gamma_4$ are the part of boundary subjected to specified chloride ions and relative humidity flux, respectively. $\Gamma_1$ and $\Gamma_2$ form the complete boundary surface for the chloride diffusion problem, and $\Gamma_3$ and $\Gamma_4$ form the moisture diffusion problem [16].

### 2.2. Material Parameters

To numerically solve the chloride diffusion problem, many material parameters must be determined: the moisture capacity ($\partial w/\partial H$), chloride binding capacity ($\partial C_t/\partial C_f$), humidity diffusion coefficient ($D_H$), and chloride diffusion coefficient ($D_{Cl}$) (in Equations (5) and (6)) [15].

### 2.2.1. Moisture Capacity

The moisture capacity of the concrete was developed based on the multiphase and multiscale model [17]. Assuming that the effect of the shrinkage of concrete can be evaluated simply by the average of the moisture capacities of the aggregate and the cement paste as shown in Equation (11),

$$\frac{\partial w}{\partial H} = f_{agg}\left(\frac{\partial w}{\partial H}\right)_{agg} + f_{cp}\left(\frac{\partial w}{\partial H}\right)_{cp} \tag{11}$$

where $f_{agg}$ and $f_{cp}$ = weight percentages of the aggregate and cement paste, and $\left(\frac{\partial w}{\partial H}\right)_{agg}$ and $\left(\frac{\partial w}{\partial H}\right)_{cp}$ = moisture capacities of aggregate and cement paste, which can be calculated based on the model developed (Xi et al., 1994a, b) and Xi (1995a, b) [18–21].

2.2.2. Chloride Binding Capacity

A modified relationship between the bound chloride $C_b$, and the free chloride $C_f$, was established by Tang and Nilson (1993) based on Freundlich isotherm, and was proposed by Xi and Bazant (1999);

$$C_b = \frac{\beta_{C\text{-}S\text{-}H}}{1000} \left( \frac{C_f}{35.45\beta_{sol}} \right)^A 10^B \tag{12}$$

which is differentiated with respect to $C_f$ and then one would obtain the chloride binding capacity of concrete as defined by Xi and Bazant (1999) to be;

$$\frac{\partial C_f}{\partial C_t} = \frac{1}{1 + \frac{A \cdot 10^B \cdot \beta_{C\text{-}S\text{-}H}}{35450\beta_{sol}} \left( \frac{C_f}{35.45\beta_{sol}} \right)^{A-1}} \tag{13}$$

where $A$ and $B$ = chloride adsorption related constants, 0.3788 and 1.14, $\beta_{sol}$ = ratio of pore solution volume to concrete weight, L/g, and $\beta_{C\text{-}S\text{-}H}$ = weight ratio of C-S-H gel to concrete (g/g) [14,22].

Based on the definition of $\beta_{sol}$, the following Equation (14) can be easily derived;

$$\beta_{sol} = \frac{f_{cp}n_{cp} + f_{agg}n_{agg}}{\rho_{sol}} \tag{14}$$

where $f_{cp}$ and $f_{agg}$ = weight percentages of cement paste and aggregate in concrete mix (g/g), $n_{cp}$ = cement paste adsorption isotherm, $n_{agg}$ = aggregate adsorption isotherm, and $\rho_{sol}$ = density of pore solution measured in g/L.

Specific gravities of concrete and C-S-H are similar, therefore, Xi and Bazant (1999) assumed that the weight fraction of C-S-H in concrete, $\beta_{C\text{-}S\text{-}H}$, is equal to the volume fraction of C-S-H in concrete, $f_{C\text{-}S\text{-}H}$, then,

$$\beta_{C\text{-}S\text{-}H} = f_{C\text{-}S\text{-}H} = \frac{V_{total} - V_1 - V_{cp}}{V_{total}} = 1 - f_1 - f_{cp} \tag{15}$$

where $f_1$ = volume fraction of anhydrous pores of cement particles, and $f_{cp}$ = volume fraction of capillary pores of cement paste [14].

2.2.3. Humidity Diffusion Coefficient

The humidity diffusion coefficient of concrete depends on the diffusion coefficients of aggregate and cement paste. Using the composite theory (Christensen, 1979), the effective diffusion coefficient of concrete can be evaluated as following Equation (16),

$$D_H = D_{Hcp} \cdot \left( 1 + g_i \left/ \left[ \frac{(1 - g_i)}{3} + 1 \left/ \left( \frac{D_{Hagg}}{D_{Hcp}} - 1 \right) \right. \right] \right. \right) \tag{16}$$

where $g_i$ = aggregate volume fraction, $D_{Hcp}$ = humidity diffusion coefficient of the cement paste, and $D_{Hagg}$ = humidity diffusion coefficient of the aggregates [23].

The humidity diffusion coefficient of aggregates in concrete is very small due to the fact that the pores in aggregates are discontinuous and enveloped by cement paste and so it can be neglected. The humidity diffusion coefficient of cement paste can be predicted by using the empirical formula described in Xi et al. (1994b) [19].

2.2.4. Chloride Diffusion Coefficient

Chloride diffusion coefficient in saturated concrete was studied by Xi and Bazant (1999) as following Equation (17),

$$D_{cl} = f_1(w/c, t_0) \cdot f_2(g_i) \cdot f_3(H) \cdot f_4(T) \cdot f_5(C_f) \tag{17}$$

In Equation (17), the functions of the chloride diffusivity consist of the concrete curing time ($t_0$), gravel volume fraction ($g_i$), relative humidity ($H$), temperature ($T$), and free chloride concentration ($C_f$) [14].

First factor of chloride diffusion coefficient accounts for the effect of the water–cement ratio ($w/c$) and curing time ($t_0$) in Equation (18),

$$f_1(w/c, t_0) = \frac{28 - t_0}{62500} + \left(\frac{1}{4} + \frac{(28 - t_0)}{300}\right)(w/c)^{6.55} \tag{18}$$

The second influence factor is to consider the effect of composite action of the aggregate and the cement paste in Equation (19),

$$f_2(g_i) = D_{cp}\left(1 + \frac{g_i}{[1 - g_i]/3 + 1/[(D_{agg}/D_{cp}) - 1]}\right) \tag{19}$$

where $g_i$ = volume fraction of aggregate in concrete, and $D_{agg}$ and $D_{cp}$ = chloride diffusion coefficient of aggregate and cement paste.

The third factor, $f_3(H)$, in Equation (20) is to consider the effect of relative humidity level on the chloride diffusion coefficient. A model proposed by Bazant et al. (1972) can be used, which was developed initially for moisture diffusion [24].

$$f_3(H) = \left[1 + \left(\frac{1 - H}{1 - H_C}\right)^4\right]^{-1} \tag{20}$$

where $Hc$ = critical humidity level, 0.75.

Arrhenius law was used by Xi and Bazant (1999) to introduce the temperature effect of the forth factor in chloride diffusion coefficient as shown in Equation (21),

$$f_4(T) = \exp\left[\frac{U}{R}\left(\frac{1}{T_0} - \frac{1}{T}\right)\right] \tag{21}$$

where

$T_0$ and $T$ = reference and current temperatures in Kelvin, $T_0$ = 296 K;
$R$ = gas constant, 8.314 J/mol·K; and
$U$ = diffusion process activation energy, depending on $w/c$ ratio [14].

The detail description of diffusion process activation energy, $U$ can be found in the paper of Ababneh et al. (2003) [15].

The dependence of chloride diffusion coefficient on free chloride concentration $C_f$ for the fifth factor is presented in Equation (22);

$$f_5(C_f) = 1 - k_{ion}(C_f)^m \tag{22}$$

where $k_{ion}$ and $m$ = 8.333 and 0.5, respectively, according to Xi and Bazant (1999) [14].

### 3. Finite Element Formulation

In order to solve a time-dependent coupled moisture-chloride diffusion problem, a large linear system equation was derived with finite element method. The finite element formulation will be briefly introduced in this section. The continuous variables in the coupled chloride and moisture diffusion equations, free chloride ($C_f$) and relative humidity ($H_m$) are spatially discretized over the space domain, $\Omega$. The domain discretization can be described as shown in Equation (23).

$$\Omega = \overset{nel}{\underset{e=1}{U}} \Omega^e \tag{23}$$

in which *nel* is the total number of elements in space domain and $\Omega_e$ is an element. It is also defined $\partial\Omega$ as the boundary of computational domain and $\partial\Omega^e$ the boundary of subdomain.

The unknown variables in Equations (24) and (25) are defined in terms of nodal values, $\{C_f\}$ and $\{H_m\}$,

$$C_f \simeq \lfloor N \rfloor \left\{ \overset{\wedge}{C}_f \right\} \tag{24}$$

$$H_m \simeq \lfloor N \rfloor \left\{ \overset{\wedge}{H}_m \right\} \tag{25}$$

where $\lfloor N \rfloor$ is the triangle element shape function. The notations $\lfloor \ \rfloor$ and $\{\ \}$ are row and column vectors, respectively. The element shape functions are expressed as following Equation (26),

$$\lfloor \mathbf{N} \rfloor = \lfloor N_1 \quad N_2 ... \quad N_n \rfloor \tag{26}$$

in which $N_i$ is the shape function for node $i$ and $n$ is the total numbers of nodes in an element. The unknown vectors of free chloride $\{\overset{\wedge}{C}_f\}$ in Equation (27) and relative humidity $\{\overset{\wedge}{H}_m\}$ in Equation (28) can be defined as,

$$\{\overset{\wedge}{C}_f\} \equiv \{\overset{\wedge}{C}_1, \overset{\wedge}{C}_2, \overset{\wedge}{C}_3, \cdots \cdots \cdots, \overset{\wedge}{C}_n\} \tag{27}$$

$$\{\overset{\wedge}{H}_m\} \equiv \{\overset{\wedge}{H}_1, \overset{\wedge}{H}_2, \overset{\wedge}{H}_3, \cdots \cdots \cdots, \overset{\wedge}{H}_n\} \tag{28}$$

The nodal free chloride concentrations and relative humidity are solved by substituting the approximated values of Equations (24) and (25) into governing equations of Equations (5) and (6), and applying the Galerkin procedure to the weak forms, then the finite element matrix can be obtained as shown in Equation (29):

$$\frac{d}{dt}\left(\left[C_e(\overset{\wedge}{\phi})\right]\left\{\overset{\wedge}{\phi}\right\}\right) = \left[K_e(\overset{\wedge}{\phi})\right]\left\{\overset{\wedge}{\phi}\right\} \tag{29}$$

where the element matrices and vector are as following Equations (30)–(32),

$$[C_e] = \begin{bmatrix} C_c & 0 \\ 0 & C_h \end{bmatrix} \tag{30}$$

$$[K_e] = \begin{bmatrix} K_{cc} & K_{ch} \\ K_{hc} & K_{hh} \end{bmatrix} \tag{31}$$

$$\{\overset{\wedge}{\phi}\} = \left\lfloor \overset{\wedge}{C}_f \overset{\wedge}{H}_m \right\rfloor \tag{32}$$

In detail, the components in element matrices are as Equations (33)–(38),

$$[K_{cc}] = -\int_{\Omega_e} \nabla\lfloor N_c \rfloor^T D_{Cf} \nabla\lfloor N_c \rfloor d\Omega + \int_{\partial\Omega_e} \lfloor N_c \rfloor^T D_{Cf} \nabla\lfloor N_c \rfloor d\Gamma \tag{33}$$

$$[K_{ch}] = -\int_{\Omega_e} \nabla\lfloor N_c \rfloor^T D_\varepsilon \nabla\lfloor N_h \rfloor d\Omega + \int_{\partial\Omega_e} \lfloor N_c \rfloor^T D_\varepsilon \nabla\lfloor N_h \rfloor d\Gamma \tag{34}$$

$$[K_{hc}] = -\int_{\Omega_e} \nabla\lfloor N_h \rfloor^T D_\delta \nabla\lfloor N_c \rfloor d\Omega + \int_{\partial\Omega_e} \lfloor N_h \rfloor^T D_\delta \nabla\lfloor N_c \rfloor d\Gamma \tag{35}$$

$$[K_{hh}] = -\int_{\Omega_e} \nabla\lfloor N_h \rfloor^T D_{Hm} \nabla\lfloor N_h \rfloor d\Omega + \int_{\partial\Omega_e} \lfloor N_h \rfloor^T D_{Hm} \nabla\lfloor N_h \rfloor d\Gamma \tag{36}$$

$$[C_C] = \int_{\Omega_e} \lfloor N_C \rfloor^T C_C \lfloor N_C \rfloor d\Omega \tag{37}$$

$$[C_h] = \int_{\Omega_e} \lfloor N_h \rfloor^T C_h \lfloor N_h \rfloor d\Omega \tag{38}$$

Finally, Equation (39) is also discretized in time space with time interval $\triangle t = t^{\xi+1} - t^\xi$ as following,

$$\left( \left[ C_e(\hat{\phi}) - \theta \cdot \Delta t \cdot K_e(\hat{\phi}) \right] \{\hat{\phi}\} \right)^{\xi+1} = \left( \left[ C_e(\hat{\phi}) - (1-\theta) \cdot \Delta t \cdot K_e(\hat{\phi}) \right] \{\hat{\phi}\} \right)^\xi \tag{39}$$

The value of parameter $\theta$ is related to the solution method adopted in the program. Typical values of $\theta$ are 0, $1/2$ and 1 correspond to fully explicit, semi-implicit and fully implicit methods, respectively. The semi-implicit method called Crank–Nicholson method is used in this study.

Equation (39) is simplified as linear system equation.

$$[A]^{\xi+1} \{\hat{\phi}\}^{\xi+1} = \{b\}^\xi \tag{40}$$

where

$$[A]^{\xi+1} = \left[ C_e(\hat{\phi}) - \theta \cdot \Delta t \cdot K_e(\hat{\phi}) \right]^{\xi+1} \tag{41}$$

$$\{b\}^\xi = \left( \left[ C_e(\hat{\phi}) - (1-\theta) \cdot \Delta t \cdot K_e(\hat{\phi}) \right] \{\hat{\phi}\} \right)^\xi \tag{42}$$

## 4. Implementation of Parallel Finite Element Method

### 4.1. Various Programs Adapted in Parallel Finite Element Program

In order to implement the parallel finite element program for the coupled moisture and chloride problem, various programs were employed such as Triangle for mesh generation, ParMETIS, PETSc (Portable, Extensible Toolkit for Scientific Computation), and MPI (Message Passing Interface) [25–31].

In this study, Triangle was used for the mesh generation of triangle element, which was created at Carnegie Mellon University. Triangle generates exact delaunay triangulations, and are suitable for finite element analysis [25].

PETSc (3.0.0 p8) is a large and versatile package integrating distributed vectors, distributed matrices in several sparse storage formats, Krylov subspace methods, preconditioners, and Newton-like nonlinear methods with built-in trust region or linesearch strategies and continuation for robustness.

It is designed to provide the numerical infrastructure for application codes involving the implicit numerical solution of PDEs, and it uses on MPI for portability to most parallel machines. The PETSc library is written in C, but may be accessed from user codes written in C, Fortran, and C++ [32].

MPI (Message Passing Interface) is a standardized and portable message-passing system designed to function on a wide variety of parallel computers. The standard defines the syntax and semantics of library routines and allows users to write portable programs in the main scientific programming languages (Fortran, C, or C++) [27–29].

ParMETIS extends the functionality of METIS and includes routines based on a parallel graph-partitioning algorithm that are especially suited for parallel computations and large-scale numerical simulations involving unstructured meshes. In typical FEM computations, ParMETIS dramatically reduces the time spent in interprocess communication by computing mesh decompositions such that the number of interface nodes/elements is minimized [30].

### 4.2. Overlapping Domain Decomposition Method with Additive Schwarz Preconditioner

A parallel program is typically developed by dividing the program into multiple fragments that can execute simultaneously, each on its own processor. In the finite element analysis, this can be accomplished by applying a domain decomposition method. Domain decomposition method is the method usually used for solving large scale system equations and it is also suitable for parallel programming because of data locality [15]. There are two types of domain decomposition methods, overlapping and non-overlapping methods. In this study, the overlapping method was employed to solve the linear sparse matrix. That is because the advantage of overlapping domain decomposition is easier to setup in algebraic approach and faster convergence than non-overlapping domain decomposition. Furthermore, the boundaries of extended subdomains are smoother than non-overlapping subdomains.

Iterative solver must be used in the iterative domain decomposition method. In this study, for the iterative solver, GMRES (Generalized Minimal Residual method) was chosen for both global and local matrix. GMRES is mainly chosen because of its ability to solve non-symmetric linear system as in the case of our problem. To improve the convergence of this problem, the additive Schwarz method preconditioner was applied. Figure 1 shows the flow chart of parallel pre-process and FE solver.

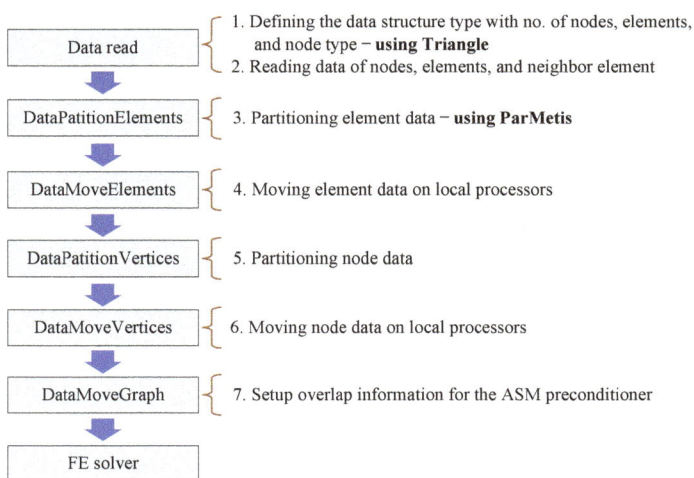

(a)

**Figure 1.** *Cont.*

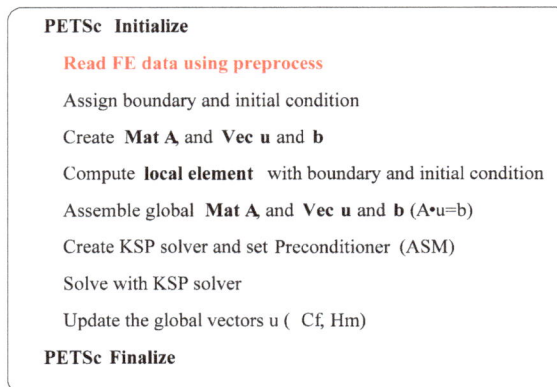

> **PETSc  Initialize**
>
>   Read FE data using preprocess
>
>   Assign boundary and initial condition
>
>   Create **Mat A** and **Vec u** and **b**
>
>   Compute **local element**  with boundary and initial condition
>
>   Assemble global **Mat A** and **Vec u** and **b** (A•u=b)
>
>   Create KSP solver and set Preconditioner (ASM)
>
>   Solve with KSP solver
>
>   Update the global vectors u (  Cf, Hm)
>
> **PETSc  Finalize**

(**b**)

**Figure 1.** Framework of parallel FE method based on PETSc: (**a**) flow chart for parallel pre-process; and (**b**) flow chart for parallel FE solver.

## 5. Numerical Results

### 5.1. Applied Bridge Overview

The Castlewood Canyon bridge is located on Highway 83 in the Black Forest of central Colorado. The original two-lane reinforced concrete arch bridge was built in 1946. The arches are 1.93 m wide by 1.78 m deep at the base with the depth tapering down to 1.0 m at the highest point. This bridge was severely dilapidated and was in need of repairing, enlarging, and strengthening. Parts of the concrete had spalled off of the deck, columns, and arches, and the steel rebar under concrete were severely rusted. In 2003, the original arch was repaired, the spandrel columns and decks were replaced and widened from about 0.9 m to 1.0 m including railings, and the overall length of the bridge was increased from 114 m to 123 m as shown in Figure 2.

**Figure 2.** Castlewood Canyon bridge, Franktown, Colorado.

### 5.2. Parallel Finite Method of Large-Scale Concrete Structure

#### 5.2.1. Modeling of Castle Wood Canyon Bridge with Large Meshes

In order to analyze the penetration of chloride and humidity in a concrete bridge, a large number of meshes are needed to capture the diffusion phenomenon within thin layer from concrete top surface to steel rebar. For depicting in detail concrete cover depth on rebar, the information of nodes and elements are created by mesh generator. Triangle as mesh generator is specialized for creating two-dimensional finite element meshes.

For castle wood canyon bridge, approximately 1.5 million nodes and three million elements were generated for input files as shown in Figure 3. To visualize and check the mesh size and shape, ParaView was employed as illustrated in Figure 4 [33]. ParaView is an open-source, multi-platform data analysis and visualization application. It is developed to analyze extremely large datasets using distributed memory computing resources and can be run on supercomputers to analyze datasets of terascale as well as on laptops for smaller data. For large scale information, VTK file format is adopted because this format is easy to read and write by hand or programmatically. This file format is automatically created when the program runs. For partitioning origin meshes, Parmetis was used to be embedded in the parallel program to automatically partition. Parmetis is an MPI-based parallel library that implements a variety of algorithms for partitioning unstructured graphs, meshes, and for computing fill-reducing orderings of sparse matrices. ParMETIS extends the functionality provided by METIS and includes routines that are especially suited for parallel AMR computations and large scale numerical simulations. The algorithms implemented in ParMETIS are based on the parallel multilevel k-way graph-partitioning, adaptive repartitioning, and parallel multi-constrained partitioning schemes developed.

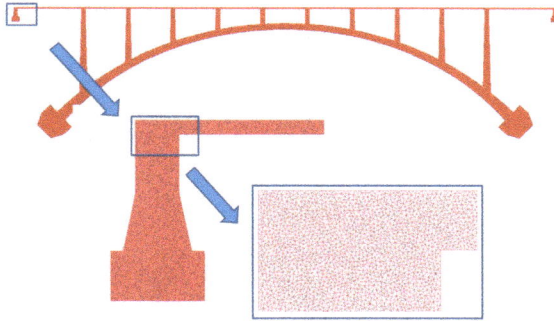

**Figure 3.** Modelling of Castlewood Canyon bridge.

**Figure 4.** Partitioning of Castlewood Canyon bridge with 16 processors.

### 5.2.2. Environmental Humidity Model

Environmental humidity model plays an important role of diffusion of chloride in concrete. The humidity model used in this study was proposed by Bazant and Xi (1993) [13]. Even though the moisture and temperature transport in concrete were coupled, humidity fluctuation was considered and temperature was assumed to be constant during analysis time. To establish a practicable and realistic humidity input model, real climate record in Chicago was employed as shown in Figure 5.

The environmental humidity model consists of three different components as shown in Equation (43) and demonstrated in Figure 6. First, $H_1$ is the mean value, standing for the stable horizontal trend. Second, $H_2$ is a random phase process corresponding to the harmonic variation of humidity. Third, $H_3$ is a random normal distribution with a specific mean and variance [13].

$$H = H_1 + H_2 + H_3 \tag{43}$$

where $H_1 = \overline{H}$, mean value of environmental model, $H_2 = A_1 \cos(2\pi t + \varphi_1) + A_{365} \cos(2\pi t + \varphi_{365})$, the random phase process of a one day period with $A_1 = 0.1$ and the random phase process of a one year period with $A_{365} = 0.08$, $\varphi_1$ and $\varphi_{365}$ have uniform distributions with constant density $1/2\pi$, $H_3 = \frac{1}{\sqrt{2\pi\sigma^2}} e^{[-\frac{(x-\mu)^2}{2\sigma^2}]}$, normally distributed random numbers are generated with specific mean and variance, 0.0046 [13].

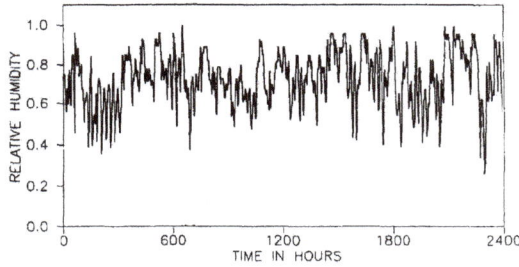

**Figure 5.** Environmental humidity Record, midway station, Chicago (Bazant et al., 1993) [13].

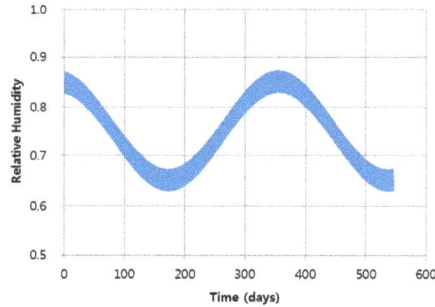

(a)

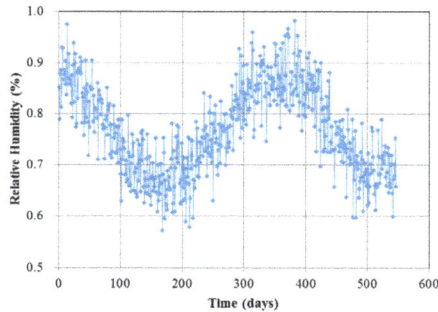

(b)

**Figure 6.** Environmental humidity model: (a) humidity model without random noise; and (b) humidity model with three components (including random noise).

For the chloride concentration on the top surface of concrete, various boundary conditions were used to actualize the realistic boundary conditions as illustrated in Figure 7. One can see that the interference between the interfaces of different concentration.

(a)                                                    (b)

**Figure 7.** Comparison of single and various boundary conditions: (**a**) single boundary condition; and (**b**) two boundary conditions.

### 5.2.3. Speed-Up for Parallel Algorithm

Speed-up is measured for the performance of the parallel implementation of finite element program. The definition of speed-up is a process for increasing in performance between two systems processing the same problem. With regard to speed-up in this study, 8–2048 processors were used in this study and the meshes with the number of about 1.5 million nodes were analyzed. When using up to eight processors, the program was not operated due to memory capacity when input data can be assigned on each memory which means the problem was too huge to solve with a couple of processors.

As the number of processors used in the analysis increased, speed-up was improved up to 512 processors because speed-up was better than ideal condition. After more than 512 processors were used, speed-up gradually increased because the communication time between the processors increased. As shown in Figure 8, one can see the optimum point of number of processors.

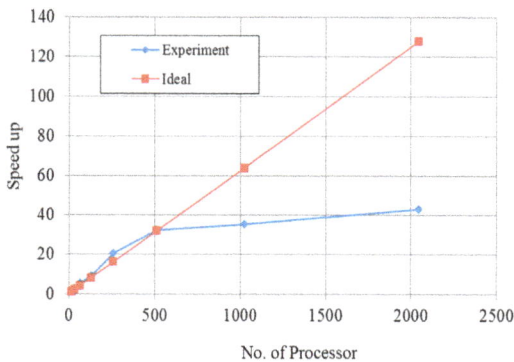

**Figure 8.** Speed up over number of processors.

### 5.2.4. Effect of Boundary Condition

(1)  Constant boundary condition

The physical and material models of the castle wood canyon bridge used in this paper were mentioned above. The validation of these models was secured by Ababneh et al. and Suwito et al. [14,15]. Ababneh et al. proved the model with alternating-direction implicit (ADI) finite-difference method and the numerical solutions were compared with the experimental results obtained by the 90-day ponding test (AASHTO T 259-80) [14]. Suwito et al. verified the accuracy of the model implemented with finite element method and numerical results agreed well with the experimental data [15].

It was assumed that the concrete bridge contained initially no chloride ions and has 60% relative humidity (RH). The top surface of concrete bridge was exposed to 3% NaCl and 100% RH. To analyze the diffusion of coupled chloride and humidity, 8 to 2048 processors and approximately 1.5 million nodes were used with parallel finite element method.

Obviously, this example is not to simulate the concrete behavior under service condition. The triangle element meshes were employed and the whole domain was partitioned into 16 sub-domains to the same as the number of processors shown in Figures 3 and 4. The sub-domains were divided with ParMetis so that each sub-mesh had almost the same number of elements.

Figures 9 and 10 show the contour of chloride and humidity distribution into the entire domain. The total and free chloride concentration at 50 mm below from the top surface of concrete is Figure 11 and the variation of relative humidity is Figure 12 over time up to 550 days. As a result of analysis, the concentration of chloride and humidity is continuously accumulated inside concrete. Specially, the variation of humidity inside concrete might not be increased in reality due to the seasonally change of the humidity distribution, so that periodic humidity model should be applied as the boundary condition.

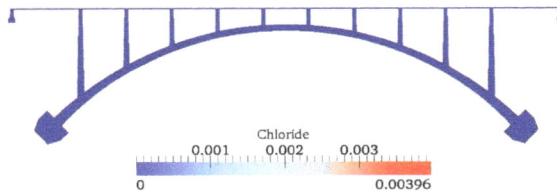

**Figure 9.** Contour of chloride distribution.

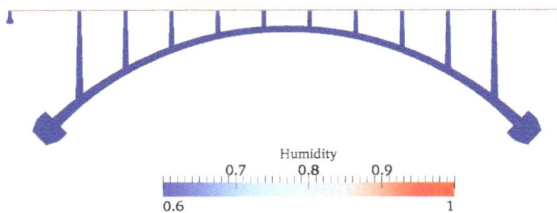

**Figure 10.** Contour of humidity distribution.

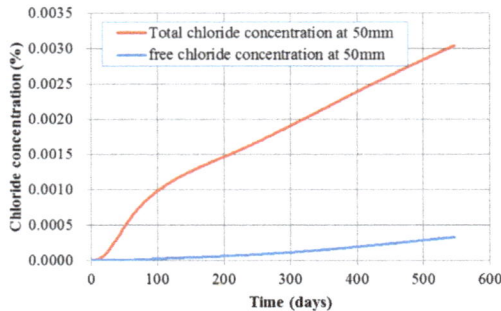

**Figure 11.** Chloride concentration over time at 50 mm depth.

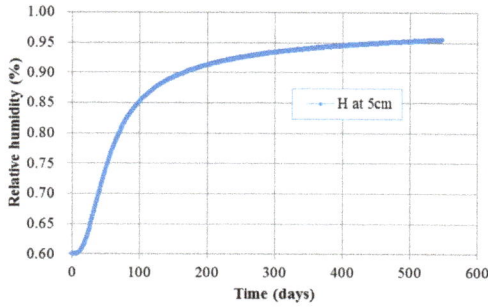

**Figure 12.** Humidity over time at 50 mm depth.

(2)   Periodic boundary condition

In order to simulate the real diffusion phenomenon in concrete, the periodic boundary condition should be required. Figures 13 and 14 describe the result of chloride concentration and relative humidity at different depth over time. As shown in Figure 13, total concentration is gradually increasing until almost one year. That means the chloride concentration can be affected by the periodic humidity boundary condition. After one year it suddenly increases because of the influence of the constant chloride initial condition. However, the total chloride concentration at 50 mm is reduced up to 15% rather than the concentration using constant boundary. In order to predict and demonstrate the realistic and seasonal chloride ingress through concrete cover, seasonal change information of chloride concentration on the top surface of concrete deck.

In Figure 13, the change of humidity distribution is influenced from the humidity model inputted as boundary condition. The short period random noise, one of random noise terms in environmental humidity model, can only reach a shallow portion near the top surface of concrete deck. As the concrete depth increased, the effect of short period random noise decreases. However, the seasonal variation of humidity concentration has an influence on 50 mm depth of concrete deck. At 50 mm from the top surface of concrete deck, the maximum value of humidity is about 77%, which is decreased up to 20% comparing with constant boundary condition. In addition, the result of about 500 days shows that the humidity trend is reversed between 10 mm and 50 mm, which means that evaporation of humidity just near concrete surface is easier than evaporation at 50 mm. Therefore, this prediction model is reasonable and effective to describe the diffusion phenomenon of chloride and humidity.

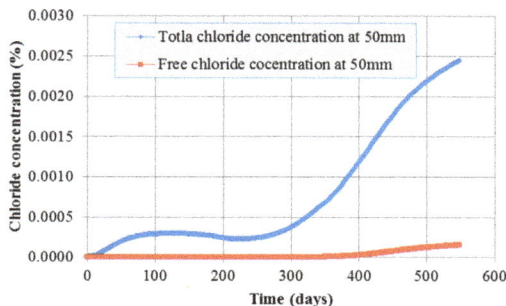

**Figure 13.** Chloride concentration over time at 50 mm depth.

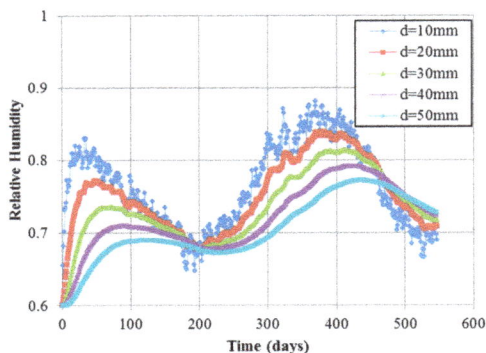

**Figure 14.** Humidity over time according to depth.

## 6. Conclusions

(1) The parallel finite element program was developed based on the robust mathematical material model. The program can be used to simulate the coupled moisture and chloride penetration into non-saturated concrete structures. Chloride ion is one of sources causing the steel corrosion in reinforced concrete structures. The material parameters related to chloride and moisture diffusion in concrete are taken into account. These parameters include chloride and moisture diffusion coefficients, moisture capacity, chloride binding capacity, and the coupling parameters reflecting the coupling effects between moisture and chloride transfer in concrete.

(2) For the implementation of parallel FE analysis, Triangle for mesh generation, ParMetis, PETSc, and MPI are employed. This program also used the overlapping domain decomposition method with additive Schwarz preconditioner. As the result of simulation, the computation time decreased until the number of processors became optimized when the number of processors increased. Then, the computational time increased because the communication between the processors increased.

(3) The present model can be used to simulate the unsaturated concrete structures subjected to other aggressive chemicals from de-icing salt. The framework of present model can be extended to simulate the multi-species de-icing salts ingress into non-saturated concrete structures in future work.

**Acknowledgments:** The authors wish to acknowledge the partial support by the US National Science Foundation under Grant CNS- 0722023 to University of Colorado at Boulder. Opinions expressed in this paper are those of the authors and do not necessarily reflect those of the sponsor.

**Author Contributions:** Okpin Na, he did all of the numerical implements, analyzed the numerical data, and wrote the manuscript. Xiao-Chuan Cai advised Okpin Na to code the parallel algorithm with C++ and PETSc and did the analysis of the numerical results. Yunping Xi directed and advised the theoretical research work, and revised the whole manuscript.

**Conflicts of Interest:** The authors declare no conflict of interest.

## References

1. Ali, H.A. Numerical simulation of chloride diffusion in RC structures and the implications of chloride binding capacities and concrete mix. *IJCEE* **2010**, *10*, 22–35.
2. Masi, M.; Colella, D.; Radaelli, G.; Bertolini, L. Simulation of chloride penetration in cement-based materials. *Cem. Concr. Res.* **1997**, *27*, 1591–1601. [CrossRef]
3. Conciatori, D.; Sadouki, H.; Bruhwiler, E. Capillary suction and diffusion model for chloride ingress into concrete. *Cem. Concr. Res.* **2008**, *38*, 1401–1408. [CrossRef]

4. Conciatori, D.; Laferriere, F.; Bruhwiler, E. Comprehensive modeling of chloride ion and water ingress into concrete considering thermal and carbonation state for real climate. *Cem. Concr. Res.* **2010**, *40*, 109–118. [CrossRef]

5. Isgor, O.B.; Razaqpur, A.G. Finite element modelling of coupled heat transfer, moisture transport and carbonation processes in concrete structures. *Cem. Concr. Compos.* **2004**, *26*, 57–73. [CrossRef]

6. Lin, G.; Liu, Y.H.; Xiang, Z.H. Numerical modeling for predicting service life of reinforced concrete structures exposed to chloride environments. *Cem. Concr. Compos.* **2010**, *32*, 571–579. [CrossRef]

7. Chernin, L.; Val, D.V.; Stewart, M.G. Prediction of cover crack propagation in RC structures caused by corrosion. *Mag. Concr. Res.* **2012**, *64*, 95–111. [CrossRef]

8. Zhang, J.; Ling, X.; Guan, Z. Finite element modeling of concrete cover crack propagation due to non-uniform corrosion of reinforcement. *Constr. Build. Mater.* **2017**, *132*, 487–499. [CrossRef]

9. Lim, S.; Akiyama, M.; Frangopol, D.M. Assessment of the structural performance of corrosion-affected RC members based on experimental study and probabilistic modeling. *Eng. Struct.* **2016**, *127*, 189–205. [CrossRef]

10. Šomodíková, M.; Lehký, D.; Doležel, J.; Novák, D. Modeling of degradation processes in concrete: Probabilistic lifetime and load-bearing capacity assessment of existing reinforced concrete bridges. *Eng. Struct.* **2016**, *119*, 49–60. [CrossRef]

11. Hu, N.; Burgueño, R.; Haider, S.W.; Sun, Y. Framework for Estimating Bridge-Deck Chloride-Induced Degradation from Local Modeling to Global Asset Assessment. *J. Bridge Eng.* **2016**, *21*, 06016005. [CrossRef]

12. Tang, L.; Nilson, L.O. *Prediction of Chloride Penetration into Concrete by Using the Computer Program CLINCON, Proceedings of the Second International Conference on Concrete under Severe Conditions 2-Environment and Loading, Tromsø, Norway, 21–24 June 1998*; Gjørv, O.E., Sakai, K., Banthia, N., Eds.; E & FN Spon: New York, NY, USA, 1998; pp. 625–634.

13. Bazant, Z.P.; Xi, Y. Stochastic Drying and Creep Effects in concrete structures. *J. Struct. Eng.* **1993**, *119*, 301–322. [CrossRef]

14. Xi, Y.; Bazant, Z.P. Modeling chloride penetration in saturated concrete. *J. Mater. Civ. Eng.* **1999**, *11*, 58–64. [CrossRef]

15. Suwito, A.; Cai, X.; Xi, Y. Parallel finite element method for coupled chloride moisture diffusion in concrete. *Int. Numer. Anal. Model.* **2006**, *3*, 481–503.

16. Ababneh, A.; Benboudjema, F.; Xi, Y. Chloride penetration in Nonsaturated concrete. *J. Mater. Civ. Eng.* **2003**, *15*, 183–191. [CrossRef]

17. Xi, Y.; Willam, K.; Frangopol, D.M. Multiscale modeling of interactive diffusion processes in concrete. *J. Eng. Mech.* **2000**, *126*, 258–265. [CrossRef]

18. Xi, Y.; Bazant, Z.P.; Molina, L.; Jennings, H.M. Moisture diffusion in cementitious materials: Adsorption isotherm. *Adv. Cem. Based Mater.* **1994**, *1*, 248–257. [CrossRef]

19. Xi, Y.; Bazant, Z.P.; Molina, L.; Jennings, H.M. Moisture diffusion in cementitious materials: Moisture capacity and diffusivity. *Adv. Cem. Based Mater.* **1994**, *1*, 258–266. [CrossRef]

20. Xi, Y. A model for moisture capacities of composite materials. I: Formulation. *Comput. Mater. Sci.* **1995**, *4*, 65–77.

21. Xi, Y. A model for moisture capacities of composite materials. II: Application to concrete. *Comput. Mater. Sci.* **1995**, *4*, 78–92. [CrossRef]

22. Tang, L.; Nilson, L.O. Chloride binding capacity and binding isotherms of OPC pastes and mortars. *Cem. Concr. Res.* **1993**, *23*, 247–253.

23. Christensen, R.M. *Mechanics of Composite Materials*, 2nd ed.; Wiley: New York, NY, USA, 1979.

24. Bazant, Z.P.; Najjar, L.J. Nonlinear water diffusion of nonsaturated concrete. *Mater. Constr.* **1972**, *5*, 3–20. [CrossRef]

25. Triangle. Available online: https://www.cs.cmu.edu/~quake/triangle.html (accessed on 27 March 2017).

26. Balay, S.; Buschelman, K.; Eijkhout, V.; Gropp, W.D.; Kaushik, D.; Knepley, M.G.; McInnes, L.C.; Smith, B.F.; Zhang, H. PETSc-users manual. In *Technical Report ANL-95/11–Revision 2.1.5*; Argonne National Laboratory: Argonne, IL, USA, 2008.

27. MPI Forum. MPI: A message passing interface standard. *Int. J. Supercomput. Appl.* **1994**, *8*, 159–416.

28. MPI Forum. MPI2: A message passing interface standard. *High Perform. Comput. Appl.* **1998**, *12*, 1–299.

29. MPICH Team. MPICH: A Portable Implementation of MPI. 1996–2005. Available online: http://www.mpich. org/documentation/guides/ (accessed on 27 March 2017).

30. Karypis, G. ParMETIS: Parallel Graph Partitioning and Sparse Matrix Ordering. 1996–2005. Available online: http://glaros.dtc.umn.edu/gkhome/metis/parmetis/download (accessed on 27 March 2017).

31. Dalcín, L.D.; Paz, R.R.; Anca, A.A.; Storti, M.A.; D'Elía, J. Parallel FEM application development in Python. *Mec. Comput.* **2005**, *24*, 1823–1838.

32. PETSc. Available online: http://www.mcs.anl.gov/petsc/documentation/index.html (accessed on 27 March 2017).

33. Paraview. Available online: http://www.paraview.org/ (accessed on 27 March 2017).

## Article

# New Insights in the Long-Term Atmospheric Corrosion Mechanisms of Low Alloy Steel Reinforcements of Cultural Heritage Buildings

**Marie Bouchar [1,2], Philippe Dillmann [2,*] and Delphine Neff [2]**

[1]   Saint-Gobain Recherche, 39 quai Lucien Lefranc, 93303 Aubervilliers CEDEX, France; marie.bouchar@saint-gobain.com
[2]   LAPA-IRAMAT, NIMBE, CEA, CNRS, Université Paris-Saclay, CEA Saclay, 91191 Gif-sur-Yvette, France; delphine.neff@cea.fr
*   Correspondence: philippe.dillmann@cea.fr; Tel.: +33-169-081-469

Received: 3 April 2017; Accepted: 12 June 2017; Published: 19 June 2017

**Abstract:** Reinforcing clamps made of low alloy steel from the Metz cathedral and corroded outdoors during 500 years were studied by OM, FESEM/EDS, and micro-Raman spectroscopy. The corrosion product layer is constituted of a dual structure. The outer layer is mainly constituted of goethite and lepidocrocite embedding exogenous elements such as Ca and P. The inner layer is mainly constituted of ferrihydrite. The behaviour of the inner layer under conditions simulating the wetting stage of the RH wet/dry atmospheric corrosion cycle was observed by in situ micro-Raman spectroscopy. The disappearance of ferrihydrite near the metal/oxide interface strongly suggests a mechanism of reductive dissolution caused by the oxidation of the metallic substrate and was observed for the first time in situ on an archaeological system.

**Keywords:** cultural heritage metals; iron; low alloy steel; atmospheric corrosion; in-situ measurement; micro-Raman

## 1. Introduction

Since Antiquity and the Middle Ages, stone buildings have been reinforced by metallic clamps, rods, and ties. Recent studies have demonstrated that thousands of kilograms of ferrous alloys have been used since their inclusion in buildings in monuments of the Middle Ages, such as cathedrals [1–4]. Today, these reinforcements are considered as part of heritage buildings and must be preserved as testimonies of the skills of ancient builders. Most of these artefacts are exposed to atmospheric corrosion, either indoor or outdoor [5]. Besides, numerous archaeological artefacts are conserved in museum storage rooms, and in some cases, are submitted to uncontrolled environmental conditions. For that reasons, for several years, the very long term atmospheric corrosion of iron and steel has been studied, especially on these monuments. Another reason for the study of these kind of corrosion systems is that they can be considered as analogues for the prediction of the corrosion behaviours of today's materials that are going to be used for very long periods in civil engineering or the nuclear industry [6,7].

Former characterisation studies performed on ferrous metals coming from historical monuments [8–11] observed that the layers developed in such conditions over several hundred years were a mix of Fe-containing phases such as goethite ($\alpha$-FeOOH), lepidocrocite ($\gamma$-FeOOH), akaganeite (($\beta$-FeO$_{1-x}$(OH)$_{1+x}$,Cl$_x$), feroxyhyte ($\delta$-FeOOH), ferrihydrite (Fe$_2$O$_3$,1.8H$_2$O), maghemite ($\gamma$-Fe$_2$O$_3$), magnetite (Fe$_3$O$_4$), and hematite (Fe$_2$O$_3$), whose proportions depend on the initial state of the alloy (presence of a former scale due to the elaboration process) and the corrosion conditions (wet/dry cycle, presence of pollutants, ... ). Nevertheless, the exact corrosion mechanisms involved in these long-term systems are still under discussion. Most authors admit that it is based on the well-known RH (relative humidity) "wet/dry" cycle that occurs in the case of atmospheric corrosion, but the different processes involved during this stage are not completely deciphered.

The wet/dry cycle is divided into three steps: (i) the wetting stage, when the electrolyte progressively covers the surface of the material (here, the corrosion product); (ii) the wet step, when a continuous layer is covering the surface; and (iii) the drying step, when the electrolyte evaporates from the surface. In the first step, it is assumed by some authors [12–15] that the Fe(III)-phases constituting the corrosion product layer (CPL) participate in the oxidation of the metal. This was suggested for short-term [16,17] and long-term corrosion systems [18–21]. These Fe(III)-phases are lepidocrocite [22,23], ferrihydrite [9,24] and, to a lesser extent, ferroxyhite and akaganeite. Some authors made the hypothesis that these phases could be reduced into conductive ones. A different hypothesis was suggested for the reduced phase that is obtained: solid state transformation into an Fe(II)-phase $\gamma$-Fe·OH·OH, with a structure similar to a hydrogel [23], dissolution, reduction, and re-precipitation into Fe(II)-containing species [20]. Monnier et al. [21] demonstrated that the nature of the final phase (a mixture of magnetite and Fe(II)-hydroxide) depends on the reduction mode (current or potential imposed) and on the electrolyte pH. Other studies [25,26] suggest that the anodic and cathodic reactions are decoupled in thick CPLs (the latter one being located at the outer part of the CPL), supposing the existence of transient conductive phases formed during the wetting stage by the reduction of electrochemically reactive phases. This transient phase is re-oxidised at the end of the cycle (see below).

Nevertheless, despite all these different studies, the reduction of reactive phases inside ancient corrosion thick layers was never observed directly. During the following steps (wet step) of the wet/dry cycle, it is admitted that dissolved $O_2$ is transported in the pores of the CPL and reduces when it meets species that can be oxidised ($Fe^{2+}$ ion dissolved after the reduction of an electrochemically reactive phase during the first stage, Fe(II) conductive phase, or even the metallic substrate). In the case that conductive species are present, the behaviour of the system would completely change, allowing a decoupling of anodic and cathodic oxygen reduction reactions. At the end of the cycle, the massive supply of oxygen and the possible drying of the pores lead to the oxidation of all Fe(II)-species present in the system. At least part of the pores of the CPL are empty of electrolyte.

Considering this short state of the art, the aim of the present study is twofold. First, a supplementary long-term corrosion system is studied: the iron clamps corroded outdoors on the Metz cathedral, since the 15th centuries. Because of the potential variability of the ancient corrosion layers, it is of primary importance to extend the set of ancient systems analysed with fine characterisation methods. One important issue is to estimate the proportion of potentially reactive phases inside the layer and their location. The other aim of the study presented in this paper is to perform an in situ experiment on the corrosion system to simulate the wetting stage and to follow the possible reduction of reactive phases by micro-Raman spectroscopy.

## 2. Materials and Methods

### 2.1. Set of Samples

Samples were taken on iron clamps removed from the belfry tower of the Metz Cathedral (France) during a restoration campaign (2010–2014) (Figure 1a). These clamps were put in the monument during the building (15th centuries AD), as demonstrated by archaeological science studies [1,3]. The clamps are about 40 cm long and have a rectangular section of 3 cm by 2 cm (Figure 1c). Because of the heterogeneity of ancient iron and steel [4], the average chemical composition is a value of low significance. Investigations made by EPMA on the metallic matrix reveal that except P and C, all of the elements have contents below 100 ppms. The metal was studied by metallographic observations on cross sections (see Supplementary Materials). It is composed of low alloy steel (wt. % C < 0.3) containing several thousand ppm of phosphorus. Some non-metallic slag inclusions reaching several 100 μm in size are also observed in the metallic matrix. These clamps are located outdoors on walls, in a vertical position (Figure 1b). Meteorological data collected daily by the Metz-Frescaty weather station from 1981 to 2010 give an accurate estimate of the environmental conditions. The RH ranges between 70% and 90%, and the temperature ranges between 1 °C and 19 °C.

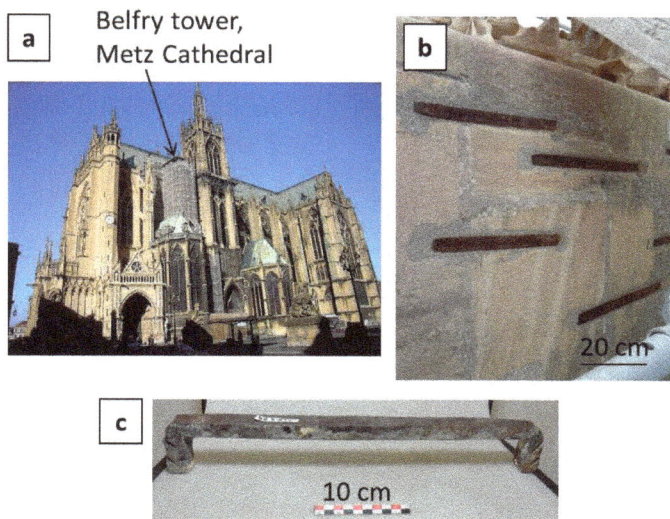

**Figure 1.** (**a**) Metz Cathedral; (**b**) iron reinforcement clamps located outdoors on walls of the La Mutte tower; (**c**) one of the iron clamps studied at the laboratory.

Complete cross sections of the clamps were performed, using a slow speed diamond saw. These samples were mounted in epoxy resin, and prepared by grinding (SiC, grade 80–4000) and polishing (diamond paste, 3 and 1 μm) under ethanol.

### 2.2. Characterisation Methods

The entire corrosion system corresponding to the outer face of the clamp, exposed to atmospheric corrosion, was observed on transverse sections by an Optical Microscope (OM) and by a Field Emission gun Scanning Electron Microscope (FESEM) JEOL 1200. The chemical composition of the CPL was analysed by EDS (IdFix™ and Maxview™ softwares, Fondis Electronic, France) coupled to FESEM (accelerating voltage of 15 kV).

The structure of the CPL was investigated by micro-Raman Spectroscopy (μRS) on the cross sections using a Renishaw InVia spectrometer equipped with a frequency doubled Nd:YAG laser at 532 nm. The laser was focused on the sample thanks to a Leica DM/LM microscope. With the ×50 focus used for the acquisitions, the beam diameter is 1.5 μm and the penetration length is about 2 μm. The spectral resolution given by the CCD detector is 2 cm$^{-1}$. As some iron oxides are very sensitive to laser exposure, density filters are used to control the laser power on the sample under 100 μW. Spectral acquisitions are performed with the Renishaw Wire 3.2 software in an ultra-fast mode (StreamLine™, Renishaw, UK).

The iron oxides and oxyhydroxides constituting the rust layer were quantified on spectra extracted from maps or acquired locally on the corrosion product layer (CPL) using a program developed in our laboratory, inspired from the CorATmos program developed by Monnier et al. [9], and adapted to process data of maps with a very large number of spectra. This program, based on linear combinations of reference spectra of pure phases, fits each experimental spectrum between 200 and 800 cm$^{-1}$ with a Levenberg-Marquardt algorithm in order to reach the proportion of the phases at each point of a map.

The reference spectra are those used by Monnier et al. The error has been estimated by the latter authors as about 5% for each quantified phase. More details on the procedure can be found in [11].

*2.3. In Situ Experimental Setup*

The in situ cell (Figure 2) is designed to allow, at the same time, the circulation of deoxygenated water on the surface of the CPL and μRS analyses on a cross section of the CPL. A transverse section of the clamp of about 1 cm$^3$ is mounted inside the cell and then polished. After polishing, an adhesive film (HD ClearTM Crystal Tape, Duck® Brand, USA) transparent to visible light and a laser beam is put on the cross section, avoiding the solution to bypass the corrosion layer and to directly reach the bare metal. The Raman spectrum of the adhesive film only shows two bands (810 cm$^{-1}$ and 398 cm$^{-1}$) in the region of the corrosion product signals (200–850 cm$^{-1}$), so its contribution can be easily removed from the Raman spectra obtained on the CPL.

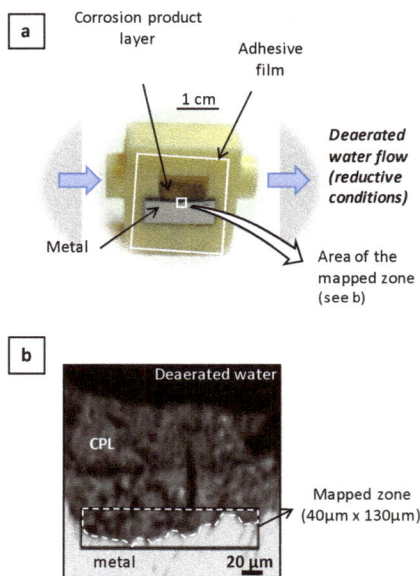

**Figure 2.** (**a**) In situ experimental cell; (**b**) Optical microphotograph of the transverse section. Black rectangle: zone observed by μRS maps. Dotted line: zone where the phase quantification is performed. CPL: Corrosion Product Layer.

## 3. Results

*3.1. Characterisation of the Corrosion Products*

The average thickness of the CPL has been estimated on a cross section from 154 measurements performed with a regular step covering a CPL length of 12 mm along the metal/CPL interface. The average value is 110 ± 50 μm. The local values are dispersed and vary between 20 and 260 μm. The CPL is crossed by many cracks with variable orientations relative to the metal/CPL interface. The crack aperture ranges from less than 1 μm to a few micrometers (Figure 3a). These cracks may have formed by the effect of mechanical strengths generated by corrosion on the metal, leading to a volume increase when iron oxides replace metallic iron. Moreover, the CPL presents a network of micrometric and submicrometric pores (Figure 3b). As stated by other authors [8,10] on ancient corrosion layers, the CPL seems to be bi-layered with an internal zone much more dense (despite presenting micro-cracks) than an outer and more porous layer.

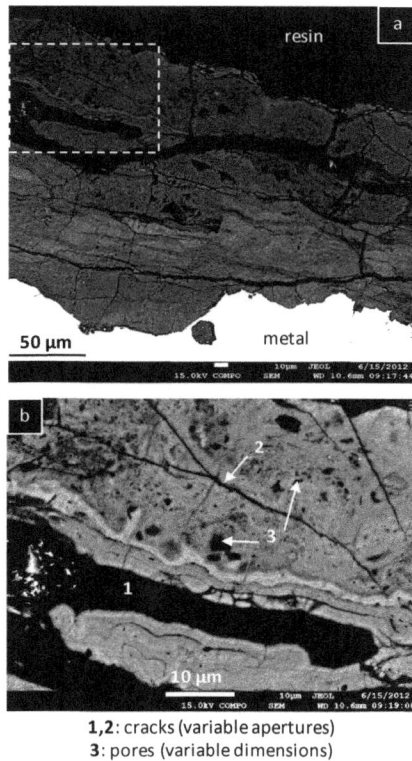

**1,2**: cracks (variable apertures)
**3**: pores (variable dimensions)

**Figure 3.** FESEM images of the corrosion product layer in the backscattered electron mode. Image (**b**) corresponds to a magnification in the white dot rectangle drawn on the image (**a**). 1&2: cracks with various apertures; 3: pores with various sizes.

The FESEM/EDS maps performed on cross section samples confirm the structuration of the CPL into two sublayers with different elemental compositions (Figure 4). In both the inner and outer sublayers, the major elements are iron and oxygen. The sublayers differ from each other by their average contents in minor elements. Their distribution is quite inhomogeneous inside each sublayer. Calcium is mainly located along the cracks and pores in both sublayers, with the contents reaching 10 wt. % locally. This minor element corresponds to an exogenous pollution, as calcium is found in high amounts in stone and mortar dusts, as well as particles from monument walls in which iron clamps were sealed. Silicon is essentially located in the most external part of the CPL. As calcium, it is assumed that silicon comes from building walls. Some spots can nevertheless be identified in the inner sublayer. They correspond to non-corroded slag inclusions initially present in metallic matrix (see [3] and "Materials and methods" section). Chlorine is only present in rare zones in the outer sublayer, with higher contents reaching 1 to 2 wt. %. Even far from the seacoast, as in the present case, low quantities of this element can be found in the atmosphere, due to anthropogenic sources, leading to its accumulation in the corrosion product. Almeida et al. [27] showed in a corrosion study on 19 urban and rural sites conducted in 2000, that even with low deposition rates of chlorides ($<3$ mg·m$^2$·d$^{-1}$), Cl- can be detected in the corrosion products at levels higher than 308 mg·m$^2$, with most of the values being comprised between 34 and 66 mg·m$^2$. Sulphur is detected only in the extreme outer sublayer (sometimes reaching about 1 wt. %). This element comes from the SO$_2$ of urban and industrial atmospheres [28]. The level of this compound is controlled for several decades and the annual mean concentration is now well below 100 µg·m$^{-3}$; however, at the end of the XIXth centuries and during the XXth centuries, this value could reach several 100 µg·m$^{-3}$ [29]. The phosphorus content may locally reach several weight percents (2 to 10 wt. %) in the outer sublayer. The zones rich in phosphorus correspond to one or several strips parallel to the metal/CPL interface. The presence of phosphorus in the metallic substrate may not explain the high contents in the CPL alone. The most probable hypothesis to explain the origin of phosphorus in the outer sublayer is the contribution of an exogenous source. Animal excrement residues rich in phosphates could be one of these sources [30]. The phosphorus content in the inner sublayer is lower than 0.5 wt. %.

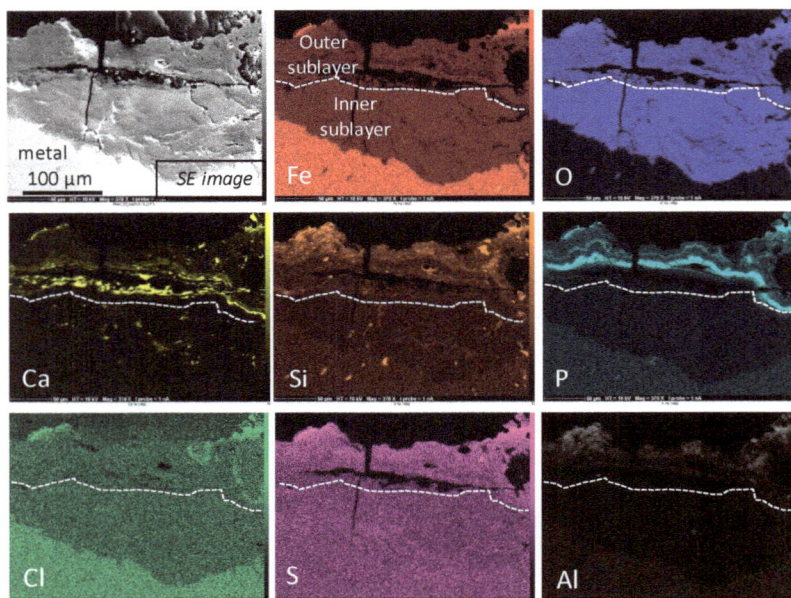

**Figure 4.** FESEM image of the corrosion product layer in the backscattered electron mode, and related EDS maps (Kα lines) of detected chemical elements. The white dot line corresponds to the limit between the inner and outer sublayers of the corrosion product layer.

The CPL phases have been identified on a cross section by µRS. At the scale of the laser probe, most of the spectra present bands corresponding to a mixture of several phases. Figure 5 shows a set of spectra testifying the presence of four oxyhydroxides in the CPL:

- goethite α-FeOOH (bands at 300 and 390 cm$^{-1}$, spectra N° 1–2);
- lepidocrocite γ-FeOOH (intense peak at 250 cm$^{-1}$, spectrum N° 2);
- akaganeite β-FeO$_{1-x}$(OH)$_{1+x}$Cl$_x$) (combination of asymmetric bands at 309, 390, and 724 cm$^{-1}$, spectrum N° 3), only present in rare zones;
- ferrihydrite Fe$_2$O$_3$,1.8H$_2$O [31], and more generally hydrated oxyhydroxides (broad and symmetric signal, without shoulder, around 710 cm$^{-1}$, spectrum N° 4). In fact, feroxyhite (δ-FeOOH) also presents a broad band in the same region as ferrihydrite, but at 677 cm$^{-1}$ [32]. The Raman spectra obtained on the Metz cathedral samples are mainly in agreement with ferrihydrite, but the presence of small amounts of feroxyhite cannot be completely excluded.

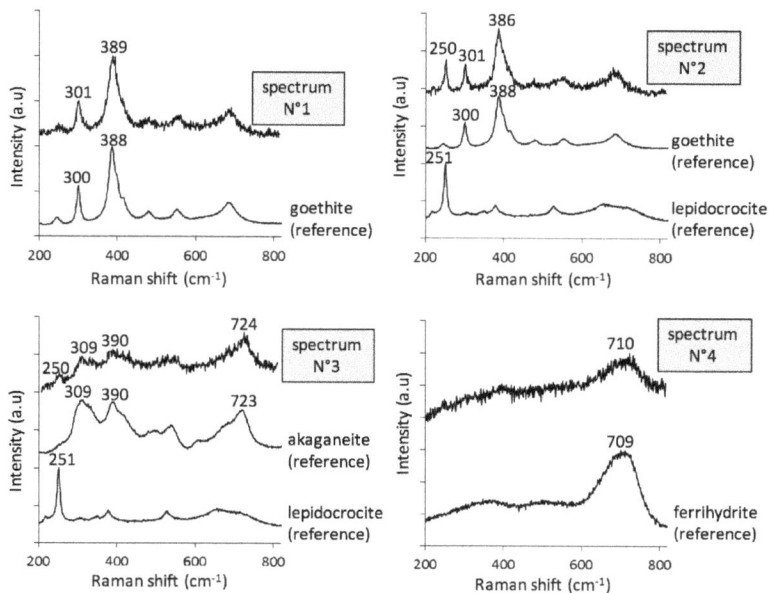

**Figure 5.** Set of Raman spectra acquired on the corrosion product layer. Each spectrum is plotted together with the corresponding 'reference' spectra. (spectrum N°1): goethite; (spectrum N°2): blended goethite/lepidocrocite; (spectrum N°3): blended akaganeite/lepidocrocite; (spectrum N°4): ferrihydrite.

Maghemite ($\gamma$-Fe$_2$O$_3$) that presents a sharp double peak at 670 and 720 cm$^{-1}$, is very different from the broad band observed on the experimental spectra at this location, whilst magnetite (Fe$_3$O$_4$) and hematite ($\alpha$-Fe$_2$O$_3$) were not detected in the CPL.

µRS quantitative maps (see "Materials and methods" section) also highlight the CPL structuration into two sublayers. Figure 6 presents a map acquired in a representative zone of the CPL. The limit between the inner and outer sublayers is shown by a black dot line on the schematic view of the map. The phase quantifications in both zones delimited by this line (zones a and b) are given in Table 1. The inner sublayer contains mainly ferrihydrite—contents higher than 70%—and goethite—about 5 to 25%. Lepidocrocite is present in small amounts (generally <10%), but only in a few islets. The outer sublayer also contains goethite in similar quantities as the inner one. However, this layer is much poorer in ferrihydrite—with local contents ranging between 0 and about 50%—and much richer in lepidocrocite—with contents of about 25%, locally reaching 70%. In rare limited zones of the outer sublayer, akaganeite is detected in quantities lower than 10%. Lastly, it was not possible to detect by µRS a specific phosphorus-containing phase in the most phosphorous zones of the outer layer. This is probably due to the fact that the phosphorous products are not crystallised as stated by other authors for ancient corrosion products [33,34]. All the results are summarised in Table 2.

**Table 1.** Average phase contents in zone a (inner sublayer) and zone b (outer sublayer) of Figure 6. n.d.: not detected.

| | Average % | | | |
|---|---|---|---|---|
| | **Ferrihydrite** | **Goethite** | **Lepidocrocite** | **Akaganeite** |
| Inner sublayer (zone a) | 77 | 20 | 3 | <1 |
| Outer sublayer (zone b) | 44 | 21 | 35 | n.d. |

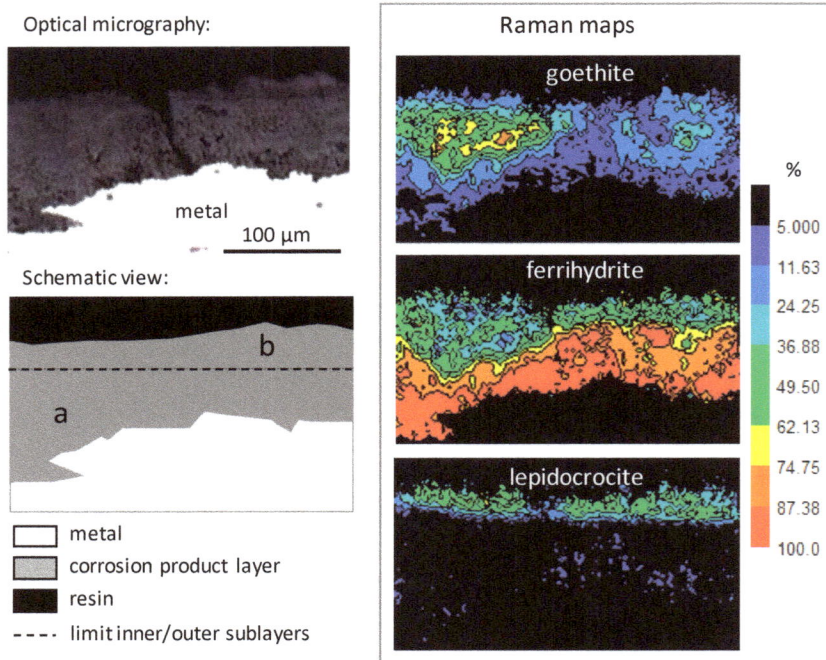

**Figure 6.** Quantitative μRS maps on a corrosion product layer containing goethite, ferrihydrite, and lepidocrocite. The dot line on the schematic view corresponds to the average limit between the inner sublayer (a) and outer sublayer (b).

**Table 2.** Summing up of the characterization results.

| Layer Part | Morphology | Minor Elements | Phases % |
|---|---|---|---|
| Inner | Cracks (micro and sub-micro) Multiphased Relatively dense | Ca (exogenous) along cracks (up to 10%) Si (endogenous) from former slag inclusions P (endogenous) from the metal (about 0.5%) | Ferrihydrite: 77 Goethite: 20 Lepidocrocite: 3 |
| Outer | Cracks (micro and sub-micro) Porous | Ca (exogenous) along cracks (up to 10%) Si (exogenous) Cl (exogenous) in rare zones S (exogenous) along outer surface P (exogenous) up to 10% | Ferrihydrite: 44 Goethite: 21 Lepidocrocite: 35 |

## 3.2. In Situ Reactivity Test of the Corrosion Layer

A specific sample, with the same corrosion pattern as the one evidenced in Section 3.1, was selected to perform a reduction test on the CPL. This test aims to show that reactive phases (here, mainly ferrihydrite) in contact with the metal are susceptible to playing the role of an oxidizer in the absence of dioxygen and thus to being reduced. The experimental setup and method used for the test are described in the "Materials and methods" section.

The evolution of the CPL of a clamp sample, set in a cell, has been observed by μRS at the metal/CPL interface over nine days (Figure 2). Spectral maps of the same zone have been acquired during the experiment: (i) before the introduction of deaerated water in the cell; (ii) after 1, 2, 6, and 9 days of deaerated water circulation; (iii) after stopping the flow of deaerated water, then exposure to ambient air, for one day—see Figures 7 and 8. The mapped zone near the metal/CPL interface is 130 μm long and 40 μm wide—i.e., 52 × 16 points—with a step of 2.5 μm in length and width

(Figure 2b). It is worth noting that this zone does not present any crack detectable by OM and FESEM, which could link directly the metal/CPL interface to the external environment. From each quantitative map acquired during the experiment, an average phase content of the entire zone can be deduced. The evolution of this average content versus time is presented in Figure 9.

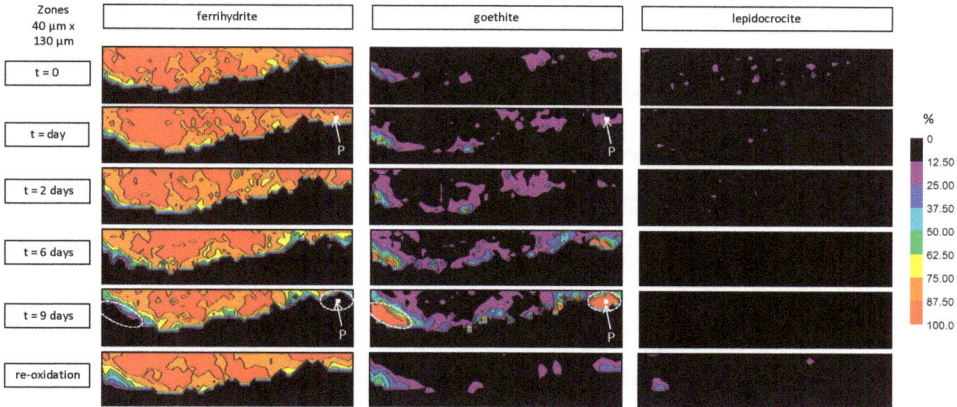

**Figure 7.** Quantitative μRS maps in the zone identified by the dot line on Figure 2b: before the test (*t* = 0), after 1, 2, 6, and 9 days of deaerated water circulation, and after re-exposure to ambient air during one day (re-oxidation). The regions circled by a dot line correspond to those where ferrihydrite has disappeared after nine days of deaerated water exposure. The spectra corresponding to point P, for one and nine days, are presented in Figure 8.

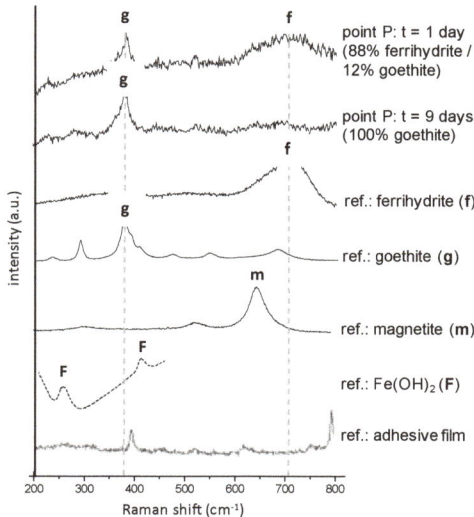

**Figure 8.** Raman spectra at point P identified on Figure 7, after one and nine days of exposure to deaerated water. The 'reference' spectra of the pure phases are also presented (ferrihydrite, goethite, and magnetite from [9]; Fe(OH)$_2$ from [35]; adhesive film). The contribution of the adhesive film of the cell was substracted from the experimental spectra.

At $t = 0$, before the beginning of deaerated water circulation through the experimental cell, the mapped zone mainly contains ferrihydrite (about 90% average spread over the entire zone, and thus in contact with the metal) (Figures 7 and 9, $t = 0$). Goethite (about 7% on average) and lepidocrocite (about 3% on average) are minor phases and are present as localized islets. After one day of deaerated water circulation in the cell, quantitative maps do not present any significant difference with those obtained at time $t = 0$. However, for longer treatment times, the obtained μRS maps are quite different. After two days, goethite islets have expanded spatially and lepidocrocite ones have disappeared. Between two and six days, the average contents of ferrihydrite and lepidocrocite decrease progressively, whereas the average goethite content increases (Figure 9). After six days in deaerated water, overall content variations are significant compared to the maximal quantification error, which is estimated to be 5% (see "Materials and methods" section). The ferrihydrite content has decreased by more than 10%, whereas the goethite content has increased by more than 10%. Then, the CPL composition stabilizes. After nine days in deaerated water, the quantification results are similar to those obtained at time $t = 6$ days. Quantitative maps of phases at time $t = 9$ days show that variations of ferrihydrite and goethite contents are localised at some specific places in the CPL. The highest variations—up to 90%—are mainly concentrated in two regions, both in contact with the metal/CPL interface. In both regions, ferrihydrite has disappeared and the goethite relative content has increased. Figure 8 compares the spectra acquired at times $t = 1$ day and $t = 9$ days at the same point in one of these two regions (labelled "P" on Figure 7). The broad band around 700 cm$^{-1}$, corresponding to the presence of ferrihydrite and visible on the spectrum at $t = 1$ day, is no longer detectable on the spectrum at $t = 9$ days.

No other new phases than the ones present at time $t = 0$ have been detected during the test. Specifically, magnetite and iron hydroxide (II), detected by Monnier et al. by X-ray absorption spectroscopy and X-ray diffraction during ferrihydrite reduction induced by an electric current or potential [21], are not identified in this test on spectra from the map at $t = 9$ days. The decrease in the local ferrihydrite content is only balanced by an increase in the local goethite content.

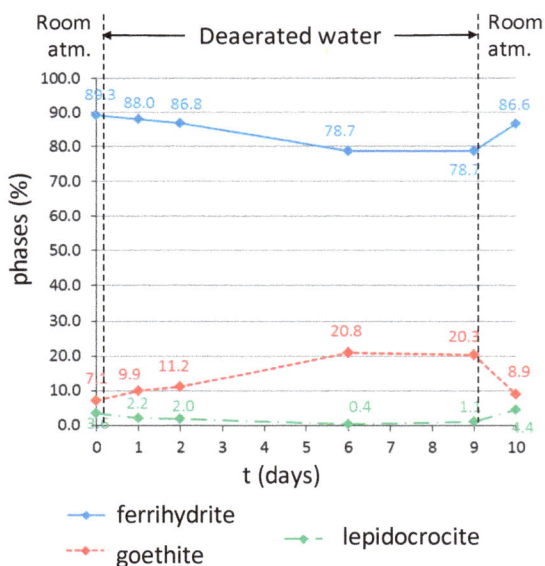

**Figure 9.** Global average contents (%) of the phases detected in the zone marked with a black rectangle on Figure 2b, calculated from the maps presented in Figure 7.

## 4. Discussion

The chemical and structural features of the CPL from the Metz cathedral are relatively similar to the one measured by precedent studies on a reinforcement made of ferrous alloy from monuments of several hundred years old. Figure 10 compares the thicknesses measured on the samples of the Metz cathedral with the ones found on low alloy steel reinforcements of cultural heritage monuments with different ages. Values found in the present studies correspond to the lower ones measured on monuments of the same age (Amiens cathedral). An important difference to note here is that contrary to all of the other studied monuments, the clamps of the Metz cathedral were not corroded indoors. Consequently, these relatively low values could partly be due to the fact that a non negligible part of the outer zone of the layer may have been eliminated by leaching processes under the falling rain.

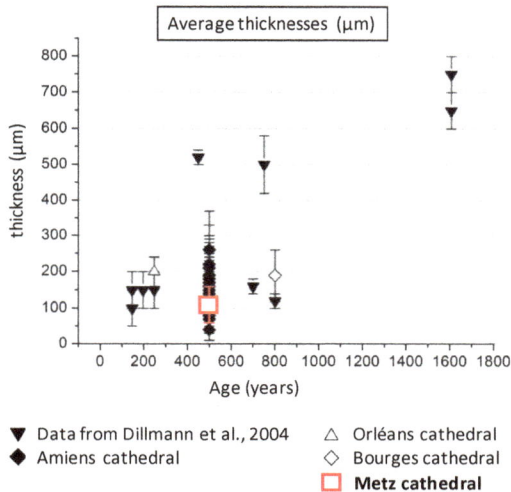

**Figure 10.** Average thicknesses and standard deviations of CPL measured on ancient low alloy steels in monuments v time. Black down triangle: from [8], losanges: Amiens cathedral [10], up open triangle: Orléans Cathedral [36], open losange: Bourges cathedral [11], red square: Metz cathedral, this study.

Nevertheless, despite this possible leaching, the outer layer is at least partly preserved, as demonstrated by the presence of the typical bi-layered structure evidenced on the cross sections. This morphology has already been observed by several authors on low alloy steels [8,11] and weathering steels [37,38] corroded during decades and centuries under comparable conditions. The outer layer contains typical exogenous elements coming from the atmosphere (S, Cl) or environment (Ca, Si, ... ). Some hypotheses about the formation of the bi-layered structure have been made by Burger et al. [39]. These authors examined low alloy steel coupons preliminary covered with a thin gold layer in order to mark the original surface and corroded in a climatic chamber during several months. After the corrosion test, the gold layer was located exactly between the two sublayers; the inner one mainly constituted of ferrihydrite, the outer one of lepidocrocite, as for the Metz samples. The authors deduced from these observations that the growth of the corrosion layers happened not only in the detriment of the metallic substrate (forming the inner layer), but was also governed by the dissolution of some iron cations and subsequent precipitation at the external interface of the rust layer. This dissolution/precipitation mechanism would explain the porous nature of the outer layer and the fact that important quantities of exogenous elements are entrapped in it.

Concerning the constitutive phases, the CPL of the Metz cathedral contain the compounds usually reported by characterisation studies on such systems: goethite ($\alpha$-FeOOH), lepidocrocite ($\gamma$-FeOOH), akaganeite ($\beta$-FeO$_{1-x}$(OH)$_{1+x}$,Cl$_x$), and ferrihydrite (Fe$_2$O$_3$,1.8H$_2$O) were identified. Nevertheless, some differences must be noted. The CPL of the Metz cathedral clamps contain significantly less akaganeite (1%) than those of the reinforcements coming from other monuments, such as the Amiens cathedral (20% for some samples) [10]. However, Metz is very far from the seacoast (350 km as crow flies) compared to Amiens (only 60 km as crow flies). The marine atmosphere must favour the formation of akaganeite in the case of Amiens, but has no influence in the Metz case. The presence of chlorine-containing phases in the CPL of Metz should more probably be due to anthropogenic polluting compounds like de-icing salts or chlorofluorocarbons (CFCs) released in the atmosphere by industries [40].

Another difference concerning the structure of the Metz corrosion pattern compared to former studies is the high content of ferrihydrite detected in the inner sublayer (up to 80% of the corrosion products for some samples). As a comparison, the ferrihydrite (and feroxyhite) contents in the CPL of Amiens and Bourges samples reached up to 65% and 43%, respectively. The identification of ferrihydrite on the Raman spectra is sometimes not obvious in the case of mixed phases. Specifically, its distinction with feroxyhyte, which also presents a broad band at 677 cm$^{-1}$, may be hard [32]. However, the typical band observed on the spectra around 700 cm$^{-1}$ in the case of the Metz samples CPL tends to favour the hypothesis of the presence of ferrihydrite instead of the one of feroxyhyte. Despite the fact that the presence of feroxyhyte cannot be completely excluded, it seems to be the minority and, for simplification reasons, only ferrihydrite will be considered in the following. This choice is reinforced by the fact that Monnier et al. also preferentially identified by XAS studies ferrihydrite rather than feroxyhyte in corrosion products formed on phosphoric iron [41].

The presence of amorphous or low-crystallised phases (such as ferrihydrite) is often reported in publications concerning the long-term atmospheric corrosion of iron and weathering steel [10,11,42]. Nevertheless, the present observation of massive quantities of ferrihydrite on 500 year-old samples is not in agreement with the hypothesis of several authors [39,43], who proposed that ferrihydrite that is one of the major phases in the first stage of the atmospheric corrosion process, progressively transforms into more crystallised phases as goethite because of reduction/dissolution/precipitation mechanisms (see below) during the wet/dry cycles. This is clearly not the case here. This confirms that the age of the sample is not the predominant parameter that influences the proportion of phases in the presence of long-term atmospheric CPL. A hypothesis to explain the major presence of this phase after several hundred years could be the fact that it contains certain amounts of phosphorus, an element that is well-known to reduce the reactivity of the corrosion products [31,44–48]. This point is detailed below.

Contrary to former studies on long-term atmospheric corrosion, ferrihydrite is not present in the CPL under the form of marblings surrounded by goethite (as for example in samples from the Amiens and Bourges cathedrals). Here, it is present under the form of a homogeneous layer mixed at a microscopic scale with other minor phases. As said before, several thousand ppm of P were locally detected in the inner layer. Some authors have proposed that this element could be adsorbed at the surface of ferrihydrite [33], lowering the electrochemical reactivity. Nevertheless, ferrihydrite is reported in the literature as one of the most electrochemically reactive substances during the wet/dry cycle of atmospheric corrosion [18]. As explained in the introduction part, oxygen does not reach the metal/oxide interface during the wetting stage of the RH cycle. During this stage, iron is susceptible to oxydize by reducing ferrihydrite, provided that this reactive phase is electrically connected to the metallic substrate [21]. In this case, the reduction equation would be the following:

$$Fe_2O_3,1.8H_2O_{(CPL)} + H^+ + e^- \Leftrightarrow Fe(OH)_2 + 1.8H_2O \qquad (1)$$

The disappearance of ferrihydrite during the in-situ test can be explained by its transformation either into a non-detected phase or into a soluble phase. Then, the relative contribution of the remaining phase (goethite) to the Raman spectra would increase. Ferrihydrite has a very low solubility (K$_s$ = 10$^{-39}$ at 25 °C) [48,49]. Consequently, the quantity of dissolved Fe$^{3+}$ ions should be very low (<10$^{-6}$ mol/L for 2 < pH < 12). In reducing media, such as in the present case, the solubility may increase by several orders of magnitude [50]. This is mainly due to the reduction of the structural Fe(III)-species into Fe(II)-species which are much more soluble than the Fe(III) ones. The reductive dissolution of ferrihydrite could thus be envisaged. In that case, the fact that the disappearance of this phase is observed by µRS, mainly in zones neighbouring the metal, strongly suggests that the main reduction process is due to the oxidation of the metal and not to the one of Fe species dissolved in water. In that latter case, the disappearance of the ferrihydrite should have been detected in the entire layer and not only at the metal/CPL interface. It has been noted by several authors that ferrihydrite especially shows a high elechtrochemical reactivity [21,51]. It was observed that at pH = 7.5 and under a potential of −1 V/SCE, a synthetic ferrihydrite is reduced into a mix of 80% magnetite ($\alpha$-$Fe_2O_3$) and 20% $Fe(OH)_2$ [21]. At the same pH but under an imposed current of −100 µA/mg, the same phase was mainly reduced under the form of Fe(II)-hydroxide ($Fe(OH)_2$). This latter phase, less crystallised, was identified only by X-Ray absorption techniques under Synchrotron Radiation. With the µRS used in the present study, $Fe(OH)_2$ is hardly detectable. Its spectra present large and weak Raman bands. Thus, it must be present in very high quantities to be detected in a mix of phases, which seems not to be the case here. On the contrary, magnetite produces intense and very characteristic Raman bands. If present, even in low quantities, this phase should have been detected. Lastly, the hypothesis of the direct re-precipitation of goethite from species formed by ferrihydrite dissolution is not likely because of the reducing conditions.

## 5. Conclusions

A corrosion pattern formed under the outdoor atmospheric corrosion of reinforcing clamp on mediaeval monuments for several hundred years is reported. The bi-layered corrosion product layer contains exogenous minor elements (Cl, Ca, Si, S, and P) in the outer layer and about 0.5% of endogenous P in its inner layer. The latter is mainly constituted of Ferrihydrite (77%) and Goethite (20%). The outer layer, in addition to these two phases (respectively, 44 and 21%), contains higher quantities of lepidocrocite (35%). Compared to other corrosion patterns of long-term atmospheric corrosion, the quantity of ferrihydrite is particularly high.

A simulation of the wetting stage of the wet/dry cycles of atmospheric corrosion coupled to in situ µRaman observation allowed us to observe the disappearance of ferrihydrite, particularly at the metal/CPL interface. No neo-precipitated compounds could be evidenced. The disappearance of ferrihydrite along the metal/CPL interface is caused by a reduction of this phase (before or after dissolution), either under the form of dissolved $Fe^{2+}$ ions or under the form of a phase which is non detectable by µRS. In the first case, the dissolved $Fe^{2+}$ ions may migrate in the pore water and be re-oxidised by the supply of solvated dioxygen during the following stages of the wet/dry cycle. In the case of the reduction of ferrihydrite under the form of a solid phase, as proposed in the literature, the "reduced ferrihydrite" could be partially conductive [18,23]. Provided that ferrihydrite is significantly reduced, the conductivity of the reduced species could lead to a decoupling of the anodic (Fe(0) oxidation) and cathodic corrosion (dissolved dioxygen reduction) reactions in the next step of the wet/dry cycle.

The fact that the disappearance of ferrihydrite is only detected after one day suggests that the process is quite slow. This brings us to raise the question of the influence of the duration and the nature of the wet/dry cycle on this process. In the case of a short wetting stage compared to the characteristic reduction time, the quantity of reduced ferrihydrite could be negligible and consequently have only a small impact on the mechanism. Nevertheless, one should keep in mind that µRS—especially in the conditions used here (through a polymer film in an in situ cell)—could not be sensitive to small variations of phase contents. Thus, it is difficult to evaluate the intensity of the reduction process and its predominance in the case of short wetting stages. No further evolution of the system is observed after nine days of the experiment. It seems that the electrical connection with the metallic substrate is not preserved after a certain time, forbidding the reduction of Fe(III)-phases farther in the CPL. This strongly questions the hypothesis of the formation of a reduced conductive phase, allowing the decoupling of anodic and cathodic reactions. This observation more probably suggests the reductive dissolution of ferrihydrite under the form of $Fe^{2+}$ ions. With the supply of solvated dioxygen at the end of the in situ experiment, the ferrihydrite content at the interface increases again, probably corresponding to the precipitation of this phase after the re-oxidation of $Fe^{2+}$ ions dissolved in the pore solution. It is interesting to note that this new Fe(III)-phase is mainly ferrihydrite and not goethite.

This in-situ approach, by μRaman analyses, allowed us for the first time to observe the behaviour of a thick corrosion layer during the first stages of atmospheric corrosion. It would be very interesting in further studies to confront these observations with electrochemical approaches that would allow one to assess quantitative aspects of the reduction process and to refine the mechanisms.

**Supplementary Materials:** The following are available online at www.mdpi.com/1996-1944/10/6/670/s1, Figure S1: metallographic study of the clamp.

**Acknowledgments:** We would like to thank Jean-Paul Gallien for the design of the cell.

**Author Contributions:** Marie Bouchar, conceived the experiments and implemented the characterization methodology, Delphine Neff supervised the characterization experiments, Philippe Dillmann supervised the project and coordinated the writing of the paper. The three authors were involved at equal part in the discussion of the results.

**Conflicts of Interest:** The authors declare no conflict of interest.

## References

1. Disser, A.; Dillmann, P.; Leroy, M.; L'Héritier, M.; Bauvais, S.; Fluzin, P. Iron supply for the building of Metz cathedral: New methodological development for provenance studies and historical considerations. *Archaeometry* **2016**, *59*, 493–510. [CrossRef]
2. Leroy, S.; L'Héritier, M.; Delqué-Kolic, E.; Dumoulin, J.-P.; Moreau, C.; Dillmann, P. Consolidation or initial design? Radiocarbon dating of ancient iron alloys sheds light on the reinforcements of French Gothic Cathedrals. *J. Archaeol. Sci.* **2015**, *53*, 190–201. [CrossRef]
3. Disser, A.; Dillmann, P.; Bourgain, C.; L'Héritier, M.; Vega, E.; Bauvais, S.; Leroy, M. Iron reinforcements in Beauvais and Metz Cathedrals: From bloomery or finery? The use of logistic regression for differentiating smelting processes. *J. Archaeol. Sci.* **2014**, *42*, 315–333. [CrossRef]
4. L'Héritier, M.; Dillmann, P.; Aumard, S.; Fluzin, P. Iron? Wich iron? Methodologies for metallographic and slag inclusion studies applied to ferrous reinforcements from Auxerre Cathedral. In *The World of Iron*; Humphris, J., Rehren, T., Eds.; Archetype Publications: London, UK, 2013; pp. 409–420.
5. Leygraf, C.; Graedel, T.E. *Atmospheric Corrosion*; John Wiley & Sons: New York, NY, USA, 2000; p. 354.
6. Dillmann, P.; Neff, D.; Féron, D. Archaeological analogues and corrosion prediction: From past to future. A review. *Corros. Eng. Sci. Technol.* **2014**, *49*, 567–576. [CrossRef]
7. Feron, D.; Crusset, D.; Gras, J.M. Corrosion Issues in the French High-Level Nuclear Waste Program (Reprinted from Proceedings of the CORROSION/2008 Research Topical Symposium). *Corrosion* **2009**, *65*, 213–223. [CrossRef]
8. Dillmann, P.; Mazaudier, F.; Hoerle, S. Advances in understanding atmospheric corrosion of iron I—Rust characterisation of ancient ferrous artefacts exposed to indoor atmospheric corrosion. *Corros. Sci.* **2004**, *46*, 1401–1429. [CrossRef]
9. Monnier, J.; Bellot-Gurlet, L.; Baron, D.; Neff, D.; Guillot, I.; Dillmann, P. A methodology for Raman structural quantification imaging and its application to iron indoor atmospheric corrosion products. *J. Raman Spectrosc.* **2011**, *42*, 773–781. [CrossRef]
10. Monnier, J.; Neff, D.; Reguer, S.; Dillmann, P.; Bellot-Gurlet, L.; Leroy, E.; Foy, E.; Legrand, L.; Guillot, I. A corrosion study of the ferrous medieval reinforcement of the Amiens cathedral. Phase characterisation and localisation by various microprobes techniques. *Corros. Sci.* **2010**, *52*, 695–710. [CrossRef]
11. Bouchar, M.; Foy, E.; Neff, D.; Dillmann, P. The complex corrosion system of a medieval iron rebar from the Bourges' Cathedral. Characterization and reactivity studies. *Corros. Sci.* **2013**, *76*, 361–372. [CrossRef]

12. Suzuki, I.; Masuko, N.; Hisamatsu, Y. Electrochemical properties of iron rust. *Corros. Sci.* **1979**, *19*, 521–535. [CrossRef]
13. Evans, U.R. Electrochemical Mechanism of Atmospheric Rusting. *Nature* **1965**, *206*, 980–982. [CrossRef]
14. Matsushima, I.; Ueno, T. On the protective nature of atmospheric rust on low-alloy steel. *Corros. Sci.* **1971**, *11*, 129–140. [CrossRef]
15. Stratmann, M. The atmospheric corrosion of iron—A discussion of the physico-chemical fundamentals of this omnipresent corrosion process. Invited Review. *Ber. Bunsenges. Phys. Chem.* **1990**, *94*, 626–639. [CrossRef]
16. Strattmann, M.; Streckel, H. On the atmospheric corrosion of metals which are covered with thin electrolyte layers-I. verification of the experimental technique. *Corros. Sci.* **1990**, *30*, 681–696. [CrossRef]
17. Strattmann, M.; Streckel, H.; Kim, K.T.; Crockett, S. On the atmospheric corrosion of metals which are covered with thin electrolyte layers-III. The measurment of polaristaion curves on metal surfaces which are covered by thin electrolyte layers. *Corros. Sci.* **1990**, *30*, 715–734. [CrossRef]
18. Lair, V.; Antony, H.; Legrand, L.; Chausse, A. Electrochemical reduction of ferric corrosion products and evaluation of galvanic coupling with iron. *Corros. Sci.* **2006**, *48*, 2050–2063. [CrossRef]
19. Antony, H.; Legrand, L.; Peulon, S.; Lair, V.; Chaussé, A. *Applications of Thin Layers of Iron Oxidation Products for Corrosion Studies*; Eurocorr: Nice, France, 2004; CD ROM.
20. Antony, H.; Legrand, L.; Maréchal, L.; Perrin, S.; Dillmann, P.; Chaussé, A. Study of lepidocrocite electrochemical reduction in neutral and slightly alkaline solutions at 25 °C. *Electrochim. Acta* **2005**, *51*, 745–753. [CrossRef]
21. Monnier, J.; Reguer, S.; Foy, E.; Testemale, D.; Mirambet, F.; Saheb, M.; Dillmann, P.; Guillot, I. XAS and XRD in situ characterisation of reduction and reoxidation processes of iron corrosion products involved in atmospheric corrosion. *Corros. Sci.* **2014**, *78*, 293–303. [CrossRef]
22. Stratmann, M.; Bohnenkamp, K.; Engell, H.-J. An electrochemical study of phase-transitions in rust layers. *Corros. Sci.* **1983**, *23*, 969–985. [CrossRef]
23. Stratmann, M.; Hoffmann, K. In Situ Mössbauer spectroscopic study of reactions within rust layers. *Corros. Sci.* **1989**, *29*, 1329–1352. [CrossRef]
24. Neff, D.; Bellot-Gurlet, L.; Dillmann, P.; Reguer, S.; Legrand, L. Raman imaging of ancient rust scales on archaeological iron artefacts for long-term atmospheric corrosion mechanisms study. *J. Raman Spectrosc.* **2006**, *37*, 1228–1237. [CrossRef]
25. Burger, E.; Monnier, J.; Berger, P.; Neff, D.; L'Hostis, V.; Perrin, S.; Dillmann, P. The long-term corrosion of mild steel in depassivated concrete: Localizing the oxygen reduction sites in corrosion products by isotopic tracer method. *J. Mater. Res.* **2011**, *26*, 3107–3115. [CrossRef]
26. Monnier, J.; Burger, E.; Berger, P.; Neff, D.; Guillot, I.; Dillmann, P. Localisation of oxygen reduction sites in the case of iron long term atmospheric corrosion. *Corros. Sci.* **2011**, *53*, 2468–2473. [CrossRef]
27. Almeida, E.; Morcillo, M.; Rosales, B.; Marrocos, M. Atmospheric corrosion of mild steel. Part I—Rural and urban atmospheres. *Mater. Corros.* **2000**, *51*, 859–864. [CrossRef]
28. Graedel, T.E.; Frankenthal, R.P. Corrosion Mechanism for Iron and Low Alloy Steels Exposed to the Atmosphere. *J. Electrochem. Soc.* **1990**, *137*, 2385–2394. [CrossRef]
29. Tidblad, J. Atmospheric corrosion of heritage metallic artefacts: Processes and prevention. In *Corrosion and Conservation of Cultural Heritage Metallic Artefacts*; Woodhead Publishing: Oxford, UK, 2013; pp. 37–52.
30. Shahack-Gross, R.; Berna, F.; Karkanas, P.; Weiner, S. Bat guano and preservation of archaeological remains in cave sites. *J. Archaeol. Sci.* **2004**, *31*, 1259–1272. [CrossRef]
31. Schwertmann, U.; Cornell, R.M. *Iron Oxides in the Laboratory*; Wiley-VCH: Weinheim, Germany, 2000; p. 137.
32. Neff, D.; Reguer, S.; Bellot-Gurlet, L.; Dillmann, P.; Bertholon, R. Structural characterization of corrosion products on archaeological iron. An integrated analytical approach to establish corrosion forms. *J. Raman Spectrosc.* **2004**, *35*, 739–745. [CrossRef]
33. Monnier, J.; Réguer, S.; Vantelon, D.; Dillmann, P.; Neff, D.; Guillot, I. X-rays absorption study on medieval corrosion layers for the understanding of very long-term indoor atmospheric iron corrosion. *Appl. Phys. A Mater. Sci. Process.* **2010**, *99*, 399–406. [CrossRef]
34. Balasubramaniam, R.; Kumar, A.V.R.; Dillmann, P. Characterization of rust on ancient indian iron. *Curr. Sci.* **2003**, *85*, 101–110.

35. Lutz, H.D.; Möller, H.; Schmidt, M. Lattice vibration spectra. Part LXXXII. Brucite-type hydroxides M(OH)$_2$ (M = Ca, Mn, Co, Fe, Cd)—IR and Raman spectra, neutron diffraction of Fe(OH)$_2$. *J. Mol. Struct.* **1994**, *328*, 121–132. [CrossRef]

36. Provent, E. *Etude de la Corrosion Atmosphérique et dans la Pierre des Armatures Métalliques du 18ème siècle de la Cathédrale d'Orléans*; LAPA/CEA; CEA Saclay: Gif Sur Yvette, France, 2010.

37. Keiser, J.T.; Brown, C.W.; Heidersbach, R.H. Characterization of the passive film formed on weathering steels. *Corros. Sci.* **1983**, *23*, 251–259. [CrossRef]

38. Kashima, K.; Hara, S.; Kishikawa, H.; Miyuki, H. Evaluation of protective ability of rust layers on weathering steels by potential measurment. *Corros. Eng.* **2000**, *49*, 25–37. [CrossRef]

39. Burger, E.; Fénart, M.; Perrin, S.; Neff, D.; Dillmann, P. Use of the gold markers method to predict the mechanisms of iron atmospheric corrosion. *Corros. Sci.* **2011**, *53*, 2122–2130. [CrossRef]

40. Seinfeld, J.H.; Pandis, S.N.; Noone, K. *Atmospheric Chemistry and Physics: From Air Pollution to Climate Change*; AIP: University Park, MD, USA, 1998.

41. Monnier, J.; Vantelon, D.; Reguer, S.; Dillmann, P. X-ray absorption spectroscopy study of the various forms of phosphorus in ancient iron samples. *J. Anal. At. Spectrom.* **2011**, *26*, 885–891. [CrossRef]

42. Morcillo, M.; Chico, B.; Díaz, I.; Cano, H.; de la Fuente, D. Atmospheric corrosion data of weathering steels. A review. *Corros. Sci.* **2013**, *77*, 6–24. [CrossRef]

43. Yamashita, M.; Miyuki, H.; Matsuda, Y.; Nagano, H.; Misawa, T. The long term growth of the protective rust layer formed on weathering steel by atmospheric corrosion during a quarter of a century. *Corros. Sci.* **1994**, *36*, 283–299. [CrossRef]

44. Misawa, T.; Hashimoto, K.; Shimodaira, S. The mecanism of formation of iron oxide and oxyhydroxides in aquaeous solutions at room temperature. *Corros. Sci.* **1974**, *14*, 131–149. [CrossRef]

45. Misawa, T.; Kyuno, T.; Suetaka, W.; Shimodai, S. Mechanism of atmospheric rusting and effect of Cu and P on rust formation of low alloyed steels. *Corros. Sci.* **1971**, *11*, 35–48. [CrossRef]

46. Arai, Y.; Sparks, D.L. ATR–FTIR Spectroscopic Investigation on Phosphate Adsorption Mechanisms at the Ferrihydrite–Water Interface. *J. Colloid Interface Sci.* **2001**, *241*, 317–326. [CrossRef]

47. Torrent, J.; Barron, V.; Schwertmann, U. Phosphate adsorption and desorption by goethites differing in crystal morphology. *Soil Sci. Soc. Am. J.* **1990**, *54*, 1007–1012. [CrossRef]

48. Cornell, R.; Schwertmann, U. *The Iron Oxides—Structure, Properties, Occurrences and Uses*, 2nd ed.; Wiley-VCH Verlag: Weinheim, Germany, 2003; p. 664.

49. Fox, L.E. The solubility of colloidal ferric hydroxide and its relevance to iron concentrations in river water. *Geochim. Cosmochim. Acta* **1988**, *52*, 771–777. [CrossRef]

50. Postma, D. The reactivity of iron oxides in sediments: A kinetic approach. *Geochim. Cosmochim. Acta* **1993**, *57*, 5027–5034. [CrossRef]

51. Antony, H.; Perrin, S.; Dillmann, P.; Legrand, L.; Chaussé, A. Electrochemical study of indoor atmospheric corrosion layers formed on ancient iron artefacts. *Electrochim. Acta* **2007**, *52*, 7754–7759. [CrossRef]

![materials logo] *materials*

MDPI

*Article*

# Analysis of Historic Copper Patinas. Influence of Inclusions on Patina Uniformity

Tingru Chang [1], Inger Odnevall Wallinder [1], Daniel de la Fuente [2], Belen Chico [2], Manuel Morcillo [2], Jean-Marie Welter [3] and Christofer Leygraf [1],*

[1] KTH Royal Institute of Technology, Div. Surface and Corrosion Science, School of Chemical Science and Engineering, SE 10044 Stockholm, Sweden; tingru@kth.se (T.C.); ingero@kth.se (I.O.W.)
[2] National Centre for Metallurgical Research (CENIM-CSIC), 28040 Madrid, Spain; delafuente@cenim.csic.es (D.d.l.F.); bchico@cenim.csic.es (B.C.); morcillo@cenim.csic.es (M.M.)
[3] Independent scholar, Luxembourg-1361, Luxembourg; jean-marie.welter@pt.lu
* Correspondence: chrisl@kth.se; Tel.: +46-8790-6468

Academic Editor: Yong-Cheng Lin
Received: 21 February 2017; Accepted: 13 March 2017; Published: 16 March 2017

**Abstract:** The morphology and elemental composition of cross sections of eight historic copper materials have been explored. The materials were taken from copper roofs installed in different middle and northern European environments from the 16th to the 19th century. All copper substrates contain inclusions of varying size, number and composition, reflecting different copper ores and production methods. The largest inclusions have a size of up to 40 $\mu$m, with most inclusions in the size ranging between 2 and 10 $\mu$m. The most common element in the inclusions is O, followed by Pb, Sb and As. Minor elements include Ni, Sn and Fe. All historic patinas exhibit quite fragmentized bilayer structures, with a thin inner layer of cuprite ($Cu_2O$) and a thicker outer one consisting mainly of brochantite ($Cu_4SO_4(OH)_6$). The extent of patina fragmentation seems to depend on the size of the inclusions, rather than on their number and elemental composition. The larger inclusions are electrochemically nobler than the surrounding copper matrix. This creates micro-galvanic effects resulting both in a profound influence on the homogeneity and morphology of historic copper patinas and in a significantly increased ratio of the thicknesses of the brochantite and cuprite layers. The results suggest that copper patinas formed during different centuries exhibit variations in uniformity and corrosion protection ability.

**Keywords:** atmospheric corrosion; historic copper; patina; bilayer; cuprite; brochantite; antlerite; inclusions; Volta potential; micro-galvanic effect

---

## 1. Introduction

The production and use of copper and copper alloys is many thousands of years old and has been based on several different production technologies over the millennia [1]. In early times, native copper and copper extracted from oxide ores was obtained through a reduction with charcoal. These ores were rapidly exhausted in Europe, and ancient metallurgists had to learn to obtain the copper from a large variety of copper sulfide ores. They range from rather pure chalcopyrite ($CuFeS_2$) to the *Fahlerz* ore with high antimony and arsenic contents. Well-known *Fahlerz* deposits were exploited until the 19th century in Schwaz (Austrian Tirol) and Neusohl (Slovakia), which belonged to the Austro-Hungarian Empire. Often the copper ore was a minority component in an ore body. In the mine of Falun (Sweden), for instance, lead and zinc sulfides were the main constituents. After this, oxygen became the agent that helped to reduce sulfur and other major and minor elements contained in the ore. This explains why old copper samples contain larger amounts of oxygen (ranging from 400 to 4000 ppm by weight) and other metallic impurities. The most important one is lead, with contents of up to 1% by weight, followed by arsenic and antimony, often up to 0.5% by weight. Further minor elements included nickel, silver, tin, zinc, iron, and bismuth. It is only at the end of the 19th century that progress in electro-refining resulted in a total impurity content well below 1000 ppm by weight. A few decades later, the emergence of acetylene torches led to an almost complete desoxidation of copper intended to be joined by brazing and welding. Nowadays, phosphor is the preferred desoxidant and some 200 to 300 ppm by weight is left in copper used to fabricate e.g., roofing sheets and water tubes. This brand is named *Desoxidized high phosphor* copper (DHP-Cu). During its annealing—even at reduced oxygen partial pressures—some oxygen diffuses into the bulk and forms with phosphor a thin sub-surface layer of copper meta-phosphate, which will modify the corrosion behavior of the copper [2]. Sheets of copper were manufactured until the end of the 18th century by hammering, after which rolling was made possible by the development of large and strong mills [3]. A great deal of our knowledge of early production methods originates from sophisticated characterizations of copper and copper alloy artifacts [4–6]. It also originates from detailed analysis of their inclusions [7,8], which to some extent can reveal the nature of the copper ores from which the copper was extracted. Combined with knowledge about the metallurgy of copper or copper alloys, such studies have contributed to a better understanding of historic metallurgical and foundry processes and also of the ores and, hence, the minerals from which the objects emanate.

This study of selected historic copper materials aims to explore how such objects can reveal information about their history, by analysis not only of the copper substrate, but also of the corrosion product layers that have grown on the surface over extended exposure times. These layers are commonly designated as (natural) patina, although the term also can refer to man-made, chemically induced modifications of the surface appearance. Thus, can an in-depth analysis reveal anything about the environment in which the layers were formed? Chemical characteristics of patina formed on copper alloys as an environmental indicator has been reported before: for instance, the detailed analysis of sulfur isotopes in the sulfur-containing corrosion products suggested that the element could be traced to sulfur contained in air-polluting species, despite large geographic variations in the analyzed samples [9]. An important study to be mentioned in this context is the extensive patina study that was performed as part of the restoration of the Statue of Liberty [10,11]. Two families of inclusions were identified, but until now they have not been related to the structure of the corrosion product layers [12].

The present study includes copper materials that originate from roofs on different historical buildings, eight historic surfaces and one modern commercially available sheet for comparison. The focus of this paper is on the inclusions embedded in the copper matrix. They turn out to vary substantially between the different materials with respect to size, frequency and chemical composition. The underlying question to be explored herein concerns if and how the characteristics of these inclusions may influence the composition and homogeneity of the copper patina formed.

## 2. Materials and Methods

### 2.1. Materials and Sample Preparation

The examined historic copper materials have been exposed for roofing on different historic buildings in variety of environments and exposure times, the oldest at the Royal Summer Palace (Belvedere) in Prague, Czech Republic (completed between 1538 and 1560) and the youngest at a church in the Old Town of Stockholm, Sweden (exposure period around 100 years, starting in the 19th century). Other exposure sites include Drottningholm Castle outside Stockholm, Sweden, the Mausoleum in Graz, Austria, Basilica Maria Dreieichen in Lower Austria, Helsinki Cathedral, Finland, Otto Wagner Church in Vienna, Austria, and Kronborgs Castle, Elsinore, Denmark. Information on exposure site and estimated length of exposure period is given in Table 1.

**Table 1.** Exposure site, location (city, country) and approximate exposure period of investigated copper surfaces.

| Exposure Site | City (Country) | Exposure Period (Years) |
|---|---|---|
| Royal Summer Palace (Belvedere) | Prague (Czech Republic) | ~425 |
| Mausoleum | Graz (Austria) | ~390 |
| Drottningholm Castle | Stockholm (Sweden) | ~300 |
| Basilica | Maria Dreieichen (Austria) | ~160 |
| Helsinki Cathedral | Helsinki (Finland) | ~150 |
| Kronborgs Castle | Elsinore (Denmark) | ~120 |
| Otto Wagner Church | Vienna (Austria) | ~110 |
| Church in Old Town | Stockholm (Sweden) | ~100 |

Cross-sections of the historic copper patinas were prepared by embedding the materials into a conductive polymer followed by polishing using 0.25 μm diamond paste to obtain a near mirror-like cross-sectional surface. Prior to the investigations, all samples were ultrasonically cleaned in analytical-grade ethanol for 10 min and subsequently dried by cold nitrogen gas.

## 2.2. SEM/EDS (Scanning Electron Microscopy and Energy Dispersive Spectroscopy)

SEM/EDS-analysis (scanning electron microscopy and energy dispersive spectroscopy) was conducted to obtain morphology and elemental information. Cross-sections were analyzed using a LEO 1530 instrument (Zeiss, Oberkochen, Germany) with a Gemini column, upgraded to a Zeiss Supra 55 (Zeiss) and an EDS X-Max SDD (Silicon Drift Detector) 50 mm × 50 mm detector from Oxford Instruments (Oxford, UK). All analyses were performed by means of a FEI-XL 30 Series instrument (Oxford, UK), equipped with an EDS system (EDAX Phoenix) with an ultra-thin windows Si-Li detector. All images were obtained using backscattered electrons at an accelerating voltage of 15 kV. The data of EDS elemental composition was collected by Oxford Instruments INCA 5.04.

## 2.3. AFM/SKPFM (Atomic Force Microscopy and Scanning Kelvin Probe Force Microscopy)

AFM-analysis (Nanoscope Icon AFM from Bruker, Karlsruhe, Germany) was employed to image the surface topography and to map Volta potential variations along the cross-section of selected historic copper substrates. The Volta potential mapping was carried out using the Kelvin probe technique [13,14]. For this purpose, a SCM-PIT probe from Bruker and a HQ:NSC18/PT probe from MicroMash (Wetzlar, Germany) were used. The data were processed with Gwyddion (Czech Metrology Institute, Brno, Czech Republic), which is a free modular software program for SPM data visualization and analysis [15].

## 2.4. EIS (Electrochemical Impedance Spectroscopy)

EIS measurements were performed at the open circuit potential (OCP) to estimate the polarization resistance of one historic and one modern, commercially available, copper sheet (DHP-Cu, >99.9 wt % Cu) by using a Multi Autolab (Metrohm Autolab B.V., Utrecht, The Netherlands) instrument. 0.1 M $Na_2SO_4$ was used as the electrolyte with an Ag/AgCl reference electrode and platinum mesh as counter electrode. The applied perturbation amplitude was 10 mV with the measured frequency range of $10^4$–$10^{-2}$ Hz with 75 measuring points. To perform the experiments at oxygen-free conditions, the solution was purged with $N_2$ gas for 30 min prior to the experiment. The EIS data was analyzed by using the Nova 1.8 software (Metrohm Autolab B.V., Utrecht, The Netherlands).

## 3. Results

Figure 1 displays SEM images at lower (a) and higher (b) magnification of the copper sample collected from the Royal Summer Palace in Prague, the oldest of the historic samples investigated. It is evident that the substrate contains numerous inclusions, seen as white spots (lower part in light grey,

Figure 1a,b at higher magnification). Most of them exhibit an elongated shape in the hammering or rolling direction of the sheet, which was also observed in historic copper materials from the Middle Ages [16]. The inclusions are also seen within the patina (upper part in dark grey, Figure 1a) and suggest that the inclusions, at least partially, remain intact during the atmospheric corrosion process when the outermost surface of the copper substrate is oxidized and transformed to copper patina. Similar inclusion distributions were previously observed in historical French roofs [17].

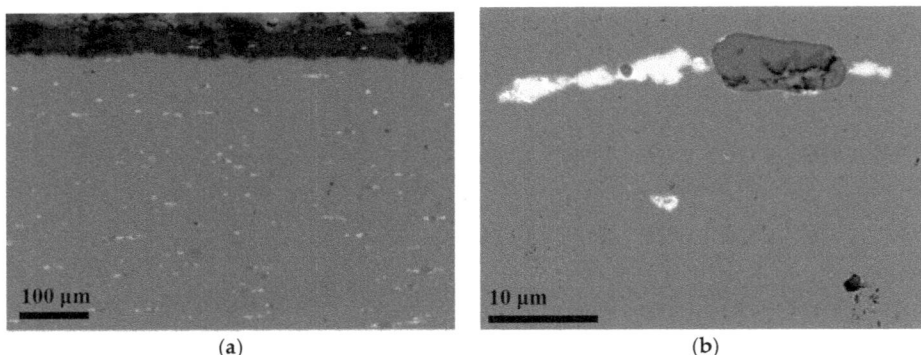

(a)                                                            (b)

**Figure 1.** SEM images of an overview cross-section of the historic copper material from the Royal Summer Palace (Belvedere) in Prague showing the copper substrate (lower part in light grey) and the patina (upper part in dark grey) (**a**); Different inclusions in the substrate at higher magnification (**b**).

Figure 2 is another example of a copper sample (Drottningholm Castle outside Stockholm) with inclusions embedded both in the substrate (lower part in figure, light grey) and in the patina (upper part, darker grey). As in all other samples, the patina consists of two irregularly shaped layers, one closer to the substrate and another, slightly darker, on the outside. It can be seen that the inner layer consists of cuprite ($Cu_2O$) and the outer mainly of brochantite ($Cu_4SO_4(OH)_6$), sometimes partially also of antlerite ($Cu_3SO_4(OH)_4$).

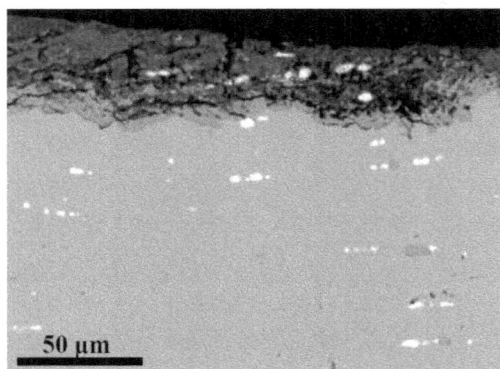

**Figure 2.** Overview SEM image of a cross-section of the historic copper material from the Drottningholm Castle outside Stockholm showing the copper substrate (lower part, light grey) and the patina (upper part, darker grey). The thin cuprite layer next to the substrate is slightly lighter grey than the outer brochantite layer in darker grey.

As seen in Figure 2, the cuprite layer thickness (inner layer) varies considerably along the copper substrate. Moreover, the thickness is generally thinner where the brochantite/antlerite layer contains inclusions (the central part of Figure 2) and thicker where the same layer lacks any evident presence of inclusions. The same tendency is also seen in other historic materials in this study, and suggests that the presence of inclusions may influence the homogeneity of the patina. To explore if the nature of the inclusions can have an impact on the patina formation, an investigation based on SEM/EDS was made of the density, size range and composition of the inclusions observed in all the historic materials. For each copper sample, at least three areas, sized 300 μm × 300 μm, were selected for detailed investigations. The results have been summarized in Table 2.

**Table 2.** Density, size range in different directions, and elemental composition of selected inclusions in substrates of all investigated historic copper materials. bdl: below detection limit, around 0.5 atomic % or lower.

| Exposure Site | Number of Inclusions/ 300 × 300 μm$^2$ | Size Range/μm | Elemental Composition/Atomic % | | | | | | |
| --- | --- | --- | --- | --- | --- | --- | --- | --- | --- |
| | | | O | Cu | Pb | As | Sb | Ni | Sn |
| Royal Summer Palace, Prague | 48 ± 9 | 2.1–38.2 | 66.8 | 4.0 | 10.7 | bdl | 18.5 | bdl | bdl |
| Mausoleum, Graz | 39 ± 11 | 2.5–14.8 | 52.9 | 26.4 | bdl | bdl | 7.3 | 13.4 | bdl |
| Drottningholm Castle, Stockholm | 73 ± 12 | 2.7–8.9 | 63.5 | 3.0 | 25.2 | 5.1 | 3.2 | bdn | bdn |
| Basilica, Maria Dreieichen | 62 ± 20 | 0.3–3.1 | 46.8 | 26.1 | 23.1 | 4.0 | bdl | bdl | bdl |
| Helsinki Cathedral | 18 ± 3 | 6.2–25.4 | 59.6 | 17.1 | 14.2 | 3.9 | 5.2 | bdl | bdl |
| Kronborgs Castle, Elsinore | 39 ± 11 | 1.5–4.8 | 34.8 | 65.2 | bdl | bdl | bdl | bdl | bdl |
| Otto Wagner Church, Vienna | 57 ± 18 | 0.5–8.6 | 67.7 | 1.5 | 14.6 | 0.7 | 6.3 | bdl | 9.2 |
| Church in Old Town, Stockholm | 35 ± 5 | 3.1–15.5 | 65.7 | 1.4 | 23.9 | 7.8 | 1.2 | bdl | bdl |

From Table 2 it is evident that the inclusion characteristics vary considerably between the samples. For the density and size range the intervals shown reflect variations between different investigated areas in the substrate. The density is highest for Drottningholm Castle and Basilica Maria Dreieichen, and lowest for Helsinki Cathedral. The largest inclusion sizes, on the other hand, are seen for the Royal Summer Palace and Helsinki Cathedral, and the smallest for Basilica Maria Dreieichen and Kronborgs Castle.

Regarding the elemental composition, the information in Table 2 is given for a representative selected single larger inclusion rather than for a mixture of inclusions. The most common element in the inclusions is oxygen, suggesting that most inclusions are oxides of different mixtures of metals. The elemental ratio between copper and oxygen in the Kronborgs Castle sample suggests that cuprite ($Cu_2O$) is the most probable compound in the inclusion analyzed. Similarly, the results from the Royal Summer Palace sample suggest that the inclusion most likely contains rosiaite ($PbSb_2O_6$), sometimes with smaller cuprite inclusions adjacent to rosiaite inclusions. Separate XRD measurements of inclusions in the patina formed on the copper roof of the Royal Summer Palace confirm that the inclusions, indeed, consist of cuprite and rosiaite [18]. The marked variation in elemental composition of the inclusions analyzed in the different copper samples suggests that several other single or mixed oxides of primarily lead, antimony, copper and tin may be present in the inclusions. A third type of inclusion, different from the others and present in the Graz Mausoleum sample, is a mixture of oxides containing copper, nickel and antimony. Due to their small size, however, it is hard to reveal their exact identity by techniques such as X-ray diffraction, which is able to identify crystalline phases.

The origin of the precipitates is clear. Oxygen is soluble to some extent in liquid copper, but not in the solid phase. When the melt solidifies, oxygen precipitates by forming an oxide. In pure copper—such as modern *Electrolytic tough pitch* copper (ETP copper), containing some 200 to 300 ppm by weight of oxygen—only cuprite is formed. In old copper, the impurities present in large amounts compete with copper to form more or less complex oxides depending on their respective chemical potentials. A special element is lead: like oxygen, it is not soluble in solid copper and precipitates either as a metallic nodule or as a (multi-element) oxide.

We consider next the thickness of the patina formed, in particular the thickness of the cuprite and brochantite/antlerite sub-layers. Figure 3 exhibits the cross-section of the patina and adjacent substrate (a) and the corresponding elemental distribution of oxygen and sulfur relative to the sum of (copper + oxygen + sulfur) (in atomic %) based on EDS analysis (b) of the copper patina formed at the Royal Summer Palace in Prague. It is evident that the relative sulfur- and oxygen-distributions show two levels. An inner sub-layer close to the substrate is characterized by an insignificant sulfur-level and a lower oxygen-level, while an outer sub-layer is characterized by both higher sulfur and oxygen levels relative to the inner sub-layer. There is substantial evidence that this inner layer consists of cuprite and the outer layer of brochantite, antlerite or both, where their formation rate strongly depends on prevailing environmental and pollutant conditions [19,20]. Detailed studies of the evolution of copper patina in different exposure sites have provided great evidence that the inner cuprite layer forms instantaneously when copper is exposed to air, and then continues to grow during extended exposure time. In more sulfur-rich atmospheres, the growth of the patina results in antlerite ($Cu_3SO_4(OH)_4$) as the end product, with strandbergite ($Cu_{2.5}SO_4(OH)_3 \cdot 2H_2O$) as a precursor. In less sulfur-rich atmospheres the end product of patina growth is brochantite ($Cu_4SO_4(OH)_6$), with posnjakite ($Cu_4SO_4(OH)_6 \cdot H_2O$) as a precursor [20]. In order to distinguish the end products brochantite and antlerite from each other it is necessary to use X-ray diffraction. In the current investigation, which is based on SEM/EDS alone, we denote the outer layer as brochantite/antlerite with no further distinction between the two sulfur-containing phases. However, a more detailed analysis of the patina formed at the Royal Summer Palace in Prague and based on X-ray diffraction is the subject of a later publication [18].

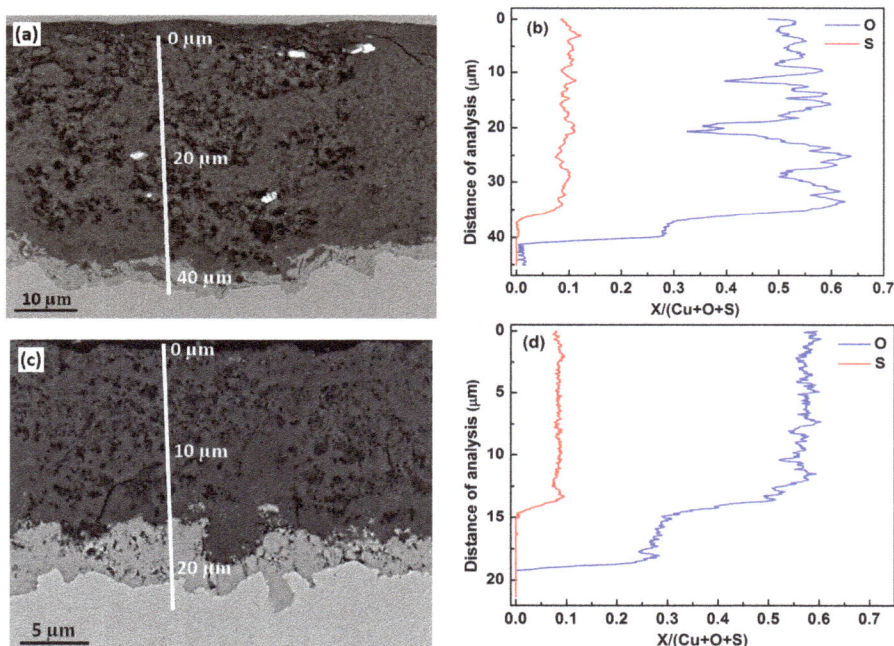

**Figure 3.** Cross-sections of the patina and adjacent substrate (**a,c**) and the corresponding elemental distribution of oxygen and sulfur relative to the sum of (copper+oxygen+sulfur) (in atomic %) based on Energy Dispersive Spectroscopy (EDS) analysis (**b,d**) of copper patina formed at the Royal Summer Palace in Prague (**a,b**) and at Kronborgs Castle (**c,d**), respectively.

In an analogous way, the results in Figure 3c,d provide evidence of the existence of two sub-layers, cuprite and brochantite/antlerite, for the patina formed at Kronborgs Castle. An interesting difference is that the cuprite layer thickness on average is higher on Kronborgs Castle material than on the material from the Royal Summer Palace.

Table 3 compiles thicknesses of the cuprite and brochantite/antlerite sub-layers determined for the patinas of all historic materials. The average thickness was obtained by measuring the area of each patina layer of five SEM cross-section images ($\geq$90 µm $\times$ 65 µm) of each material. A large variation in thicknesses of both sub-layers can be seen. The average cuprite thickness ranges from around 1 µm (Helsinki Cathedral) to around 10 µm (Basilica Maria Dreieichen), while the average brochantite/antlerite thickness varies from less than 10 µm (Helsinki Cathedral) to around 42 µm (Royal Summer Palace). The thickness ratio ((brochantite/antlerite)/cuprite) between the sub-layers also varies with an order of magnitude from 10.6 (Royal Summer Palace) to 1.1 (Otto Wagner Church).

**Table 3.** Approximate average thicknesses with standard deviation (µm) of the two sub-layers and the corresponding data for the ratio of brochantite/antlerite sub-layer and cuprite sub-layer. Data in brackets show corresponding minimum and minimum thickness values (µm). The data were obtained through measurements of five different cross-sections of the patina using SEM images ($\geq$90 µm $\times$ 65 µm) for each material.

| Exposure Site | Thickness of Cuprite ($t_1$) | Thickness of Brochantite/ Antlerite ($t_2$) | $t_2/t_1$ |
|---|---|---|---|
| Royal Summer Palace, Prague | 4.2 ± 3.3 (1.2, 8.2) | 42.5 ± 11.8 (30.7, 52.5) | 10.6 ± 6.1 |
| Mausoleum, Graz | 9.1 ± 4.4 (3.3, 13.7) | 13.3 ± 14.5 (7.8, 33.8) | 1.5 ± 1.4 |
| Drottningholm Castle, Stockholm | 9.5 ± 6.1 (3.4, 17.4) | 23.4 ± 11.3 (11.7, 30.2) | 2.5 ± 1.5 |
| Basilica, Maria Dreieichen | 9.9 ± 6.9 (3.1, 15.5) | 11.2 ± 5.2 (6.1, 17.5) | 1.2 ± 0.9 |
| Helsinki Cathedral | 1.1 ± 0.7 (0.8, 1.8) | 8.7 ± 2.1 (6.7, 9.1) | 8.1 ± 3.2 |
| Kronborgs Castle, Elsinore | 4.1 ± 3.9 (1.0, 8.1) | 15.4 ± 6.2 (9.5, 22.9) | 3.7 ± 2.6 |
| Otto Wagner Church, Vienna | 9.7 ± 4.8 (4.4, 16.1) | 10.6 ± 6.8 (5.5, 17.9) | 1.1 ± 0.9 |
| Church in Old Town, Stockholm | 10.3 ± 6.6 (3.6, 13.2) | 17.4 ± 5.9 (14.4, 24.3) | 1.6 ± 1.5 |

It is interesting to compare these ratios with available information from patinas formed on commercial sheet copper during contemporary periods rather than historic ones. Reference [20] contains a rich set of data from sheet copper (DHP-Cu, >99.9 wt % Cu) exposed at 39 exposure sites and exposure times of up to eight years, where the mass of the brochantite, cuprite and other sub-layers have been determined. The brochantite/cuprite thickness ratio from Prague after eight years, for instance, shows a thickness ratio of around 1.25, and corresponding thickness ratios for all other sites also give values of around 1.25 or lower. Therefore the cuprite layer is much more dominating in patinas formed on DHP-Cu than in historic patinas, and they also exhibit more uniform sub-layers than historic patinas. Figure 4a is an example of such modern sheet copper (DHP-Cu, >99.9% Cu, 0.02%–0.1% P, by weight), which is basically free of inclusions (if present < 1 µm) due to its purity. Figure 4b shows the substrate after exposure in a marine test site (Brest, France) for five years. It is evident that the inner cuprite layer has a thickness that clearly exceeds the thickness of the outer layer, in this case atacamite ($Cu_2Cl(OH)_3$). The formation of atacamite is due to the dominance of chlorides in the marine test site of Brest, as opposed to all other sites investigated herein, where the chlorides obviously were not present in sufficiently high concentrations to detect any atacamite.

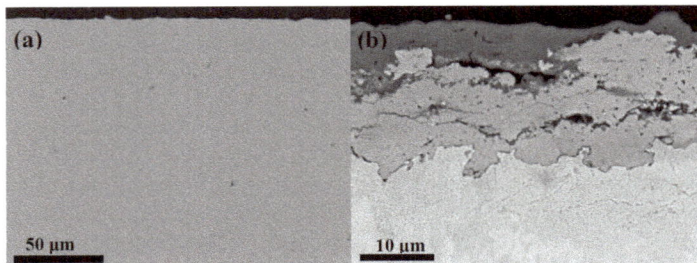

**Figure 4.** Overview SEM-image of a cross-section of modern sheet copper (*Desoxidized high phosphor copper (DHP-Cu)*) with patina (**a**); More detailed view of cross-section of patina after 5 years in a marine test site of Brest, France (**b**). The inner layer consists of cuprite, the outer layer of atacamite.

In all, the results so far show that patinas formed in historic times to a much greater extent are dominated by a brochantite/antlerite layer and by more inhomogeneous and fragmented sub-layers. Patinas formed on commercial DHP-copper, on the other hand, possess a more dominating cuprite layer whereby both sub-layers are characterized by a much higher homogeneity. It should be added that the thickness ratio brochantite/cuprite may change over exposure time for a given exposure situation. This was demonstrated by Fitzgerald et al. for copper exposed in Brisbane, Australia, over a time period of 140 years. The study suggested different growth mechanisms for the two layers: the growth of the inner cuprite layer is controlled by diffusion of cuprous ions through cuprite, while the growth of the outer brochantite layer is controlled mainly by precipitation from the aqueous ad-layer of the ionic species, involved [21].

Nevertheless, the question is to what extent the brochantite/cuprite thickness ratio may depend on the copper substrate characteristics, in particular those of the inclusions. By comparing the thickness ratio (right column of Table 3) with data on inclusion properties (Table 2), it can be concluded that neither the inclusion density nor their elemental composition can be used to explain the brochantite/cuprite thickness ratio or the patina homogeneity. However, when plotting the maximum size of the inclusions (Table 2) in the different historic copper materials against the observed brochantite/cuprite thickness ratio, a surprisingly good correlation is obtained (Figure 5).

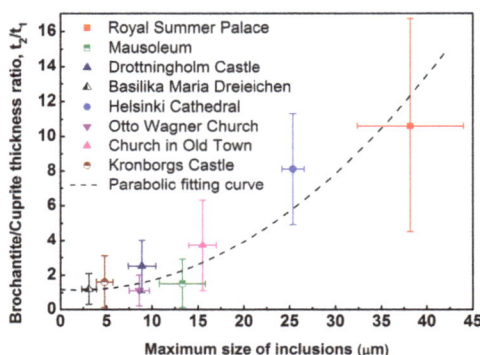

**Figure 5.** Relation between the thickness ratio of the brochantite/cuprite layer, $(t_2/t_1)$, and the observed maximum size of inclusions within the copper substrates of the historic copper materials from the various exposure sites. The dotted parabolic line is only included for guidance–with no physical meaning.

It is clearly seen that the brochantite/cuprite thickness ratio increases continuously with the maximum size of the inclusions. The error bar for the maximum size is based on the different observed maximum values of inclusion sizes obtained on each investigated 300 μm × 300 μm area, and presented as average values with standard deviation. The figure also shows a dashed line which is the result of a parabolic fit (with no underlying physical meaning) of all data points. When extrapolating the data to a maximum inclusion size of 0 μm, a thickness ratio of around 1.1 is obtained. This value is in good agreement with observations based on patinas formed on sheet DHP-Cu, as mentioned above (see further reference [20]).

## 4. Discussion

In what follows, a mechanism is proposed to explain why the inclusion size may be of importance for the reduced uniformity of the patina and the increase in thickness ratio between the brochantite/antlerite sub-layer and the cuprite sub-layer, as indicated in Figure 5. The discussion starts by analyzing the electrochemical nobility of some larger inclusions found in the historic copper materials. From the elemental analysis of the inclusions analyzed (see Table 3), it is evident that at least three types of inclusions can be identified: lead-antimony-oxygen-rich inclusions (such as rosiaite) seen in, e.g., the Royal Summer Palace sample; cuprite-rich inclusions seen in the Kronborgs Castle substrate; and copper-antimony-nickel-oxygen-rich inclusions seen in the Mausoleum Graz substrate. These three types of inclusions were analyzed by AFM-based Scanning Kelvin Probe Force Microscopy to reveal the Volta potential of the inclusions relative to the surrounding copper matrix. The results are displayed in Figure 6.

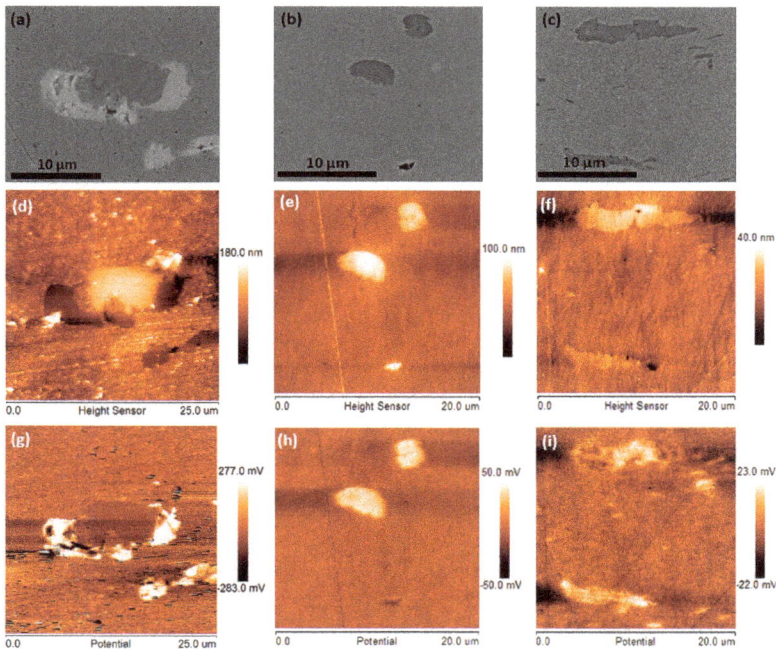

**Figure 6.** SEM-images (**a–c**), the corresponding atomic force microscopy (AFM)-based topographies (**d–f**) and AFM-based Volta potential images (**g–i**) of three types of inclusions: lead-antimony-oxygen-rich inclusions (such as rosiaite) observed in the substrate of the Royal Summer Palace sample (**a,d,g**); $Cu_2O$ (cuprite)-rich inclusions in the substrate of the Kronborgs Castle material (**b,e,h**); copper-antimony-nickel-oxygen-rich inclusion in the substrate of the Mausoleum in Graz material (**c,f,i**).

The SEM-images, Figure 6a (lead-antimony-oxygen), Figure 6b (cuprite) and Figure 6c (copper-antimony-nickel-oxygen) show three representative inclusions. Their elemental composition has been analyzed by EDS. Each sample has then been transferred to an AFM-microscope where the same inclusions have been identified again (see Figure 6d–f). AFM-based Scanning Kelvin Probe Force Microscopy [22] was subsequently used to obtain Volta potential scans of each inclusion (see Figure 6g–i). As clearly seen, all inclusions show white areas in the Volta potential images indicating a relative nobility of those inclusion areas that is higher than the surrounding copper matrix [23].

With a higher Volta potential it can be concluded that all three types of inclusions act as cathodes relative to the copper matrix. An immediate implication of this is that the inclusions during the atmospheric corrosion process trigger micro-galvanic corrosion effects in their immediate vicinity. This may cause not only accelerated corrosion kinetics but also a different sequence of corrosion product evolution than what would be expected without micro-galvanic effects [24], as illustrated in Figure 7. The figure displays a cross-section of the interfacial region between the copper substrate (light grey, lower part) and the patina (dark grey, upper part). As already shown in Figures 1 and 2, the inclusions remain relatively intact during the oxidation of the copper metal surface forming a patina.

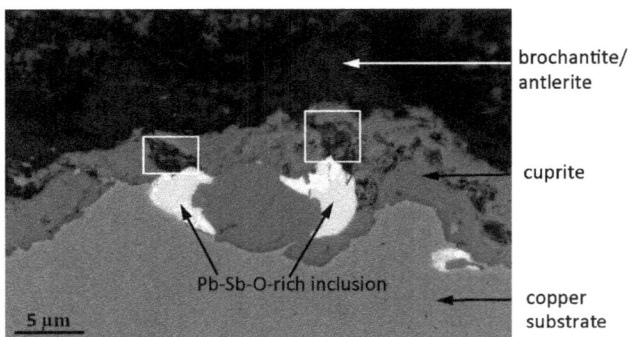

**Figure 7.** SEM cross-sectional image of the interfacial region between the copper substrate (lower part) and the patina (upper part) of the historic copper material from the Royal Summer Palace in Prague. White areas represent lead-antimony-oxygen-rich inclusions, light grey areas in the middle represent cuprite, and dark areas in the upper part brochantite/antlerite.

Figure 7 has captured the moment when two electrochemically more noble inclusions than the copper substrate (in this case two lead-antimony-oxygen-rich inclusions, white areas in the figure) have reached the copper matrix/patina interface. The grey areas in the figure represent cuprite, and the darker areas brochantite/antlerite. Near both lead-antimony-oxygen-rich inclusions the brochantite/antlerite layer is in direct contact with the inclusions (as marked by two squares in Figure 7), which has resulted in disrupture of the cuprite layer. A possible reason for this is the higher relative nobility of all inclusions relative to the copper matrix, which results in more oxidizing conditions and which therefore locally favors the formation of $Cu^{2+}$-containing corrosion products (such as brochantite or antlerite) at the expense of $Cu^+$-containing corrosion products (cuprite). It is well established that galvanic effects are favored by cathodic/anodic area ratios, and therefore not surprising that the largest inclusions result in the strongest micro-galvanic effects.

Therefore, the overall impact of the larger noble inclusions, such as those shown in Figure 7, is to disrupt the patina bi-layer structure, resulting in a more fragmentized patina and a brochantite/antlerite sub-layer, which, at least partially, grows at the expense of the cuprite sub-layer. This electrochemically-based influence of large inclusions on patina structure may not exclude other possible influences of the inclusions, e.g., the mechanical disrupture of the patina sub-layers. However, exploring this issue further requires complementary investigations.

Considering the difference in microstructure between modern high-purity copper (e.g., DHP-Cu) and historic copper materials, and also the fact that the protective properties of the patina layer mainly have been attributed to the inner cuprite layer [21], the results suggests that high-purity copper has a superior corrosion protective ability over historic ones. A demonstration of this was performed by exposing DHP-Cu, which was diamond polished to 0.25 μm, and one of the historic samples (Helsinki Cathedral), which was also diamond polished to 0.25 μm after removal of the old patina layer, to an oxygen-free 0.1 M $Na_2SO_4$-solution. The results, as obtained with EIS, are displayed in Figure 8a (Nyquist plot) and Figure 8b (impedance and phase angle). The spectra all reveal two time constants, and an impedance modulus that is slightly higher for the modern copper material compared to the historic one throughout the whole frequency region. The results suggest a higher protective ability of the corrosion products formed on modern copper under current exposure conditions. However, considering the long lifetime of historical copper roofs, even "lower quality" copper exhibits an outstanding resistance to outdoor corrosion.

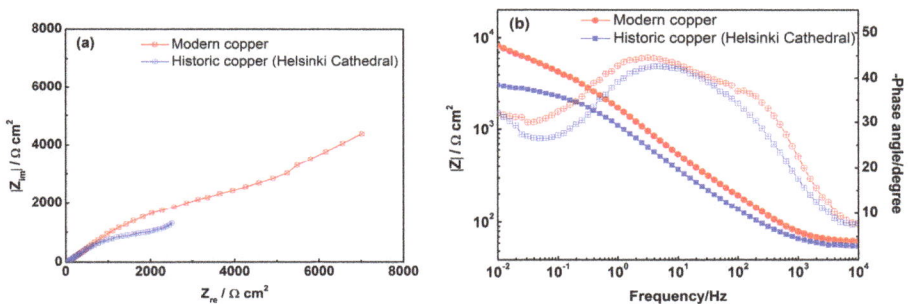

**Figure 8.** EIS spectra of Nyquist plot (**a**), impedance modulus (filled legend, (**b**)) and phase angle (unfilled legend, (**b**)) for the polished modern (DHP-Cu) and historic (Helsinki Cathedral) copper material during exposure in 0.1 M oxygen-free $Na_2SO_4$.

## 5. Conclusions

The historic naturally patinated sheet copper investigated herein shows an inhomogeneous patina with an outer layer of brochantite (or brochantite/antlerite) that is much thicker than the inner layer of cuprite. In modern sheet copper of higher purity (>99.9 wt %), on the other hand, the patina is more homogenous with comparable thicknesses of the two sub-layers.

Oxidic inclusions in the size range of up to 40 μm were observed. Three types of inclusions were identified: cuprite ($Cu_2O$)-rich; rosiaite ($PbSb_2O_6$)-rich with a smaller fraction of cuprite; and oxides of a mixture of mainly copper, antimony and nickel.

The thickness ratio between the brochantite/cuprite sub-layers varied from 1.1 to 10.6 between the investigated materials of different age and exposure conditions. This thickness ratio seems to be directly related to the maximum size of the inclusions found in each historic material. The observation can be attributed to the higher relative nobility of all inclusions found, which creates micro-galvanic effects when the inclusions reach the substrate/patina interfacial region.

**Acknowledgments:** We gratefully acknowledge Rolf Sundberg and Lennart Engström (previously at Outokumpu), and Pia Voutilainen at the Scandinavian Copper Development Association, for providing the historic Scandinavian copper materials. SEM help of Gunilla Herting is highly appreciated. The authors also express sincere thanks to the China Scholarship Council for supporting Tingru Chang's stay at KTH Royal Institute of Technology, Sweden.

**Author Contributions:** Tingru Chang, Inger Odnevall Wallinder, Manuel Morcillo and Christofer Leygraf conceived and designed the experiments; Tingru Chang performed the experiments; Belen Chico and Daniel De La Fuente contributed to SEM/EDS analysis, Manuel Morcillo, Inger Odnevall Wallinder and Jean-Marie Welter supplied the copper sheets, Jean-Marie Welter contributed to part of the text, Tingru Chang and Christofer Leygraf analyzed the data and wrote the paper.

**Conflicts of Interest:** The authors declare no conflict of interest.

## References

1. Craddock, P.T. *Early Metal Mining and Production*; Edinburgh University Press: Edinburgh, Scotland; Smithsonian University Press: Washington, DC, USA, 1995.
2. Wu, J.X.; Pedocchi, L.; Billi, A.; Rovida, G. Segregation of phosphorous at the surface of oxidized DHP copper. In Proceedings of the International Symposium on Control of Copper and Copper Alloys Oxidation, Rouen, France, 6–9 July 1992; p. 27.
3. Welter, J.-M. Du laminage à la coulée continue: Le regard de l'industriel. In *Quatre Mille ans D'histoire du Cuivre*; Pernot, M., Ed.; Presses Universitaires: Bordeaux, France, 2017.
4. Garbacz-Klempka, S. Rzadkosz, Metallurgy of copper in the context of metallographic analysis of archeological materials excavated at the Market Square in Krakow. *Arch. Metall. Mater.* **2009**, *54*, 281–288.
5. Figueiredo, E.; Araújo, M.F.; Silva, R.J.; Vilaça, R. Characterisation of a Proto-historic bronze collection by micro-EDXRF. *Nucl. Instrum. Methods Phys. Res. Sect. B* **2013**, *296*, 26–31. [CrossRef]
6. Garbacz-Klempka, A.; Kwak, Z.; Perek-Nowak, M.; Starski, M. The metallographic characterization of metal artifacts based on late medieval examples. *Arch. Foundry Eng.* **2015**, *15*, 29–34.
7. Bachmann, H.G. Prehistoric and early historic copper smelting slags. *Z. Erzbergbau Metallhuettenw.* **1968**, *21*, 419–424.
8. Eggers, T.; Ruppert, H.; Kronz, A. Change of copper smelting techniques during medieval times in the Harz-Mountains (Germany). In Proceedings of the 6th International Congress on Applied Mineralogy in Research, Economy, Technology, Ecology and Culture, Göttingen, Germany, 13–21 July 2000; pp. 971–974.
9. Nord, A.; Tronner, K.; Boyce, A. Atmospheric bronze and copper corrosion as an environmental indicator. A study based on chemical and sulphur isotope data. *Water Air Soil Pollut.* **2001**, *127*, 193–204. [CrossRef]
10. Graedel, T.E.; Nassau, K.; Franey, J.P. Copper patinas formed in the atmosphere—I. Introduction. *Corros. Sci.* **1987**, *27*, 639–657. [CrossRef]
11. Franey, J.; Davis, M. Metallographic studies of the copper patina formed in the atmosphere. *Corros. Sci.* **1987**, *27*, 659–668. [CrossRef]
12. Welter, J.-M. Understanding the copper of the Statue of Liberty. *J. Min. Metals Mater. Soc.* **2006**, *58*, 30–33. [CrossRef]
13. Nonnenmacher, M.; O'Boyle, M.; Wickramasinghe, H. Kelvin probe force microscopy. *Appl. Phys. Lett.* **1991**, *58*, 2921–2923. [CrossRef]
14. Femenia, M.; Canalias, C.; Pan, J.; Leygraf, C. Scanning Kelvin probe force microscopy and magnetic force microscopy for characterization of duplex stainless steels. *J. Electrochem. Soc.* **2003**, *150*, B274–B281. [CrossRef]
15. Nečas, D.; Klapetek, P. Gwyddion: An open-source software for SPM data analysis. *Open Phys.* **2012**, *10*, 181–188. [CrossRef]
16. Ynsa, M.D.; Chamón, J.; Gutiérrez, P.C.; Gomez-Morilla, I.; Enguita, O.; Pardo, A.I.; Arroyo, M.; Barrio, J.; Ferretti, M.; Climent-Font, A. Study of ancient Islamic gilded pieces combining PIXE-RBS on external microprobe with SEM images. *Appl. Phys. A* **2008**, *92*, 235–241. [CrossRef]
17. Welter, J.-M. La couverture en cuivre en France: Une promenade à travers les siècles. In *Monumental*; 1168-4534Direction du patrimoine: Paris, France, 2007; Semestriel 2, pp. 104–111; Bibliogr. p. 112.

18. Morcillo, M.; Chico, B.; de la Fuente, D.; Chang, T.; Wallinder, I.O.; Jiménez, J.A.; Leygraf, C. Characterization of centuries-old patinated copper roof tile from Queen Anne's Summer Palace in Prague. 2017; in preparation.

19. Leygraf, A.; Odnevall Wallinder, I.; Tidblad, J.; Graedel, T. *Atmospheric Corrosion*; John Wiley & Sons: Hoboken, NJ, USA, 2016.

20. Krätschmer, A.; Odnevall Wallinder, I.; Leygraf, C. The evolution of outdoor copper patina. *Corros. Sci.* **2002**, *44*, 425–450. [CrossRef]

21. FitzGerald, K.P.; Nairn, J.; Skennerton, G.; Atrens, A. Atmospheric corrosion of copper and the colour, structure and composition of natural patinas on copper. *Corros. Sci.* **2006**, *48*, 2480–2509. [CrossRef]

22. Schmutz, P.; Frankel, G. Characterization of AA2024-T3 by scanning Kelvin probe force microscopy. *J. Electrochem. Soc.* **1998**, *145*, 2285–2295. [CrossRef]

23. Sathirachinda, N.; Gubner, R.; Pan, J.; Kivisäkk, U. Characterization of phases in duplex stainless steel by magnetic force microscopy/scanning Kelvin probe force microscopy. *Electrochem. Solid-State Lett.* **2008**, *11*, C41–C45. [CrossRef]

24. Forslund, M.; Leygraf, C.; Claesson, P.M.; Lin, C.; Pan, J. Micro-galvanic corrosion effects on patterned copper-zinc samples during exposure in humidified air containing formic acid. *J. Electrochem. Soc.* **2013**, *160*, C423–C431. [CrossRef]

*materials*

MDPI

*Review*

# Vibrational Spectroscopy in Studies of Atmospheric Corrosion

**Saman Hosseinpour [1,†] and Magnus Johnson [2,*]**

[1]   Max Planck Institute for Polymer Research, Ackermannweg 10, 55128 Mainz, Germany;
     hosseinpour@mpip-mainz.mpg.de
[2]   Department of Chemistry, Division of Surface and Corrosion Science, KTH Royal Institute of Technology,
     SE-100 44 Stockholm, Sweden
*    Correspondence: magnusj@kth.se; Tel.: +46-8-790-9911
†    Current Address: Institute of Particle Technology (LFG), FAU Erlangen-Nürnberg, Erlangen, Germany

Academic Editor: Manuel Morcillo
Received: 28 February 2017; Accepted: 28 March 2017; Published: 18 April 2017

**Abstract:** Vibrational spectroscopy has been successfully used for decades in studies of the atmospheric corrosion processes, mainly to identify the nature of corrosion products but also to quantify their amounts. In this review article, a summary of the main achievements is presented with focus on how the techniques infrared spectroscopy, Raman spectroscopy, and vibrational sum frequency spectroscopy can be used in the field. Several different studies have been discussed where these instruments have been used to assess both the nature of corrosion products as well as the properties of corrosion inhibitors. Some of these techniques offer the valuable possibility to perform in-situ measurements in real time on ongoing corrosion processes, which allows the kinetics of formation of corrosion products to be studied, and also minimizes the risk of changing the surface properties which may occur during ex-situ experiments. Since corrosion processes often occur heterogeneously over a surface, it is of great importance to obtain a deeper knowledge about atmospheric corrosion phenomena on the nano scale, and this review also discusses novel vibrational microscopy techniques allowing spectra to be acquired with a spatial resolution of 20 nm.

**Keywords:** atmospheric corrosion; infrared spectroscopy; Raman spectroscopy; vibrational sum frequency spectroscopy; VSFS; SFG

## 1. Introduction

The main aim of this review article is to illustrate how different vibrational spectroscopy techniques are used in the field of atmospheric corrosion. Since the focus of this article is on the ability of the various techniques, the main divisions of the sections have been based on experimental technique rather than the metal or alloy studied, which is presented in subsection.

A significant number of the studies discussed here are model studies of metal surfaces exposed to one or a few corrosive gases, thus in significantly less complex atmospheres than in natural environments. Hence, the number of corrosion products formed will be lower, allowing for simpler spectral assignments. Several of the studies discussed below focus on investigating indoor atmospheric corrosion processes, which significantly differ from corrosion occurring outdoors due to the different nature of the corrosive species present among other parameters. A summary of important gases found in indoor environments is found in Table 1.

Two important properties of some of the vibrational spectroscopy techniques (e.g., infrared reflection/absorption spectroscopy) are the ability to perform experiments in-situ and in real time, as well as obtaining chemical information with a spatial resolution of 20 nm with the novel technique nano FTIR (Fourier transform infrared spectroscopy) microscopy. These two abilities, are crucial parameters in revealing the mechanisms of atmospheric corrosion processes.

**Table 1.** Important corrosive gases in indoor environments [1,2]. Reproduced from [1] by permission of The Electrochemical Society.

| Species | Typical indoor Concentration (ppb) | $H$ [b] | Equilibrium Solution Concentration (μM) |
|---|---|---|---|
| $O_2$ | 2.1 (8) [a] | 1.7 (−3) | 3.6 (2) |
| $O_3$ | 18 | 1.8 (−2) | 3.2 (−4) |
| $H_2O_2$ | 5 | 2.4 (5) | 1.2 (3) |
| $H_2S$ | 0.3 | 1.5 (−1) | 4.5 (−5) |
| COS | 0.6 | 3.7 (−2) | 2.2 (−5) |
| $SO_2$ | 30 | 1.4 | 4.2 (−2) |
| HCl | 0.4 | 2.0 (1) | 8.0 (−3) |
| $Cl_2$ | 0 | 6.2 (−2) | 0 |
| $NH_3$ | 10 | 1.0 (1) | 1.0 (−1) |
| $NO_2$ | 4 | 7.0 (−3) | 2.8 (−5) |
| $HNO_3$ | 3 | 9.1 (4) | 2.7 (2) |
| $CO_2$ | 6.0 (5) | 3.4 (−2) | 2.0 (1) |
| HCHO | 10 | 1.4 (4) | 1.4 (2) |
| HCOOH | 20 | 3.7 (3) | 7.4 (1) |
| $CH_3COOH$ | 20 | 8.8 (3) | 8.8 (1) |

[a] 2.1 (8) means $2.1 \times 10^8$; [b] Henry's law constant.

In several of the studies discussed below, complementary techniques such as scanning electron microscopy (SEM), X-ray photoelectron spectroscopy (XPS), quartz crystal microbalance (QCM), cathodic reduction (CR), and atomic force microscopy (AFM), have been used to characterize the corrosion products, with the obvious advantage that the corrosion process can be examined from different viewpoints. Although here the results obtained by these other techniques have not been discussed in detail, the complementary techniques employed have been mentioned to demonstrate the added value of combining various analytical techniques.

## 2. Results and Discussion

### 2.1. Overview

The vibrational spectroscopy techniques that are described below are infrared, Raman, and vibrational sum frequency spectroscopy, and for each of these main techniques different subtypes are described. Each section starts with a brief description of the theory behind the technique, and is followed by a discussion how the technique has been used to study atmospheric corrosion of different metals and alloys.

IR and Raman spectroscopy have different advantages and drawbacks when applied to studies of atmospheric corrosion, and before summarizing specific studies a brief comparison between the techniques is provided.

The spectral range covered by IR and Raman spectroscopy commonly differs, and usually Raman spectroscopy can offer a wider accessible range in the fingerprint and group frequency regions. In Raman spectroscopy the spectral region ~100–4000 cm$^{-1}$ is commonly covered, whereas in IR spectroscopy the lowest reachable wavenumber is usually around 600 cm$^{-1}$ due to the cutoff frequency of MCT detectors. With other IR detectors such as DTGS it is however possible to reach down to around 80 cm$^{-1}$, thus similar to Raman spectroscopy. However, even with other detectors than MCT, the accessible wavenumber range with IR can be limited at lower wavenumbers by the transparency of the windows possibly used in the system.

In studies of local corrosion, confocal Raman microscopy has been used in a number of atmospheric corrosion studies, and provides a spatial resolution of some hundreds of nm, which is significantly better than conventional IR microscopy which at the best has a spatial resolution of around 5 μm due to diffraction limitation. However, with novel techniques such as nano FTIR microscopy it is possible to study the distribution of corrosion products with a spatial resolution of 20 nm.

Infrared reflection/absorption spectroscopy (IRRAS) has commonly been used to perform in-situ studies during an ongoing atmospheric corrosion process, and although such investigations would be possible with Raman spectroscopy as well, such studies are rare.

IR spectroscopy has the advantage that the measured absorbance is directly proportional to the amount of corrosion products formed, and hence quantitative measurements can be performed. Raman spectroscopy is more of a qualitative technique, since the surface topography (e.g., surface scratches and chunks of corrosion products) affect the scattered Raman signal.

Raman spectroscopy and some IR techniques such as IRRAS do not require any sample preparation, which however is required for IR transmission and IR attenuated total reflection (ATR) studies of corrosion products that have been scraped off the sample. In Raman spectroscopy the corroded surface can be either shiny or dull, whereas in IRRAS a shiny sample is desired in order to have as much IR radiation as possible reaching the detector. Further, IR spectroscopy requires the acquisition of a background spectrum from, e.g., a gold surface or an uncorroded surface (frequently used in in-situ IRRAS measurements).

A limiting parameter in performing Raman spectroscopy on corrosion products is the fluorescence that usually arises from corrosion products. For instance, colorful corrosion products on copper (Cu I or II oxide) cause a large fluorescence background, which sometimes overwhelms the vibrational signature of the corrosion products. However, in some cases such fluorescence generation has been used as an indication of the formation of specific corrosion products.

To conclude, the discussion above reveals that IR and Raman spectroscopy have different advantages in examinations of corrosion products, and thus depending on the specific system that is of interest, the most favorable technique should be chosen. In many situations these two spectroscopic techniques can provide complementary information and could thus be combined.

### 2.1.1. Infrared Spectroscopy in General

Infrared (IR) spectroscopy owing to its abilities in providing chemical information is a widely used technique in numerous research fields, and has also found great use in the field of atmospheric corrosion, with the main aim of identifying the corrosion products formed. However, with certain input parameters IR spectroscopy can also be used to quantify the amount of corrosion products. IR spectroscopy has been described in a large amount of sources [3,4], and hence only a brief description is provided here.

In infrared spectroscopy molecular vibrations are probed, and since different functional groups absorb IR radiation of different wavelengths, the technique can be used to assist in the identification of unknown compounds. The selection rules for IR spectrocopy are that the dipole moment has to change during the vibration, and that the IR radiation must have a wavelength that corresponds to the energy difference between vibrational states. In infrared reflection/absorption spectroscopy (IRRAS) and additional selection rule for metal surfaces is that the molecular vibration must have a component that is perpendicular to the surface plane.

IR spectroscopy can be performed in various modes depending on the type of sample that is to be studied [5]. In the transmission mode, the corrosion products are scraped off from the surface, mixed with for example KBr, and pressed into a pellet. The transmitted IR radiation through the pellet is then measured. In order to use this technique, a fairly large amount of corrosion products is required, since it is necessary to scrape them off from the surface and mix them into the pellet. In the attenuated total reflection (ATR) mode (ATR), corrosion products also need to be scraped off from the sample, and they are placed on a crystal (e.g., diamond), which guides the IR beam to the sample in order to acquire a spectrum. An advantage of ATR in comparison to transmission IR is that the corrosion products can be studied directly, without the need of making a pellet. In diffuse reflectance (DRIFTS) the diffusively reflected light from the powder of corrosion products is detected. When using infrared reflection/absorption spectroscopy (IRRAS), an IR beam incident at a grazing angle of incidence (~80°) from the surface normal is reflected off the metal surface and detected. The advantage of IRRAS is that ultrathin layers of corrosion products can be examined in-situ, and hence the technique is well suited for studies of atmospheric corrosion processes. It also has the ability to provide information about thin films of corrosion inhibitors. An advantage of IRRAS is that no sample preparation is required (e.g., scraping off corrosion products from the sample). Instead, the sample is placed directly in the spectrometer and a spectrum can be acquired.

When comparing IR spectra obtained by the different sub-techniques described above, it is necessary to be aware of that spectral difference may appear both concerning relative peak intensities as well as central wavelength. Comparing transmission and ATR spectra there may be differences in relative peak intensities since the penetration depth in the ATR mode depends on the wavelength of the IR radiation. Moreover, frequently a blue shift in IRRAS spectra of a peak position is observed when comparing to transmission spectra, which is due to the fact that the shape of IRRAS spectra not only depends on the absorption constant but also the real part of the refractive index [6]. For example, for the common corrosion product $Cu_2O$ on copper, a shift from 623 (transmission) to 655 (IRRAS) $cm^{-1}$ was observed [6].

### 2.1.2. IR Transmission Spectroscopy

IR transmission spectroscopy has been used quite rarely to identify atmospheric corrosion products, probably due to the difficulties in the sample preparation, which requires that the corrosion products are scraped off from the surface and pressed into a pellet together with for example KBr as discussed above. Thus, tiny amounts of corrosion products are difficult to study with this technique. In some examinations the IR studies have been complemented by for example XRD, in order to confirm the nature of the corrosion products.

#### 2.1.2.1. Steel

A number of studies on various steels have been performed, and the IR spectra have been used to identify the different phases of FeOOH, which have different characteristic infrared absorption bands. Additionally, the presence of bound water in the film of corrosion products has been concluded in some studies. Wang et al., studied the atmospheric corrosion of carbon steel during 12 months near Qinghai salt lake in China. Based on the results obtained from IR transmission spectroscopy it was concluded that $\beta$-FeOOH (450, 690, and 840 (weak)·$cm^{-1}$), $\gamma$-FeOOH (1020 $cm^{-1}$), and $\delta$-FeOOH (790, 880, and 1110 $cm^{-1}$) were formed, where the latter peak could be used to qualitatively assess the crystallinity since less crystalline samples exhibited a broader band [7,8]. In addition, water in the samples was revealed by a peak at 1630 $cm^{-1}$ (bending vibration). The corrosion product $Fe_8(O,OH)_{16}Cl_{1.3}$ was observed by XRD but those results could not be verified by IR spectroscopy due to the lack of reference spectra.

In a study of carbon steel by Allam et al., IR spectroscopy and X-ray diffraction were used to characterize the corrosion products formed during a 12-month exposure in an industrial atmosphere near the sea. The results indicated that the corrosion started by the formation of blisters containing iron chlorides, oxyhydroxides, sulphates, oxides, and possibly hydroxides [9]. Jaen et al., used IR spectroscopy in combination with Mössbauer spectroscopy to study the atmospheric corrosion of mild steel (A-36) and weathering steel (A-558 and COR 420) during three-month exposure to a tropical marine climate in Panama. They could identify the phases $\alpha$-FeOOH and $\gamma$-FeOOH on all steels, but no $\delta$-FeOOH was observed. In a similar way (although based on different peaks) as Wang et al., above concluded that the rust was of low crystallinity, the peaks in the study by Jaen at 790 and 884 $cm^{-1}$

were weak and broad for the A-36 and A-558 samples. Hence, this indicated low crystallinity of the corrosion products. In contrast, the COR 420 showed highly crystalline $\alpha$-FeOOH and $\gamma$-FeOOH. In an atmospheric corrosion study of pre-corroded (in $SO_2$, $NO_2$, and $H_2S$) carbon steel at various locations in China, IR spectroscopy was used to identify $\alpha$-FeOOH, $\gamma$-FeOOH, and $\delta$-FeOOH in both rust flakes and powder-like rust. Strong OH stretching vibrations at 3120 and 3380 $cm^{-1}$ were characteristic of $\gamma$-FeOOH and $\delta$-FeOOH, respectively. Further, the corrosion products contained large amounts of water, indicated by the peak at 1640 $cm^{-1}$, similarly to the rust studied by Wang et al., above.

Raman et al., used transmission IR to identify different kinds of rust on samples from outdoor environments, and present spectra from several different rust types as well as tables with characteristic infrared absorption bands [10]. A general conclusion is that crystalline corrosion products exhibit well defined peaks, whereas more amorphous products yielded weaker and broader bands.

Pacheco et al., examined the rust formation in chloride rich environments with different relative humidities, and observed the formation of oxides and oxyhydroxides during early corrosion [11].

### 2.1.2.2. Copper

Morcillo et al., identified several corrosion products formed during exposure of copper samples in Ibero-America, the main product being $Cu_2O$ [12], as well as the corrosion products formed on zinc in rural and urban atmospheres [13]. The atmospheric corrosion products of the seven meter long iron cannon in Tanjore (India) were characterized by the use of IR spectroscopy [14].

### 2.1.2.3. Zinc

Transmission IR spectroscopy of corrosion products and reference samples was used to identify the corrosion products formed during dry/wet cycles of zinc first exposed to a NaCl solution and then to humid air and $CO_2(g)$ [15]. The details of the studies are found in the IRRAS section.

### 2.1.2.4. Aluminum and Magnesium Alloys

Ke et al., investigated the atmospheric corrosion of the aluminum alloy AA2024 in the presence of magnesium chloride-based multicomponent salts [16]. IR transmission studies revealed, in agreement with XRD examinations, that longer laboratory exposures resulted in corrosion products with a higher crystallinity than those formed in shorter exposure times, with the main corrosion products being $[Mg_{1-x}Al_x(OH)_2]^{x+} Cl^-_x \cdot mH_2O$. The presence of the salts could induce corrosion also below 30% RH. IR transmission was used to characterize the atmospheric corrosion products formed on the magnesium alloy AZ91D exposed in a polluted environment [17]. FTIR in combination with XRD enabled the identification of $Mg(OH)_2$, $MgCO_3$, and $Mg_2Al_2(SO_4)_5 \cdot 39H_2O$ as corrosion products, and it was further concluded that the corrosion rate was higher for an ingot sample in comparison to a die-cast sample. The corrosion was initiated in the $\alpha$ phase, being less noble than the $\beta$ phase, which remained and constituted a barrier to further corrosion.

### 2.1.2.5. Corrosion Inhibitors

In one study, transmission IR coupled with thermogravimetric analysis was used to examine model compounds that simulate amine-carboxylic acid-based volatile corrosion inhibitors, which for example have been used in the protection of mild steel against atmospheric corrosion [18]. IR measurements of the vapor evaporated from mixtures of two-component equimolar solutions were acquired, and it was concluded that initially the vapor contained essentially only free amine. However, the vapor phase composition changed over time and it was thus concluded that great care must be taken when using this type of mixtures as corrosion inhibitors.

### 2.1.3. Diffuse Reflectance (DRIFTS)

Nickel

The atmospheric corrosion of nickel surfaces exposed outdoors at 39 test sites in European and North American countries was studied by DRIFTS [19]. The measurements were complemented by investigations where XPS, X-ray powder diffraction, and elemental analysis were used, and an agreement was found between the different methods. The identified corrosion products were sulfates characterized by strong bands at 600 and 1100 $cm^{-1}$, water and hydroxides in the region 3000–3600 $cm^{-1}$, and indications of water was also seen through the bending vibration at 1595 and 1600 $cm^{-1}$, as well as libration modes in the region 660–900 $cm^{-1}$. Two types of samples were identified, where the first type was non-crystalline and exhibited a lower protective ability, and showed distinct antisymmetric sulfate stretching bands at 1080, 115, and 1150 $cm^{-1}$ as well as three hydroxyl peaks around 3600 $cm^{-1}$. The high wavenumbers indicate that these hydroxyl groups possess very weak hydrogen bonds. The other type of samples had broader and unresolved sulfate and hydroxyl bands, showed a higher resistance towards corrosion, and were formed at longer exposure times. It was suggested that the spectral differences were due to different particle sizes, which is known to affect the band widths in DRIFTS or that the broader bands signified a more disordered product [20]. Also the water bending mode displayed different properties, with the band at 1660 $cm^{-1}$ being strongest for the first type and the band at 1959 $cm^{-1}$ for the second type.

### 2.1.4. Infrared Reflection/Absorption Spectroscopy (IRRAS)

IRRAS is the infrared technique that most frequently has been utilized to study corrosion products formed under atmospheric conditions. The technique is well suited to identify the nature of corrosion products and their kinetics of formation. Preferably the metal surface should be shiny, since a dull and rough surface results in a diffuse reflection of the IR beam from the surface and hence little light that reaches the detector. If calibrated to a mass sensitive instrument such as quartz crystal microbalance (QCM), IRRAS can additionally provide information about the absolute mass of the corrosion products, as discussed below.

A problem frequently encountered when performing long time in-situ exposures is a change in the background signal, for example by drift or a varying relative humidity. Such changes can cause broad water bands to overlap with the peaks of interest, and hence make the interpretations difficult. A solution to this problem is to use polarization modulated (PM) IRRAS, in which the polarization is modulated between s and p-polarized light (s denotes light polarized perpendicular and p parallel to the plane of incidence), where s-polarized does not give rise to any surface signal due to a cancellation of the electric field in the surface region, whereas the p-polarized is enhanced in the surface region. A further advantage of PM-IRRAS is the improved signal to noise ratio. Some PM-IRRAS studies are discussed below. IRRAS has also been employed to study the properties of ultrathin organic films used as corrosion inhibitors, as elaborated in this section.

#### 2.1.4.1. Iron

Weissenrider at al. studied the atmospheric corrosion of iron induced by relative humidity, $SO_2(g)$, $NO_2(g)$, and $O_3(g)$ in-situ by IRRAS, QCM, and AFM [21]. It was concluded that $NO_2(g)$ or $O_3(g)$ in addition to $SO_2(g)$ were necessary to form sulfate nests on the iron surface. By the use of $H_2O$ as well as $D_2O$, a band at 1100 $cm^{-1}$ in the IRRAS spectra was identified as an OH vibration of some nature.

Kotenev et al., employed IRRAS, resistometry, and gravimetry to study the oxidation of metal oxide nanocomposite layers on sensors [22]. By identifying the characteristic peaks of different phases, it was concluded that $Fe_3O_4$ (412 $cm^{-1}$), $\alpha$-$Fe_2O_3$ (555 and 602 $cm^{-1}$), and $\gamma$-$Fe_2O_3$ (652 $cm^{-1}$) were initially formed, whereas after longer time $\alpha$-$Fe_2O_3$ dominated. Thus, at the initial stages (5–30 min), the metal oxide structure contains $Fe_3O_4$, $\alpha$-$Fe_2O_3$, and $\gamma$-$Fe_2O_3$, while at 2–5 h of oxidation, $\alpha$-$Fe_2O_3$ is mainly present.

Aramaki prepared iron surfaces covered with a monolayer of 11-mercapto-1-undecanol, and on top a monolayer of bis(triethoxysilyl)ethane and octadecyltriethoxysilane to study the protection against atmospheric corrosion [23]. The film was characterized by XPS, contact angle, and IRRAS, and the conclusion was that the thiol formed a polymeric structure with the silanes, with siloxane bridges in the lateral direction. The nature of the corrosion products was characterized by XPS after exposure to air.

Stratmann and coworkers used IRRAS and quartz crystal microbalance to study the formation of ultrathin polysolixane layers on iron surfaces, as well as the protection ability of the films [24]. They found that the films significantly improved the corrosion resistance in a humid atmosphere containing $SO_2(g)$, and that the corrosion started at local defects in the film, but no changes in the polymer film itself were observed.

### 2.1.4.2. Steel

Person et al., used IRRAS to study the atmospheric corrosion of Zn-Al-Mg coated, electrogalvanized, and hot dipped galvanized steel with deposited NaCl and exposed to 805 RH and 350 ppm $CO_2$ [25]. In complement, XRD and SEM were used. During exposure NaCl(s) adsorbs water and a thin electrolyte layer is formed, which promotes the corrosion. With IRRAS several bands originating from water and hydroxide were observed at early stages ($H_2O$ and $OH^-$ stretching vibrations in the region 3000–3700 $cm^{-1}$), the water bending vibration at 1635 $cm^{-1}$, and several bands in the region 400–1000 $cm^{-1}$. Later the antisymmetric stretch of the carbonate ion was observed as a symmetric peak centered at ~1385 $cm^{-1}$, which altogether indicated that layered double hydroxide products with intercalated carbonate ions had formed. Upon introduction of dry air, the water bands got reduced, an indication of that water left the surface. Complementary ATR spectra of reference compounds were acquired to assist in the assignments of the vibrations.

Kim and coworkers studied how steel could be protected against atmospheric corrosion (10 min in air at 400 °C and at 60 °C and 100% RH for five days) by covering it with silane primers and an epoxy layer, where the silane used to improve the adhesion of the epoxy layer to the metal surface [26]. The silane reacted with the steel surface to form a Fe-O-Si bond. IRRAS was used to conclude that upon heating of the epoxy covered steel surface, the epoxide peak at 915 $cm^{-1}$ vanished, an indication that the epoxy film was completely cured. For both exposure conditions the corrosion was reduced, but the corrosion products $Fe_3O_4$, $\gamma$-FeOOH, and $Fe_2O_3$ could be observed in various amounts depending on the conditions.

### 2.1.4.3. Copper

Person et al., used ex-situ IRRAS complemented by Raman spectroscopy, cathodic reduction, and XPS to examine the atmospheric corrosion of copper exposed to 75% RH and 0.25 ppm $SO_2(g)$ and $NO_2(g)$ [27]. Copper sulfite and sulfate were identified by IRRAS and a comprehensive table summarizing the band positions for sulfite and sulfate was presented. The spectra indicated the formation of a 1 nm thick film of sulfate or nitrate. Differences in peak position in comparisons with transmission and IRRAS spectra were also discussed, and exemplified by a blue shift of 28 $cm^{-1}$ for $Cu_2O$ in the IRRAS spectra compared to transmission, as also was discussed by Greenler [6]. No Cu-O-H bending vibrations were observed, an indication of that no basic copper sulfates were formed.

Person et al., later developed a sample cell for in-situ IRRAS studies of atmospheric corrosion processes, with the ability to control the relative humidity [28]. It was shown that during exposure of a copper surface to 90% RH, $Cu_2O$ was formed as revealed by an IR band at 645 $cm^{-1}$. The kinetics of formation displayed a logarithmic behavior, indicated by a straight line when the IR absorbance of the cuprite peak was plotted against the logarithm of the exposure time, as shown in Figure 1. Hence, this demonstrates the possibility to use IRRAS to determine the kinetics of formation of corrosion products. To quantify the amount of cuprite formed, ex-situ cathodic reduction was used, and the amount of cuprite was further estimated through theoretical calculations. Further exposures to 80% RH and 0.23 ppm $SO_2$(g) revealed the formation of sulfite by a band at 1010 $cm^{-1}$, in addition to cuprite.

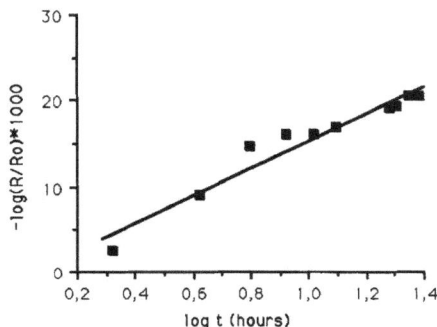

**Figure 1.** The infrared absorbance of the cuprite peak at 645 $cm^{-1}$ as a function of exposure time when a copper surface was exposed to 90% RH, revealing a logarithmic oxide growth. Reproduced by permission of [28], The Electrochemical Society. Copyright 1993.

Kleber et al., also developed an in-situ IRRAS chamber and studied the formation of cuprite on a copper surface exposed to 80% RH [29]. As a complement, simulated spectra with corrosion film thickness and angle of incidence of the IR beam as variables were obtained, and AFM measurements to study the topography were performed. By plotting the IR absorbance of the cuprite peak as a function of time, it was concluded that the early copper corrosion follows a logarithmic rate.

In a subsequent study by Person et al., copper, zinc, and nickel were exposed to 0.21 ppm $SO_2$(g) and 80% RH, and the formation of sulfite could in all three cases be followed from a few minutes after the start of the exposure to 20 h [30]. Through the IR peak positions of the sulfite bands it was possible to deduce the sulfite ion coordination on the three metal surfaces. A mechanism for the sulfur dioxide induced corrosion was also discussed, and estimations of the deposition rate of sulfur dioxide were done based on calculations and intensities of the sulfite bands. It was concluded that the deposition rate on zinc is one order of magnitude higher than that on the other metals, which indicates a surface reaction that is mass transport limited.

The same authors as above used IRRAS (complemented by XPS) to study field exposures of zinc, copper, nickel, and silver surfaces in indoor environments [2]. The locations were the Royal Palace in Stockholm, the St. Vitus cathedral in Prague, a military storage room in Karlstad (Sweden), and a computer room at a paper and pulp industry (Mönsterås, Sweden). The extreme temperatures altogether were $-3°$ and $25°$, and the relative humidity was in the range 17%–90% RH. These field exposures of course involve a significantly more complex atmosphere in comparison to the vast amount of model studies discussed in this article, and hence the peak assignments in the IR spectra become more difficult since numerous peaks are present and the nature of the atmosphere is more or less unknown. A table of important gases in indoor corrosion is found in Table 1 above. With IRRAS it was concluded that main corrosion products on zinc, copper, and nickel were carboxylates (e.g., formate and acetate), and that also tiny amounts of chloride, ammonium, nitrate, and sulfate ions were present. In contrast, silver sulfide was the main corrosion product on silver, and small amounts of sulfate, ammonium, and chloride ions were observed. Thicker layers of corrosion products were observed at the St. Vitus cathedral and in the military storage room, which was suggested to be due to a higher relative humidity and accordingly a more pronounced water layer on the metal surfaces. This study further shows the importance of organic corrosion products in indoor environments.

Aastrup et al., developed a sample cell which allowed simultaneous and in-situ studies by IRRAS and QCM of an atmospheric corrosion process of any relative humidity, hence yielding information about the nature of the corrosion products (IRRAS) and their mass (QCM) [31]. The capability of the setup was demonstrated by investigating the formation of cuprite when copper was exposed to 80% RH. By following the time evolution of the peak intensity of the cuprite band at 645 cm$^{-1}$ and changes in QCM frequency, the kinetics of formation of the corrosion process could be monitored. The detection limit for IRRAS was a film thickness of 10 Å and for QCM 2 Å. The kinetics of formation was compared with cathodic reduction results, showing that all results were in agreement with each other. Itoh et al., also developed a combined in-situ IRRAS/QCM setup and measured the initial atmospheric corrosion of copper exposed to 80% RH and 10 ppm SO$_2$(g) [32]. Sulfite (1040 cm$^{-1}$) and sulfate (1110 cm$^{-1}$) were identified as corrosion products, and complementary XPS studies showed the presence of both Cu(I) and Cu(II) oxides. Water in the film was revealed by bands at 3350 (stretches) and 1640 (bending mode)·cm$^{-1}$. Both changes in the IR absorbance and the QCM mass indicated a parabolic growth of the corrosion products.

Wadsak et al., studied the atmospheric corrosion of copper in humid air of 60% and 80% RH using a coupled setup for IRRAS and QCM, as well as AFM [33]. With IRRAS it was concluded that Cu$_2$O (645 cm$^{-1}$) formed for both relative humidities, and AFM enabled a determination of tiny corrosion products at 60% RH and a more fully covering layer at 80% RH after 80 min. With QCM it was possible to conclude that the corrosion rate initially was fastest in both cases and that the mass gain was slower at the lower relative humidity.

Gil, et al., published a series of articles where the atmospheric corrosion of copper was studied by combining IRRAS and QCM, to obtain the nature of the corrosion products as well as their mass. A combination of three analytical techniques, IRRAS, cathodic reduction, and QCM allowed a detailed study of the initial atmospheric corrosion of copper exposed to 90% RH and either 120 ppb formic, acetic [34], or propionic acid [35]. With IRRAS the corrosion products cuprite, copper carboxylates, and copper hydroxide as well as water in the film were identified. A linear correlation between the IR absorbance of the cuprite peak at 648 cm$^{-1}$, the mass of cuprite obtained from QCM, and the thickness of the cuprite layer as determined by cathodic reduction was obtained (Figure 2), thus providing a convincing conclusion about the formation of corrosion products. Correlating the mass of the corrosion products obtained by QCM with the absorbance from IRRAS allows future determinations of the mass of corrosion products to be obtained solely by the use of IRRAS, since the absorbance scales linearly with the amount of corrosion products formed. Further, the growth of the copper carboxylate was followed by the absorbance of the antisymmetric carboxylate stretch at around 1600 cm$^{-1}$. In complement AFM was used to determine the morphology of the corrosion products, and a mechanism for the reaction route was suggested. The IRRAS and QCM results were also compared with computer simulations using the GILDES model [36], which could predict the more aggressive corrosion induced by formic acid, followed by acetic acid and propionic acid [36]. The trend was explained by a higher ligand-, and proton promoted dissolution for formic acid.

Kleber et al., combined in-situ IRRAS, in-situ AFM, and SIMS to investigate the atmospheric corrosion of copper, zinc, and two brass alloys (Cu/Zn = 70/30 and 90/10 wt %) exposed to 80% RH and 250 ppb SO$_2$(g) [37]. IRRAS revealed the formation of Cu$_2$O when copper was exposed to humid air, and by integrating the cuprite peak a logarithmic rate for cuprite formation was observed. Although Zn(OH)$_2$ likely is formed in the presence of water, the bands could not be observed due to overlap with the bands from gas phase water. For copper and 90/10 brass a cuprite peak at 650 cm$^{-1}$ was observed. Its absence for 70/30 brass indicated that only zinc corroded. However, no reliable signal from ZnO at 500–550 cm$^{-1}$ could be detected due to the cutoff frequency of the MCT detector. With AFM it was concluded that the brass alloys exhibited larger corrosion products than the pure metals. By a comparison of the intensities of bands originating from metal sulfides, sulfites, and sulfates it was concluded that the brass alloy with the highest zinc content corroded faster in the presence of SO$_2$(g). As in the exposure of only humid air, Cu$_2$O was only observed for copper and 90/10 brass.

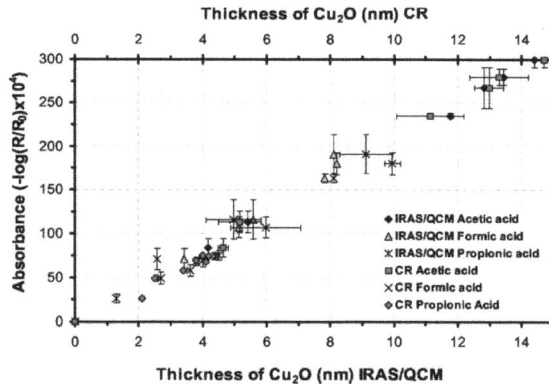

**Figure 2.** A linear relation between the results obtained from infrared reflection/absorption spectroscopy (IRRAS), quartz crystal microbalance (QCM), and cathodic reduction for the formation of $Cu_2O$ during exposure of a copper surface to 90% RH and 120 ppb carboxylic acids. Reproduced by permission of [35], The Electrochemical Society. Copyright 2007.

Faguy et al., developed a setup for in-situ PM-IRRAS measurements, and showed that their PM-IRRAS improved the signal to noise ratio in comparison with conventional IRRAS with a factor of 2.5 [38]. It was furthermore possible to obtain spectra without interference from gas phase water. The capacity of the technique was demonstrated in studies of copper exposed to $SO_2(g)$, $NO_2(g)$, and HCl(g), and nitro, nitrito, and sulfate corrosion products were identified, and the kinetics of formation could be followed.

Malvault et al., synthesized a number of basic copper salts, brochantite ($Cu_4SO_4(OH)_6$), gerhardite ($Cu_2NO_3(OH)_3$), atacamite ($Cu_2Cl(OH)_3$), aratacamite ($Cu_2Cl(OH)_3$), malachite ($Cu_2CO_3(OH)_2$), and posnjakite ($Cu_4SO_4(OH)_6 \cdot H_2O$), to use as reference compounds for interpretations of IRRAS spectra of corroded copper surfaces [39]. Additionally, cathodic reduction was used to characterize the corrosion products. Interestingly, in contrast to $Cu_2O$ [6], the band positions in the IRRAS and transmission spectra were not shifted.

The atmospheric corrosion of copper covered with sodium chloride and exposed to either $CO_2(g)$ [40] or $SO_2(g)$ [41] was studied by Chen et al., and is discussed further under the section Conventional IR microscopy, Section 2.1.6.1.

By the use of in-situ IRRAS and two-dimensional correlation analysis (2D-IR), the Osawa group determined the nature of the corrosion products as well as their kinetics of formation when a copper surface was exposed to 80% RH and 8.7 ppm $SO_2(g)$ [42]. The 2D analysis facilitated a deconvolution of the infrared bands, and was used to follow the kinetics of formation of the main corrosion products $CuSO_3Cu_2SO_3 \cdot 2H_2O$ and $CuSO_4 \cdot 5H_2O$. Initially a water adlayer was formed on the surface, and subsequently the sulfur containing corrosion products were formed.

IRRAS has been used a number of times to examine the quality of ultrathin corrosion inhibiting films, in addition to studies of corrosion products. Mekhalif and coworkers prepared self-assembled monolayers of flurothiols on copper and used IRRAS and XPS to characterize their structure and organization, and studied their protection by cyclic voltammetry [43]. Liedberg and coworkers investigated the chemisorption of the corrosion inhibitor benzotriazole on copper and cuprous oxide with IRRAS, X-ray photoelectron spectroscopy, and ultraviolet photoelectron spectroscopy [44]. IRRAS was used to study variations in the orientation of the benzotriazole molecules in the layers, and the authors further suggested a new orientation of this inhibitor on copper. The Ishida group studied the ability of imidazoles to protect copper surfaces against corrosion in a series of articles [45–47]. IRRAS was used to examine the molecular structure, orientation, and stability of the inhibitors prepared under different conditions and exposed to various heat treatments. Jang and coworkers used IRRAS and scanning electron microscopy to study the effect of heat treatments on protective films of various compositions of a copolymer consisting of vinyl imidazole and vinyl trimethoxy silane deposited on copper [48]. The authors observed that by the introduction of latter polymer, the heat resistance was improved due to the formation of disolixane linkages. Wenger and coworkers investigated the thermal stability of imidazole and 5-methylbenzimidazole films on copper surfaces in atmospheres containing nitrogen gas, air, or under vacuum conditions [49]. The authors concluded that the organic layers under heat treatment were more stable in nitrogen and vacuum in comparison to air, due to oxidation. The decomposition of the prohibiting films was monitored by plotting the IR intensity of a vibration in the organic molecules and following the change in time.

### 2.1.4.4. Silver

In an atmospheric corrosion study by Wiesinger et al., of a silver surface exposed to 90% RH and 500 ppm $CO_2$, the ability of IRRAS and PM-IRRAS to detect corrosion products was compared [50]. As shown in Figure 3, PM-IRRAS was superior in detecting weak peaks and cancelling the effect of water vapor.

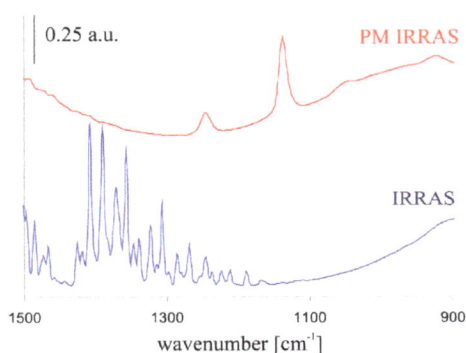

**Figure 3.** A comparison of PM-IRRAS and IRRAS to detect thin layers of corrosion products, revealing the clear benefit of using PM-IRRAS. Reprinted from [50] with permission from AIP Publishing. Copyright 2014.

The same research group used conventional IRRAS in combination with QCM to study the formation of basic carbonates on a silver surface under the influence of UV light under in-situ conditions [51,52]. IRRAS enabled the nature of the corrosion products to be determined, and QCM their mass. The bands at 922 ($[HO\text{-}CO_2]^-$ skeletal vibration), 1109 (asymmetric—C-O stretch), and 1221 (antisymmetric $CO_3^{2-}$ stretch) $cm^{-1}$ were used to identify the formation of $AgOHAg_2CO_3$. The growth of these bands with time indicated a continuous formation of basic silver carbonate. Irradiating the sample with UV light during the exposure significantly enhanced the infrared bands corresponding to the corrosion products, and accordingly an enhanced corrosion rate. Broad bands above 3000 $cm^{-1}$ indicated the presence of physisorbed water.

Further studies include investigations of the initial atmospheric corrosion of silver, iron, and copper, as investigated by IRRAS, QCM, AFM (in-situ), and XPS [53]. Under exposure to 90% RH, the IRRAS results indicate a continuous formation of cuprite (~650 $cm^{-1}$), whereas no corrosion products were observed at the iron surface, only physisorbed water. XPS indicated the formation of silver oxide and silver hydroxide. The thickness of the surface water layer, which greatly affects the corrosion rate, increased for a certain relative humidity in the order Fe < Cu < Ag, of which copper however displayed the most pronounced corrosion. Upon addition of $SO_2(g)$, $CuSO_3$ was detected as islands, but no sulfur containing species were detected on silver, only an increased corrosion rate. For iron, still no corrosion was observed, and a requirement to induce atmospheric corrosion of iron appears to be that also $NO_2(g)$ is added.

### 2.1.4.5. Zinc

Johnson et al., used IRRAS to study the atmospheric corrosion of zinc induced by the important indoor corrosion promoters formic (80 ppb) and acetic acid (100 ppb), as well as acetaldehyde (80 ppb), with a relative humidity of 90% [54–56]. Through the IRRAS data it was concluded that zinc carboxylates were formed during all three exposure conditions, and that the initial corrosion was fastest, followed by a lower rate, as indicated by comparing the absorbance of the antisymmetric carboxylate stretching vibration. The reduced corrosion rate by time is an indication that the formed corrosion products acted as a partly inhibiting layer. It was further concluded that the corrosion rate increased with an enhanced relative humidity for all three systems, since the water adlayer on the zinc surface was more pronounced at a higher humidity. This is shown in Figure 4, which clearly shows that the corrosion is faster at a higher relative humidity.

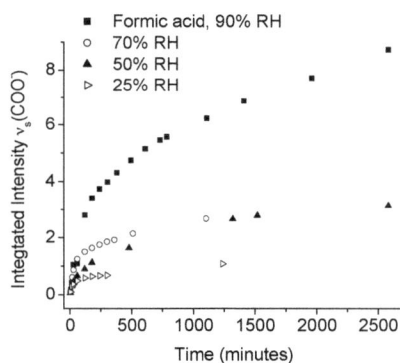

**Figure 4.** Zinc exposed to formic acid at different relative humidities. The x-axis shows the time and the y-axis the amount of zinc formate formed, as determined by the intensity of the antisymmetric formate stretch. Reproduced from [54] by permission of The Electrochemical Society. Copyright 2006.

In the exposure to formic acid, it was observed that the antisymmetric formate stretching vibration around 1630 cm$^{-1}$ was blue shifted (and broadened) when the humidity was changed from 90% to 0% RH, an indication of a change in the environment when water is removed and the interactions between zinc formate and water are reduced. The absence of a peak (C=O vibration) above approximately 1690 cm$^{-1}$ revealed that no or low amounts of physisorbed formic and acetic acid were present, but rather formate and acetate dominated. By studying the difference in wavenumbers between the symmetric and antisymmetric stretching vibrations it was concluded that the zinc acetate species formed a chelating bidentate or bridging structure. Since both the symmetric and antisymmetric carboxylate stretching vibrations possessed a significant intensity in all three systems studied, it was concluded that the zinc carboxylate species formed were randomly oriented. Complementary SEM examinations revealed that the corrosion products formed during acetic acid exposure had a radial growth, whereas acetaldehyde resulted in filiform corrosion [55]. Additionally, the role of the air/water interface in the above mentioned corrosion processes was investigated by VSFS, as discussed below.

The atmospheric corrosion of zinc was further studied by IRRAS by Qiu et al., in order to investigate the behavior of zinc exposed to 90% RH and either 120 ppb of formic, acetic, or propionic acid [57]. In order to deduce the mass of the corrosion products formed (ZnO at 570 cm$^{-1}$ and zinc carboxylates with main peaks originating from carboxylate stretching vibrations in the region

1345–1620 cm$^{-1}$), an optical model was used, and the properties of the corrosion products were further characterized by SEM and grazing incidence X-ray diffraction (GIXRD). The wavenumber difference in IR peak position between the symmetric and antisymmetric carboxylate stretching vibrations was used to conclude that the zinc formate formed a monodentate structure ($\Delta v = 275$ cm$^{-1}$), whereas exposure to acetic acid and propionic acid resulted in a chelating or bidentate structure ($\Delta v = 150$ cm$^{-1}$) [58].

Zhu et al., developed a cell for periodic dry/wet exposures, and examined the corrosion of zinc samples dipped for 1 h in a 1% aqueous NaCl solution, followed by a drying period of 23 h in 50% RH and a carbon dioxide concentration of <5 ppm or >350 ppm [15]. By following the intensity of the infrared bands (water bend at 1645 cm$^{-1}$ and water stretches at 3000–3600 cm$^{-1}$) originating from physisorbed water it was possible to follow the kinetics of the drying process, and after 6–8 h in 50% RH an equilibrium was reached. With $CO_2(g)$ concentrations below 5 ppm the dominant corrosion product was ZnO (560 cm$^{-1}$) and small amounts of $Zn_5(OH)_8Cl_2 \cdot H_2O$ (740 and 920 cm$^{-1}$) was additionally observed. In contrast, with carbon dioxide concentrations above 350 ppm, the corrosion products $Zn_5(OH)_6(CO_3)_2$ (840, 730, 1040, 1300–1600, and 3400 cm$^{-1}$) and $Zn_5(OH)_8Cl_2 \cdot H_2O$ were detected. In-situ IRRAS spectra during the first drying phase with a carbon dioxide concentration below 5 ppm is seen in Figure 5.

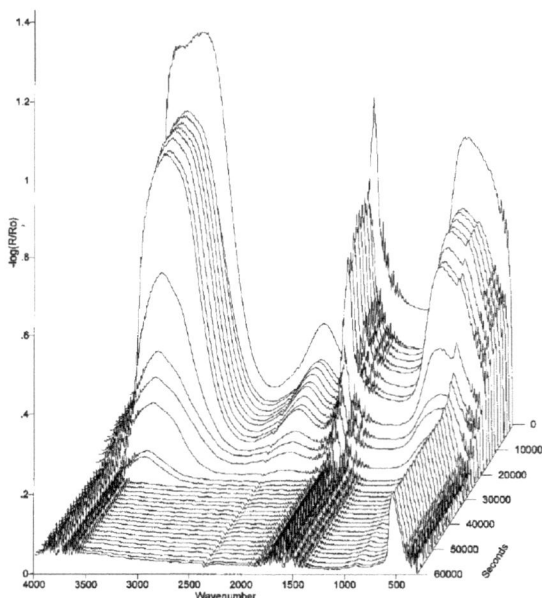

**Figure 5.** A series of IRRAS spectra acquired during the first drying phase of a zinc surface. The reduction of the water bands (~1645 and 3000–3600 cm$^{-1}$) shows the removal of physisorbed water at the surface. Reproduced from reference [15] by permission of The Electrochemical Society. Copyright 2001. The y-axis shows the absorbance given as $-\log(R/R_0)$, where R is the reflected radiation from the sample and $R_0$ from the background.

In order to obtain information about both the nature of the corrosion products formed and the changes in Volta potential, IRRAS was combined with Kelvin probe measurements by Person et al., [59]. Zinc was exposed to humid air and by IRRAS ZnO, $Zn_5(OH)_8Cl_2 \cdot H_2O$ and zinc hydroxy carbonate were observed as the corrosion products. Formation of these corrosion products was accompanied with an increase in Volta potential as a result of an enhanced inhibition of the film formed. In addition, IRRAS spectra of the corrosion products were simulated.

Kleber et al., studied the atmospheric corrosion of zinc, copper, and brass exposed to humid air and $SO_2(g)$, as described in Section 2.1.4.3 [37].

### 2.1.4.6. Brass

Qiu et al., studied the atmospheric corrosion of brass (Cu-20Zn) in 90% RH by IRRAS, and complemented the studies by SEM/EDS, AFM, cathodic reduction, confocal Raman microscopy, and scanning Kelvin probe force microscopy (SKPFM) [60]. The formation of ZnO and $Cu_2O$ was revealed by peaks at 570 and 660 $cm^{-1}$, respectively. By measuring the intensities of these bands as a function of time, it was possible to follow the kinetics of formation of the two corrosion products. As is described later, confocal Raman microscopy indicated that ZnO formed islands on a more fully covering cuprite layer. The same authors further studied the corrosion of brass exposed to formic, acetic, and propionic acid, and found by IRRAS that ZnO, $Cu_2O$, $Cu(OH)_2$ were formed [61]. To obtain more information about the corrosion products, XRD was used, and it was concluded that in the case of brass exposed to formic acid, $Cu(OH)_2 \cdot H_2O$ and $Zn(HCO_2)_2 \cdot xH_2O$ were formed, whereas for acetic acid $Zn_5(OH)_8(CH_3CO_2)_2 \cdot xH_2O$ was formed, but no crystalline oxide was observed. For propionic acid, no XRD peaks were observed, and hence the amount of crystalline products if any was below the detection limit.

As discussed in Section 2.1.4.3 on copper above, Kleber et al., studied the atmospheric corrosion of two brass alloys (Cu/Zn = 70/30 and 90/10 wt %), zinc, and copper exposed to humid air and $SO_2(g)$ [37].

### 2.1.4.7. Tin

Takeshi et al., studied the atmospheric corrosion of tin exposed to 80%–90% RH as well as $SO_2(g)$ and $NO_2(g)$ [62]. Exposure to sulfur dioxide resulted in no formation of corrosion products due to the inhibiting ability of the outermost tin oxide layer, whereas during exposure to nitrogen dioxide SnO (720 $cm^{-1}$), $SnO_2$ (610–670 $cm^{-1}$), tin nitrate (830, 1350, 1390, 1450 $cm^{-1}$), and hyponitrite (1040 $cm^{-1}$) were formed. The growth of corrosion products was observed as an increase in the IR absorbance by time. The chemical characterization was carried out in-situ by IRRAS, and complemented by Raman and XPS examinations. No synergistic effects of the two gases were observed.

### 2.1.4.8. Bronze

Wadsak et al., studied bronze exposed to 80% RH and 250 ppb $SO_2(g)$ with in-situ IRRAS, as well as AFM and XPS [63]. During exposure to only humid air, no cuprite was observed in the IRRAS spectra, but it was concluded that more water was physisorbed on bronze in comparison to copper. In fact, the corrosion on bronze was slower in comparison with copper due to the presence of lead oxide at the surface on bronze. The presence of hydroxide ions was revealed by a band at 1100 $cm^{-1}$. In exposure to humid air and sulfur dioxide, copper sulfite (1055 $cm^{-1}$) was observed and after 500 min exposure an infrared band originating from $Cu_2O$ at 650 $cm^{-1}$ appeared. The corrosion products could hence form when the protective layer of lead oxide was destroyed. IRRAS data also revealed that more water was adsorbed at the surface in presence of $SO_2(g)$ compared to its absence. With AFM it was concluded that larger corrosion products were formed in the presence of $SO_2(g)$.

### 2.1.4.9. Aluminum

The corrosion of aluminum surfaces covered with a sulfuric acid film formed from $SO_3(g)$ and $H_2O(g)$ at a pressure of 200 Torr was studied by Dai et al., [64]. The corrosion rate was observed to increase at higher relative humidities and higher deposition of sulfuric acid. From 3 to 240 min the peaks at 1103 $cm^{-1}$ ($SO_4^{2-}$) and 2510 $cm^{-1}$ ($Al_2(SO_4)_3$) increased in intensity, signifying deposition of sulfuric acid and an ongoing corrosion. The presence of the bisulfate ion was also confirmed by infrared bands at 890, 1050, and 1245 $cm^{-1}$.

### 2.1.5. Attenuated Total Reflection (ATR)

Stratmann and coworkers covered an ATR crystal with a 10 nm layer of iron, and exposed it to methyl-, and butyltrimethoxy silane at various relative humidities, resulting in a polymeric film at the surface [65]. The adsorption could be monitored by the use of a quartz crystal microbalance. At higher relative humidities thicker films were formed. Subsequently they exposed the system to 98% RH and 15 ppm $SO_2(g)$, and investigated the inhibiting properties of the polymer film. It was shown that the film increased the induction time for a corrosion process to be initiated significantly.

Person et al., used ATR as a complementary technique to IRRAS, as discussed above [25].

### 2.1.6. IR Microscopy

Infrared microscopy allows studies of the local distribution of corrosion products, which is highly important since a heterogeneous formation often occur. In conventional IR microscopy where the IR beam is focused at the surface, the spatial resolution is at the best some μm since the technique is diffraction limited. In contrast, novel IR microscopy techniques where the IR beam is focused on an AFM tip can improve the spatial resolution considerably, down to 10–20 nm. The latter type of technique has thus the potential to significantly provide more details in studies of atmospheric corrosion.

#### 2.1.6.1. Conventional IR Microscopy

The Leygraf group studied the atmospheric corrosion of copper with a single particle of sodium chloride deposited and exposed to 80% RH and different concentrations of carbon dioxide with IR microscopy (aperture size 100 × 100 μm), SEM, and Kelvin probe [66]. An electrolyte droplet formed around the salt particle and the corrosion products at different distances from the drop edge were examined by in-situ IR microscopy, as shown in Figure 6. The broad band centered at 3250 cm$^{-1}$ indicates the presence of water and hydroxide groups, the band at 1640 cm$^{-1}$ is the bending vibration of water, and the band at 1380 cm$^{-1}$ reveals the formation of carbonate (stretching vibration). A thicker electrolyte layer closer to the edge of the droplet results in larger infrared bands, and the spectra indicate that carbonate was formed in a region less than 100 μm. In similar experiments but with a carbon dioxide concentration lower than 5 ppm, an additional peak at 3572 cm$^{-1}$ appeared in certain spots, and was assigned to non-bonded OH groups, probably from a thin layer of $Cu(OH)_2$.

Similar studies were also performed on copper with deposited sodium chloride exposed to 150 ppb $SO_2(g)$ and with either <5 ppm or 350 ppm $CO_2(g)$ [41]. Similarly, a water droplet was formed at the surface, which led to galvanic corrosion. In the area of the droplet, CuCl(nantokite), $Cu_2OH_3Cl$ (paratacamite), sulfate, tiny amounts of carbonate, as well as $S_2O_6^{2-}$ (dithionate) were formed, with the latter being observed for the first time in an atmospheric corrosion process. In the area defined by the secondary spreading of the droplet, sulfite, sulfate, and dithionate were observed.

The same research group studied the atmospheric corrosion of zinc with deposited sodium chloride exposed to $SO_2(g)$, $CO_2(g)$ and 80% RH [67]. IR microscopy with an aperture of 100 × 100 μm was used to qualitatively estimate the thickness of the electrolyte layer at various positions by comparing the intensity of the water bands at 1645 and 3400 cm$^{-1}$. It was further possible to identify the corrosion products at different locations with ex-situ IR microscopy, as shown in Figure 7. In area A, the corrosion products consisted of $Zn_5(OH)_8Cl_2 \cdot H_2O$ (simonkolleite) as revealed by peaks at 3470 and 3550 cm$^{-1}$ from OH stretching vibrations, peaks at 727, 905, and 1044 cm$^{-1}$ from Zn-O-H vibrations, and ZnO was identified by a band between 477 and 575 cm$^{-1}$, whereas only a tiny peak originating from carbonate was observed at 1422 cm$^{-1}$. In area B and C the corrosion products were sodium carbonate. At <5 ppm $CO_2(g)$, the corrosion products have a local character, whereas a more general corrosion was observed for 350 ppm $CO_2(g)$. Further, a slower spreading of the droplet formed at the sodium crystal particle was observed at the higher carbon dioxide concentration. For exposures to $SO_2(g)$, no secondary spreading of the droplet was observed. Bands between 1065 and 1192 cm$^{-1}$

indicated the formation of sulfate ions, a formation that increased in the presence of sodium chloride. In contrast, no sulfite was observed.

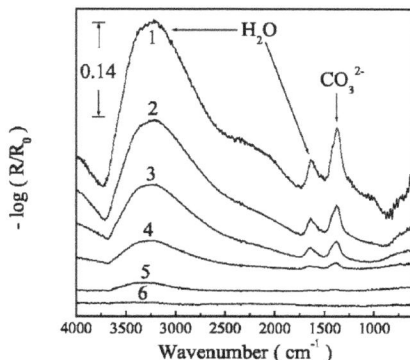

**Figure 6.** Infrared (IR) microscopy spectra taken at (1) −50; (2) −10; (3) 20; (4) 50; (5) 100; (6) 200 μm from the edge of the droplet [66]. A NaCl particle was deposited on the copper sample and it was exposed for 3 h to 350 ppm $CO_2$(g) and 80% RH. Negative values indicate that the spots are within the droplet. Reproduced from reference [66] by permission of The Electrochemical Society. Copyright 2005.

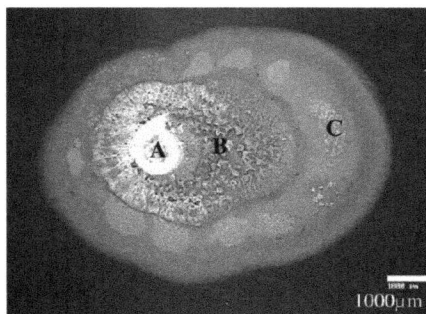

**Figure 7.** An optical image of a zinc surface with a deposited NaCl particle exposed to 90% RH for 6 h. The corrosion products identified in the three points A, B, and C are discussed in the text. Reprinted from reference [67] with permission from Elsevier. Copyright 2008.

Thierry and coworkers performed several IR microscopy studies of corroded aluminum and aluminum alloy surfaces. The AA6016 aluminum alloy coated by various methods (e.g., chromating and phosphating) was exposed to 85% RH at 25 °C for 6 weeks, and IR microscopy spectra were acquired with the aperture set to 250 × 250 μm [68]. The IR microscope allowed studies of the nature of the corrosion products both at the head and the tail of the filiform features, as shown in Figure 8. At the head (point and spectrum E), the corrosion products were identified as $Al_2(OH)_5Cl·2H_2O$ and/or $Al(OH)_2Cl$, and probably an aluminum hydroxide gel. In contrast, in the tail (B, C, D), the corrosion products were identified as an aluminum hydroxy carbonate gel. In the scratch, dawsonite (crystalline sodium aluminum hydroxyl carbonate) was observed, a compound which forms at a pH in the range 7.5–9.5, meaning that an enhanced pH value must be prevalent in the scratch. It was anticipated that the presence of hydroxide ions was due to the reduction of oxygen. The IR microscopy measurements were complemented with IR transmission spectra of reference compounds, as well as with Volta potential measurements that revealed that the head had a potential 400 mV lower than the tail.

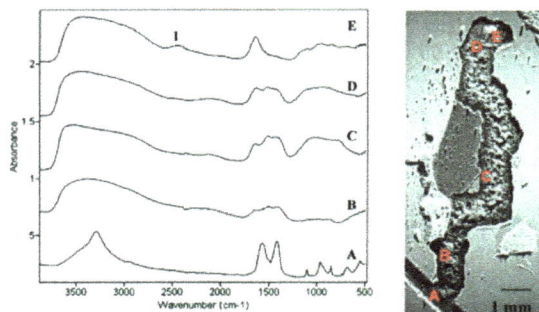

**Figure 8.** IR spectra obtained at the head and tail of the filiform corrosion products of the aluminum alloy AA6016 pretreated with Ti-Zr and covered with paint, exposed to 85% RH for six weeks [68]. Reproduced from reference [68] by permission of The Electrochemical Society. Copyright 2002.

The initiation and propagation of the filiform corrosion of the same alloy covered with a polyurethane or polyester coating was also investigated by IR microscopy and Kelvin probe [69]. The sample was exposed to 85% RH and 350 ppm $CO_2(g)$. To study the initiation of the filiform corrosion, a scratch was made in the coating, NaCl was added, 85% RH was introduced, and IR spectra were acquired as a function of time. At the initial dry conditions, $AlCl_3$ and $Al(OH)_2Cl$ were formed. The introduction of humid air results in the formation of a surface water layer, as revealed by broad bands in the region 3000–3600 cm$^{-1}$. Moreover, in the initial period (5 min) $Al(H_2O)_6^{3+}$ was identified through the peak at 2450–2500 cm$^{-1}$, which was reduced in intensity after 1 h. At the same time, filaments started to form as the adhesion of the coating was reduced, a result of a low pH value and high chloride concentration. After longer exposures the peak from $Al(H_2O)_6^{3+}$ vanished and carbonate peaks (bend at 850 cm$^{-1}$, stretches at 1090, 1430, and 1520 cm$^{-1}$) from an aluminum hydroxide gel containing carbonate appeared in the IR spectra taken in the scratch. The propagation of the filiform corrosion was further followed through the thin organic film with IR microscopy, and the movement of the head was followed by studying the peak from $Al(H_2O)_6^{3+}$ at 2500 cm$^{-1}$.

Extended investigations of the initiation and propagation of filiform corrosion on the same alloy were undertaken in order to study the effect of relative humidity, temperature, and wet-dry cycles [70]. The studies revealed that the filiform corrosion proceeded down to around 40% RH, the maximum filiform corrosion occurred at 75%–95% RH, an enhanced filiform corrosion was observed with an increasing temperature in the range 5–50 °C, and varying dry-wet cycle results were obtained depending on pretreatment.

Thierry and coworkers additionally studied the atmospheric corrosion of steel using IR microscopy. In one study the atmospheric corrosion of galvanized stainless steel covered with an epoxy resin was examined at a defect point at 90% RH [71]. Due to corrosion, de-adhesion of the coating occurred, and the corrosion products under the coating were identified by IR microscopy. It was concluded that in front of the de-adhesion simonkolleite was formed, whereas 1 mm away the main corrosion product was hydrozincite. Complementary scanning Kelvin probe measurements together with the IR microscopy measurements enabled the localization of the anode and cathode to be determined. IR microscopy was in addition used to identify anode and cathode areas on carbon steel suffering from atmospheric corrosion [72].

In a field study the distribution of corrosion products on nickel and zinc surfaces was examined by IR microscopy, and complemented studies with SEM/EDS were performed [73]. The corrosion products formed inhomogeneously over the surface and were dominated by $Zn_4CO_3(OH)_6 \cdot H_2O$, $ZnSO_3 \cdot nH_2O$, $Zn_4SO_4(OH)_6 \cdot nH_2O$, and $NiSO_4 \cdot xH_2O$.

## 2.1.6.2. Nano IR Microscopy

Although conventional IR microscopy has provided a lot of important information about the distribution of corrosion products, it suffers from the problems that the spatial resolution due to diffraction limitation at the best is around 5 μm, as well as that μm-thick films are required to be observed. With the novel technique nano FTIR microscopy both these problems are resolved, and the nature of corrosion products can be determined with a spatial resolution of 20 nm. Johnson et al., studied the distribution of the corrosion products cuprite and copper formate at a copper surface exposed to 80% RH and 100 ppb formic acid, to simulate an indoor corrosion process [74]. IRRAS was initially used to determine the nature of the corrosion products over the whole surface, and nano IR to scrutinize the spatial distribution of corrosion products. Figure 9a,b shows infrared spectra at certain positions of the corroded surface, where the spectra in Figure 9a reveal the presence of $Cu_2O$ as by the peak at 650 cm$^{-1}$, and the spectra in Figure 9b show spectra corresponding to copper formate at the points marked by "c" and "d", whereas no copper formate was present at point "e". The inset in Figure 9a is an AFM topography image that shows the positions of the points. Figure 9c shows an IR image acquired at 1600 cm$^{-1}$, which is the peak position of the strongest copper formate peak in Figure 9b. The particle denoted as "c" gives a large contrast and is hence enriched in copper formate, which agrees with the spectrum in Figure 9b where the same location results in the strongest spectrum. Figure 9d shows an IR intensity profile (1600 cm$^{-1}$) over a copper formate particle with a diameter of 150 nm, and the spatial resolution over the edge is 20 nm.

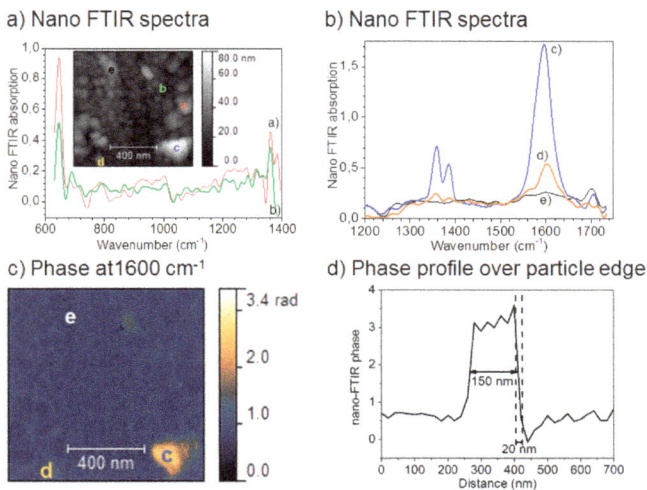

**Figure 9.** (a) Nano Fourier transform infrared spectroscopy (FTIR) spectra showing the presence of $Cu_2O$ at 650 cm$^{-1}$ with an inset displaying an AFM topography map; (b) IR spectra a point (c–e) in figure (a,c) an IR phase image at 1600 cm$^{-1}$ where the phase is related to the IR absorption; and (d) an IR profile over a particle showing a spatial resolution of 20 nm. Reprinted from reference [74] with permission from Elsevier. Copyright 2016.

Lyon and coworkers have used AFM-IR to study the heterogeneous structure in organic coatings as well as their water uptake [75–77]. In a study of a phenolic epoxy resin 100 nm thick, with a spatial resolution of 40 nm it was proven that curing of the film generated a chemically heterogeneous nanostructure, which is the origin of the nodular structure commonly found in exoxy [75]. Moreover, water uptake was investigated in epoxy-phenolic coatings and nanoscale variations were correlated with the cross-linking density at relative humidities of 35% and 65% RH [76]. The extent of the heat curing was followed by studying the intensity of the peak at 916 cm$^{-1}$, corresponding to the asymmetric oxirane ring deformation mode, and hence a lower intensity corresponds to a higher degree of local curing. The amount of water was estimated by the peak intensity at 3300 cm$^{-1}$, and the by rationing these peaks it was concluded that a high cross-linking density resulted in an enhanced local water uptake. Complementary AFM topography images were obtained simultaneously as the IR spectra, and ATR and DSC were used to further characterize the films. In another study the same research group investigated local water uptake resulting in ion channels in organic films [77]. Such studies are of importance in our understanding of why corrosion inhibiting organic films used for that appear to be unaffected suddenly do not work. By integrating the water band centered at 3416 cm$^{-1}$ in IR reflectance spectra at different times of exposure in 80% RH, it was possible to determine the absorption of water for dry and pre-soaked and then dried films. The presoaked films showed a larger water absorption, a result of the more open structure formed upon presoaking, a structure that was kept when drying. To examine how homogeneous the water uptake was, AFM-IR was used on ~500 nm thick epoxy-phenolic films on a mild steel substrate. A topography map by AFM revealed that immersion in water resulted in a rougher surface and the formation of bumps. When maps of the ratio of the IR signal at 3296 and 3420 cm$^{-1}$ (corresponding to water bands) at 60% and 30% RH were calculated, it was concluded that the bumps corresponded to a higher IR absorbance, and hence a higher local water uptake. The local water uptake in raised polymer regions is also revealed by studying images of the amplitude ratios at the wavenumbers 3420/2964 cm$^{-1}$ and 3296/2964 cm$^{-1}$, where 2964 cm$^{-1}$ is an absorption frequency of the epoxy film, 3420 cm$^{-1}$ corresponds to weakly hydrogen bonded water, and 3296 cm$^{-1}$ originates from strongly hydrogen bonded water. Figure 10 shows an AFM image in (a), which reveals raised regions, amplitude ratios acquired at the wavenumbers 3420/2964 cm$^{-1}$ and 3296/2964 cm$^{-1}$ in (b) and (c), as well as 3296/3420 cm$^{-1}$ in (d).

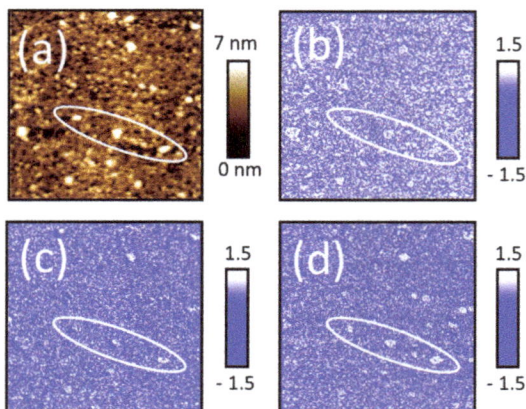

**Figure 10.** (**a**) Atomic force microscopy (AFM) topography image, amplitude images at the wavenumbers (**b**) 3420/2964; (**c**) 3296/2964; (**d**) 3296/3420. The circled regions are the ones discussed in the text. Reproduced from [77] with permission from the Royal Society of Chemistry. Copyright 2015.

### 2.1.7. Photoacoustic Infrared Spectroscopy

Palmer and coworkers developed a setup for photoacoustic IR spectroscopy, which enables in-situ studies in the field and provides information about chemical composition, thickness, quantitative analysis of corrosion products, and layering [78]. Calibration curves for brochantite ($Cu_4SO_4(OH)_6$) and antlerite ($Cu_3SO_4(OH)_4$) were obtained based on their infrared absorption frequencies and intensities, and allowed determinations of the brochantite/antlerite ratio with an accuracy of 10%.

## 2.2. Raman Spectroscopy

After the first experimental observation of Raman scattering by C.V. Raman in 1928 [79], and with the advent of laser, Raman spectroscopy has become a versatile analytical tool in many fields including surface science and corrosion. The Raman process is an inelastic scattering interaction of light with matter and in contrast to infrared processes, Raman occurs off-resonance. Upon excitation of a molecule to an intermediate virtual state, between two stationary exited states, a new photon is scattered from the virtual state accompanied by a relaxation of the molecule. As described in Figure 11, depending on the relative initial and final energy levels, Raman scattering phenomena can be defined as Stokes, Rayleigh, or anti-Stokes. Accordingly, in Raman spectroscopy, the band positions are associated with the difference between the frequencies of the exciting and scattered photon. The selection rule associated with Raman spectroscopy requires a change in the polarizability, $\alpha$, during a vibration.

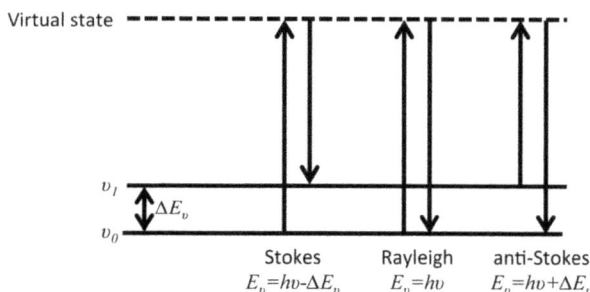

**Figure 11.** Raman scattering processes. $v_0$ and $v_1$ represent the ground and first excited vibrational energy states, respectively, with $\Delta E_v$ as the energy difference between them.

### 2.2.1. Conventional Raman Spectroscopy

#### 2.2.1.1. Steel

Atmospheric corrosion of iron and its alloys including steel has been extensively investigated using Raman spectroscopy. Li et al., characterized the rust formation on 1080 carbon steel after exposure to marine tests with a high concentration of $Cl^-$ in Hawaii [80] and utilized micro Raman spectroscopy to identify the main components of the corrosion products, lepidocrocite ($\gamma$-FeOOH) in the outer rust layer and goethite ($\alpha$-FeOOH) and akaganeite ($\beta$-FeOOH) in the inner rust layer. Complementary studies using scanning electron microscopy (SEM) and energy dispersive X-ray analyzer (EDXA) on the same point at which Raman spectra were taken enabled them to provide a schematic distribution of rust phases on different samples. They found a significant increase in corrosion rate for deposition rates of $Cl^-$ above a certain threshold (75 mg/m$^2$/day), which corresponds to the saturation of akaganeite with $Cl^-$. Below this threshold the corrosion rate of carbon steel samples was found to be independent of the $Cl^-$ deposition rate. The role of critical concentration of $Cl^-$ in the formation of akaganeite was also recently observed by Dhaiveegan et al., where the akaganeite corresponding Raman band appeared only after 2 years of exposure of 316 L and 304 stainless steels to industrial-marine-urban environment [81]. It was also showed that the characteristics of the rust layer on mild steel depend on the atmosphere salinity (chlorine ion deposition rate). At low salinity, an adherent rust layer is formed while for high salinity levels, the rust layer can easily exfoliate [82]. Raman peak positions obtained on different corrosion products of rust compounds are tabulated in reference [82]. Li et al., also investigated the very initial stages of NaCl particle induced atmospheric corrosion on 1080 carbon steel [83] combining in-situ and ex-situ Raman spectroscopy with SEM and optical microscopy. They found that the corrosion process starts with localized anodic and cathodic sites where green rust is formed in the regions close to anodic sites, lepidocrocite is mainly formed in the cathodic sites and magnetide ($Fe_3O_4$) is formed at the transition regions between anodes and cathods. The multilayer structure of the corrosion products was also observed on weathering steels with high concentration of copper, chromium, and nickel exposed to marine environments [84]. SEM-EDX analysis confirmed that nickel is distributed throughout the whole corrosion layer while the chromium concentration is higher at the inner part of the corrosion products. The innermost Cr-substitute geolite layer was believed to form the protective rust layer [85,86] limiting the penetration of the corrosive species toward the substrate. Superparamagnetic maghemite was also reported, based on Raman and Mössbauer spectroscopy, to exist in the inner layer of corrosion products and act as a protective layer [87]. Combined with X-ray diffraction (XRD) measurements it was found that lepidocrocite is the main compound of the outer corrosion product layer while the inner part was composed of ferrihydrite/low crystallized magnetite and goethite [88]. Similarly, higher amount of nickel in the composition of the weathering steels results in a greater corrosion resistance in marine environment by increasing the proportion of nanophasic or superparamagnetic goethite in the inner rust layer [89]. Hazan et al., also studied the atmospheric corrosion of AISI-4340 steel upon heat treatment in a high temperature and observed an intermediate layer between the outer wustite and the inner magnetite layers composed of small magnetite islands (bright phase) embedded in a wustite matrix (darker gray) [90].

In the presence of $SO_2$ and humidity in the atmosphere, rust layers on iron undergo a phase transition. Such a phase transition was followed using in-situ Raman spectroscopy [91]. It was found for instance that $Fe(OH)_3$ which initially is formed in the presence of several sulfur compounds is first transformed to an amorphous FeOOH, which later is crystallized by water loss. Based on these findings a minor modification to Evans model of atmospheric corrosion[92] was proposed.

Aramendia et al., used in-situ an hand held Raman spectrometer to study the rust formation and atmospheric corrosion of sculptures exposed to different conditions in the north of Spain [93]. They found goethite ($\alpha$-FeO(OH)) as the most stable phase in the corrosion products, accompanied by lepidocrocite, hematite ($\alpha$-$Fe_2O_3$), and magnetite. For the sculptures exposed to marine sites limonite ($FeO(OH)$-$nH_2O$) and akaganeita (b-FeO(OH)) were also identified. The same group used a dual laser wavelength (785 and 532 nm) portable Raman spectrometer in combination with a hand-held X-ray fluorescence spectrometer and chromatographic analysis to study the corrosion of medieval metallic artifacts from the 13th century [94]. The dual laser wavelength enabled identification of the Cu-based corrosion products phases such as cuprite, malachite, and bronchite (using 532 nm laser) together with Fe based corrosion products such as magnetite, goethite, lepidocrocte, and akaganeite (using 785 nm) on the same probe point of the sample.

Yucel et al., compared the atmospheric corrosion resistance of historic nails made of Ottoman period steels from 16th to 19th centuries and used micro-Raman spectroscopy to identify the composition of the inner and outer layer of the corrosion products [95]. They found that, a compact geolite layer is formed covering most of the inner corrosion layer resulting in enhanced corrosion protection in the Ottaman steel compared with current low alloy steels, indicating the success of iron metallurgy in that time. Although most of the atmospheric corrosion studies using Raman spectroscopy are limited to qualitative analysis, some limited works are available where Raman measurements lead to quantitative parameters. As an example, Monnier et al., analyzed ancient corroded iron samples and established quantitative composition 2D maps by mean of a home-developed software[96] based on spectral decomposition of experimental spectra by a linear combination of reference spectra. Applications of Raman spectroscopy in long term atmospheric corrosion studies on archeological iron samples and analysis of the ancient rust layers are not only important for the purpose of restoration of historical objects, but it also allows deeper understanding of long term corrosion mechanisms under atmospheric conditions. Such an understanding is important in the prediction and modeling of atmospheric corrosion of containers for storage of nuclear waste material [97].

### 2.2.1.2. Copper

Hayez et al., were the first group to explore the possibilities offered by Raman spectroscopy in studying the atmospheric corrosion of bronze statuary taking into account different types of atmospheric exposure conditions. They set up a Raman spectral database for compounds formed as the result of atmospheric corrosion of copper and its alloys in sulfur containing environments. This database includes different species of copper sulfates such as antlerite, brochantite, posnjakite, langite, and chalcanthite. The identified corrosion products on objects (specimens from sculptures) naturally exposed to an atmospheric environment were comparable to those obtained on copper (II) sulfate minerals [98]. Furthermore Hayez et al., investigated the impact of patination ingredients, the presence of additional elements like iron sulfate or chloride ions as well as the patination method on the chemical composition and nature of artificial patina using non-destructive Raman measurements [99]. The results obtained on artificial patinas were compared to the results obtained on reference products with known composition. Identification of the exact chemical component of artificial patina is a key parameter in restoration purposes, when replacement of a damaged naturally formed patina with an artificial one is required.

Bernardi et al. [100], with an innovative approach, comparatively investigated corrosion of G85 bronze in either acid rain solution collected from natural rain or in synthetic rain containing main organic components ($HCOO^-$, $CH_3COO^-$) and the aggressive inorganic components, $H^+$, $Cl^-$, $NO_3^-$, $NH_4^+$, $SO^{2-}{}_4$ mimicking the natural rain. They pointed out the effect of each alloying element in general corrosion behavior. As a result they found out that there is a slight difference between the corrosion of samples exposed to natural rain compared to those exposed to the synthetic rain. For instance, the identified corrosion products on samples exposed to natural rain were cuprite ($Cu_2O$), cerussite ($PbCO_3$), litharge ($PbO$), brochantite ($Cu_4SO_4(OH)_6$), and devillina ($CaCu_4(SO_4)^{2-}(OH)_6 \cdot 3H_2O$), whereas on samples exposed to the synthetic rain cuprite, mixed lead sulphates, and cerussite were identified in the corrosion products.

### 2.2.1.3. Silver

As a part of a larger project dealing with initial stages of atmospheric corrosion of silver, Martina et al., presented a catalogue of Raman spectra of silver compounds formed during atmospheric corrosion of pure silver using micro-Raman spectroscopy [101]. In this effort, especially for highly photosensitive compounds, micro-Raman spectroscopy with modulated laser intensity is advantageous as a nondestructive analytical tool. In this effort reference Raman spectra for the following silver compounds were obtained: Silver oxide ($Ag_2O$), silver (I) chloride ($AgCl$), silver (I) sulfide ($Ag_2S$), silver (I) sulfite ($Ag_2SO_3$), silver (I) sulfate ($Ag_2SO_4$), silver(I) carbonate ($Ag_2CO_3$), silver(I) acetate ($AgC_2H_3O_2$), and silver(I) nitrate ($AgNO_3$). These reference spectra can be used in corrosion studies of both pure silver and silver alloys. Furthermore, with the aim of understanding the role of environmental conditions promoting the atmospheric corrosion, pure silver was exposed to different controlled laboratory atmospheres (synthetic air, different relative humidity (50% and 90%), $SO_2$ (500 ppb), and $H_2S$ (500 ppb)) and formation of corrosion products with the thickness of only several monolayers was followed using micro Raman spectroscopy [102]. After even 24 h exposure of silver to $SO_2$ containing humidified air (with 50% RH or 90% RH) silver sulfate and silver sulfite were not identified as proper corrosion products. This reflects the low reactivity of silver towards $SO_2$. Comparing Raman spectra obtained on smooth silver exposed to 90% RH and those obtained on scratches on silver exposed to 50% RH revealed a secondary corrosion mechanism involving hydration reaction and chemisorption of gaseous species (oxygen, $SO_2$, and $CO_2$) in the water allayer. In contrast to these results, when silver was exposed to $H_2S$ containing humidified air a high reactivity of silver toward tarnishing (i.e., formation of $Ag_2S$) was observed, which was enhanced in the presence of higher amounts of humidity. Unlike the $SO_2$ exposure case, where the occurrence of secondary atmospheric corrosion processes was detected, after exposure of silver to $H_2S$ such a process was not observed.

2.2.1.4. Zinc

Ohtsuka et al., [103] investigated the effect of relative humidity (RH) on the formation of corrosion products on zinc in the presence of NaCl precipitations on samples using in-situ Raman spectroscopy. They identified amorphous zinc oxide as the main corrosion product when zinc was exposed to dry air. In a RH around 75%, amorphous ZnO, zinc carbonate $(ZnCO_3)_2(Zn[OH]_2)_3$ and traces of $Zn(OH)_2$ were identified. However, corrosion of zinc in presence of NaCl and high RH (>80%) was more complicated and occurred in different stages. In the first stage zinc chloride $(ZnCl_2)$ forms on the surface while zinc hydroxy chloride or simonkollite $Zn_5(OH)_8Cl_2 \cdot H_2O$ was formed as more advanced corrosion products. The formation of simonkollite included dissolution of NaCl particles, electrochemically coupled reactions and precipitation of concentrated zinc chloride.

Jayasree et al., studied the corrosion on zinc sheets exposed in marine exposure sites using FTIR and FT-Raman spectroscopy and identified two compounds, $NaZn_4Cl(OH)_6SO_4 \cdot 6H_2O$ and $Zn_4Cl_2(OH)_4SO_4 \cdot 5H_2O$ in the corrosion products [104]. They attributed the observation of multiple bands in the $\upsilon_s$ $SO_4$ vibrational mode region (433, 466, and 497 cm$^{-1}$ and its overtone at 887, 905 and 955 cm$^{-1}$) to a distorted structure for $SO_4$ anions in the corrosion products. The amount of distortion was higher in $Zn_4Cl_2(OH)_4SO_4 \cdot 5H_2O$ as a result of stronger hydrogen bonding network in its structure compared to $NaZn_4Cl(OH)_6SO_4 \cdot 6H_2O$.

Confocal Raman spectroscopy and micro spectroscopy were used to provide a laterally resolved chemical map of the localized corrosion products on zinc exposed to organic volatile solvents (formic acid and acetic acid) in dry and humidified air [105,106]. The identified corrosion products on samples exposed to acetic acid containing humid air included three dimensionally grown zinc hydroxy acetate $Zn_5(OH)_8(CH_3COO)_2 \cdot 4H_2O$ and zinc oxide distributed heterogeneously on the surface, as shown in Figure 12. Similarly, on the samples exposed to humidified air and formic acid, zinc oxide and zinc hydroxy formate were identified. However, on the sample exposed to dry air containing either acetic acid or formic acid, only zinc oxide was identified in the corrosion products. The distribution of zinc oxide was found to be more uniform compared to zinc hydroxy carboxylate.

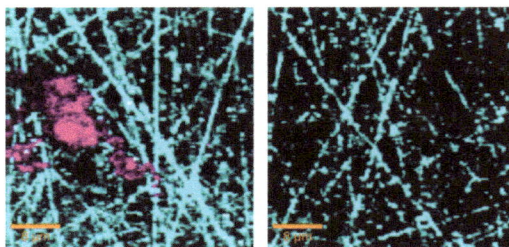

**Figure 12.** Color-coded Raman images for Zn exposed to 115 ppb acetic acid for 48 h in humid (**left**), and dry conditions (**right**). Purple color represents zinc hydroxy acetate (integrated 2850–3000 cm$^{-1}$) and blue color ZnO (300–650 cm$^{-1}$). The image size is 40 × 40 μm. Reproduced from reference [105] by permission of The Electrochemical Society. Copyright 2010.

An interesting observation from the confocal Raman microspectroscopy results obtained on zinc samples exposed to humidified air containing formic acid for 2 h is the different distribution of crystalline and amorphous zinc oxide as well as the zinc hydroxy formate (Figure 13). It was evident that more crystalline zinc oxide forms in the center of zinc hydroxy formate, while the amorphous zinc oxide is almost homogeneously distributed on the surface. After prolonged exposure (up to 48 h), the average Raman spectrum is dominated by copper hydroxy formate and Raman microspectroscopy shows that the zinc oxide and zinc hydroxy formate are still clearly separated, as seen in Figure 14.

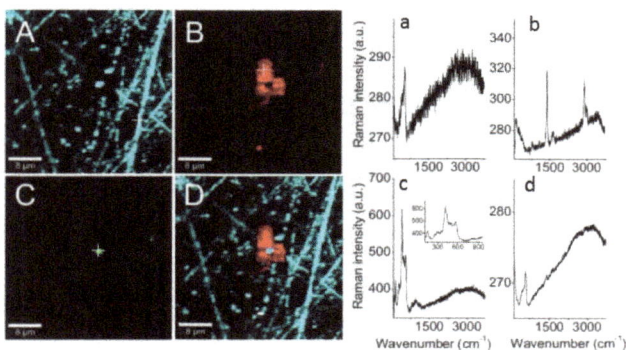

**Figure 13.** Left: Color-coded Raman images for zinc exposed to 100 ppb formic acid in humid air (90% RH) for 2 h. (**A**) ZnO (300–620 cm$^{-1}$); (**B**) CH in Zn hydroxy formate (2850–3000 cm$^{-1}$); (**C**) The region 380–480 cm$^{-1}$; (**D**) The combination of the Raman images in **A**–**C**. Right panel (**a–c**): Raman spectra corresponding to the marked spots in the color-coded images; (**d**) is the average over the whole image. Image size 40 × 40 μm. The Raman images were collected ex situ. Reproduced from reference [106] by permission of The electrochemical Society. Copyright 2010.

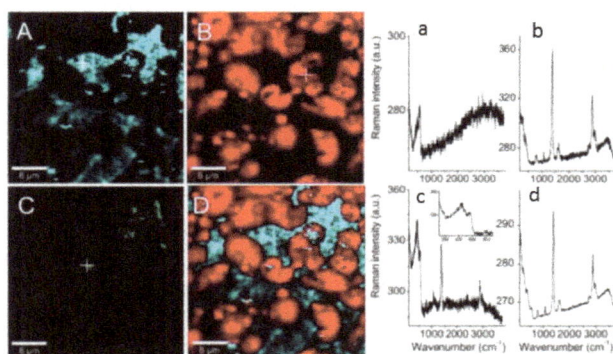

**Figure 14.** Left: Color-coded Raman images for zinc exposed to 100 ppb formic acid in humid air (90% RH) for 2 h (upper part) and 48 h (lower part). (**A**): ZnO (integrated 300–620 cm$^{-1}$); (**B**): CH in Zn hydroxy formate (2850–3000 cm$^{-1}$); (**C**): more crystalline ZnO (380–480 cm$^{-1}$); (**D**): the combination of the Raman images in **A**–**C**. Right panel (**a–c**): Raman spectra corresponding to the marked spots in the color-coded images. (**d**) is the average over the whole image. The image size is 40 × 40 microns. Reproduced from reference [106] by permission of The electrochemical Society. Copyright 2010.

When zinc samples were exposed to humidified air containing acetic acid for the period of 2 h, similar to the case of formic acid exposure, zinc oxide and zinc acetate form as corrosion products. However, unlike the formic acid exposure case where corrosion products were homogenously distributed on the surface, upon prolonged exposure (48 h) of zinc to acetic acid containing humidified air corrosion products form in two distinct morphologies: ring like disks and filaments, as shown in Figure 15. The ratio between different bands in the Raman spectra of these two aggregates indicates that zinc hydroxy acetate is formed with slightly different structures.

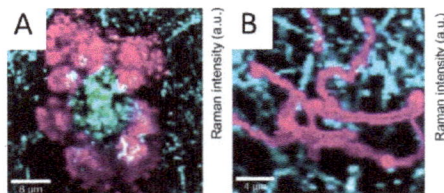

**Figure 15.** Formation of ring like (**A**) and filament-like corrosion products, characteristic of filiform corrosion (**B**). Reproduced from reference [106] by permission of The electrochemical Society. Copyright 2010.

### 2.2.1.5. Zinc Alloys

Atmospheric corrosion of duplex brass (Cu-20Zn) upon exposure to humidified air (90% RH) was studied by confocal Raman micro spectroscopy, IRRAS, scanning Kelvin probe force microscopy (SKPFM), atomic force microscopy (AFM), and scanning electron microscopy with energy dispersive X-ray analysis (SEM/EDS) [60]. Confocal Raman micro spectroscopy results demonstrated that after 3 days of brass exposure to humidified air, amorphous zinc oxide protrudes from more uniform copper (I) oxide ($Cu_2O$) on the surface. This distribution of oxides at the surface was attributed to variations in nobility along zinc rich and copper rich areas which result in galvanic effects. As a result, local growth of zinc oxide is accelerated and retards a more uniform growth of copper (I) oxide.

As an extension of the previous study, duplex brass (Cu-20Zn) samples were exposed to humidified air containing carboxylic acids (formic, acetic, and propionic acid) [61]. Raman micro spectroscopy together with IRRAS investigations revealed that the main corrosion products include copper oxide and zinc carboxylates (e.g., $Zn(HCOO)_2 \cdot xH_2O$ or $Zn_5(OH)_8(CH_3COO)_2 \cdot xH_2O$ or Zn-propionate). Investigations of the lateral distribution of the corrosion products on the samples after 3 days exposure to humidified air containing carboxylic acids indicated that the corrosion products are formed in a cell like structure, where centrally located zinc carboxylates are in general surrounded by areas of copper (I) oxide. The size of the identified corrosion cells on brass samples were in the order formic acid > acetic acid > propionic acid, which reflects the importance of ionic conductivity of the aqueous allayer (Figure 16).

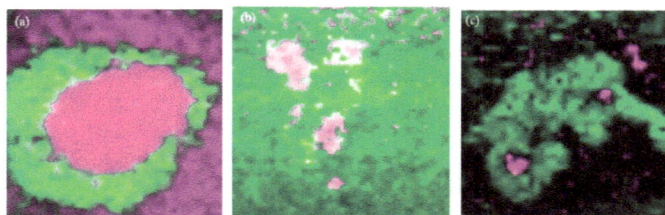

**Figure 16.** (Color online) Raman images of brass exposed 1 month to humidified air with formic acid (**a**); acetic acid (**b**); or propionic acid (**c**). The formation of Zn-carboxylate is represented by pink and of $Cu_2O$ by green. Scan size $10 \times 10$ μm. Reproduced from reference [61] by permission of The electrochemical Society. Copyright 2010.

Because of the increased need to understand the micro-galvanic effects during atmospheric corrosion of alloys, well defined patterned copper-zinc samples (25Cu-74Zn) were studied as a simple model for brass, as shown in Figure 17 [107]. After exposure of such patterned sample to humidified air (80% RH) containing formic acid for 5 days, the distribution of the corrosion products were evaluated by mean of confocal Raman microspectroscopy, displayed in Figure 18. The identified corrosion products included crystalline zinc oxide, zinc formate $Zn(HCOO)_2$, and hydrated zinc formate $Zn(HCOO)_2 \cdot 2H_2O$, as well as a very small amount of copper (II) oxide. In the corrosion products no copper (I) oxide was detected since its amount was below the detection limit. Based on the Raman peak position for the out of plane CH bending mode, which enables distinguishing between zinc formate and hydrated zinc formate [61], it was concluded that both compounds exist in the corrosion products. The combined microscopic and spectroscopic techniques in this study provided detailed information regarding the micro-galvanic coupling effect in atmospheric corrosion of brass. A clear separation between zinc oxide and zinc formate products was observed in the zinc region between copper islands where the local chemistry governs nucleation and growth of corrosion products and surface energy favors the formation of radially grown corrosion products.

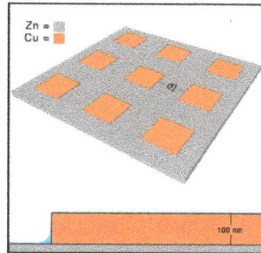

**Figure 17.** A schematic illustration of copper islands (brown) on zinc substrate (gray). Reproduced from reference [107] by permission of The Electrochemical Society. Copyright 2013.

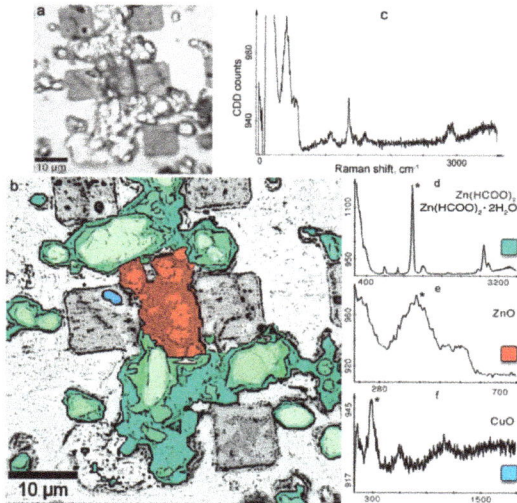

**Figure 18.** Confocal Raman image analysis of a patterned sample exposed for 5 days in 100 ppb formic acid in 80% RH. (**a**) An optical microscope image of the analyzed area; (**b**) A summary image of the local distribution of species based on interpolation of their characteristic Raman signals. The green color corresponds to interpolation between 1325 and 1435 $cm^{-1}$ (**d** \*) and denotes $Zn(HCOO)_2$ and $Zn(HCOO)_2 \cdot 2H_2O$. The red color corresponds to interpolation between 320 and 490 $cm^{-1}$ (**e** \*) and denotes poorly crystalline ZnO. The blue color corresponds to the interpolation between 250 and 320 $cm^{-1}$ (**f** \*) and denotes CuO; (**c**) An average spectrum of the red area. On the right-hand side there are average spectra (**d–f**) for the presented species in (**b**). Reproduced from reference [107] by permission of The Electrochemical Society. Copyright 2013.

When a similarly patterned sample was covered with a self-assembled monolayer of octadecanethiol (ODT) as a corrosion inhibitor before exposure to humidified air containing formic acid, the formation rate of corrosion products was initially decreased [108]. However, prolonged exposure resulted in disordering/removal of the ODT layer and an accelerated corrosion of the sample, as shown by VSFS. After 5 days of exposure Raman microspectroscopy showed a clear different distribution of corrosion products depending on the distance from the copper/zinc interface. In areas adjacent to the copper islands, hydrated zinc formate ($Zn(HCOO)_2 \cdot xH_2O$) is predominant, while hydrated zinc hydroxy formate ($Zn_5(OH)_8(HCOO)_2 \cdot xH_2O$) as well as great amounts of adsorbed water were identified in the more extended corrosion products further away from copper islands. The distribution of zinc formate over the whole distance range from copper islands was found to be more or less uniform. Such a distribution of corrosion products can be directly connected to the micro-galvanic effect as well as ion migrations during corrosion.

### 2.2.1.6. Other Elements and Alloys (Mo, Ni, Pb, SiC Fibers, U, Mg, Al, CdS, Graphene, $As_2S_3$)

Atmospheric corrosion of other metals than abovementioned alloys and metals have been also studied by Raman spectroscopy to a lesser extent and only very few examples of such studies exist in the literature on Mo, Ni, Pb, U, Mg, and Al.

The effect of formaldehyde in the indoor atmospheric corrosion of lead was studied by de Faria et al., [109]. Using a combination of SEM and Raman spectroscopy they showed that formats are produced as the results of exposure of lead coupons to formaldehyde and that oxidants such as $H_2O_2$ are not necessary. They observed significantly more complex Raman spectra for the corrosion products compared to the simple $Pb(HCO_2)_2$ spectrum.

### 2.2.2. Surface Enhanced Raman Spectroscopy (SERS) and Tip Enhanced Raman Spectroscopy (TERS)

Raman spectroscopy, despite its applications in determining the structure and composition of corrosion products suffers from a major challenge which is the very small signal to noise ratio if the number of scatterers are limited. This can be problematic especially when the intention is investigation of very thin corrosion products. However, enhancing the electric field either by roughening the sample surface (surface enhance Raman spectroscopy or SERS) or using a sharp tip as an antenna (tip enhanced Raman spectroscopy or TERS) allows huge enhancement of the Raman (by a factor of $10^6$) so that even single molecule detection becomes possible. Although there have been some efforts in using SERS and TERS in a combination with electrochemistry, very limited studies exist where enhanced Raman spectroscopy has been applied to study the atmospheric corrosion of metals. For instance, it has been shown that the atmospheric oxidation of silver, as the SERS substrate, in an ambient laboratory exposure air affects the enhancement factor dramatically such that the SERS signal intensity drops by more than 60% after only five hours exposure [110]. However, the main emphasis of this work is not studying of the corrosion of silver, but rather to emphasize the importance of substrate oxidation and contamination. Similarly, although silver is considered as a very good candidate to act as an antenna for enhancement of the Raman signal, its oxidation when used as the tip in TERS measurements in air, causes dramatic plasmonic degradation [111]. Therefore the same tip cannot be used for long term measurements and also cannot be stored.

One of the very few examples where SERS has been actually used in studying surface species upon corrosion is the work by Wang et al., [112] where a combination of in-situ, ex-situ SERS, and cathodic stripping voltammetry was used to study the effect of different exposure media including aerated 0.15 M NaCl, 0.10 M NaOH solutions, and air on the formation of corrosion product on molybdenum as well as Mo dissolution rate. The surface species were identified as molybdenum (IV) oxide, molybdate, and heptamolybdate. They also found that Mo, especially in basic solutions is not well passivated. Furthermore, Tormoen et al., [113] examined real time adsorption of volatile corrosion inhibitors on different substrates using in-situ SERS and concluded that below a relative humidity level of about 20% no adsorption of volatile corrosion inhibitors occurred regardless of the substrate.

*2.3. Vibrational Sum Frequency Spectroscopy (VSFS)*

A great challenge in studying surface reactions including oxidation and corrosion is that in most of available spectroscopic methods the signal generated from interfacial molecules is overwhelmed by the signal arising from the vast number of bulk molecules. Advances of non-linear spectroscopic tools including vibrational sum frequency spectroscopy (VSFS) and second-harmonic generation (SHG) allows probing surfaces and interfaces with no bulk signal contribution [114].

Briefly, as shown in Figure 19, in VSFS two laser beams (usually visible and IR) are spatially and temporarily overlapped on the sample surface and a third beam with its frequency equivalent to the frequency of the incoming beams is generated ($\omega_{VSF} = \omega_{Vis} + \omega_{IR}$). Under the dipole approximation, the sum frequency generation is not allowed in the centrosymmetric media including bulk materials where dipoles are randomly oriented. However, at surfaces and interfaces, due to the broken symmetry, a VSF signal is generated making this process inherently surface sensitive. Furthermore, so called non-resonant response in the VSF signal and its interference with the resonant signal can be used to probe the oxidation of the surface [115]. Besides, VSF measurements using a combination of S and P polarized beams allow an orientation analysis of the adsorbates on a surface to be performed. Therefore, VSFS is a suitable technique in studying corrosion and oxidation on a molecular level. In this respect, few studies have been specifically devoted to studying the atmospheric corrosion of metals.

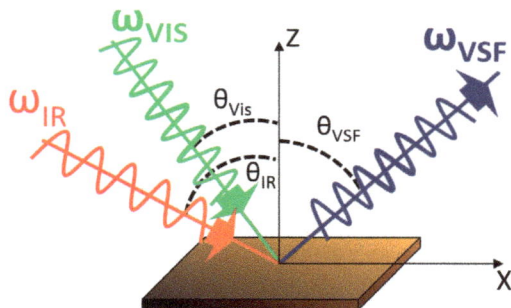

**Figure 19.** Schematic of the vibrational sum frequency spectroscopy (VSFS) geometry. The θ values represent the incident or reflected angles from the surface normal for visible, IR, and VSF beams. Each of the beams can be polarized perpendicular to the plane of incident (S-polarized) or parallel to it (P-polarized).

Atmospheric Corrosion Studies by VSFS

Hosseinpour et al., in series of studies investigated the initial stages of atmospheric corrosion of copper coated with a self-assembled monolayers (SAMs) as corrosion inhibitors, using a combination of IRAS, Raman, VSFS, quartz crystal microbalance (QCM), indirect nanoplasmonic sensing (INPS), and cathodic reduction. They adsorbed a SAM of octadecanethiol (ODT) on nonoxidized copper and observed a dramatic change in the appearance of the VSF spectra upon exposure of the sample to dry air. Qualitatively they attributed these changes to the formation of a very thin layer of copper (I) oxide underneath the ODT layer. Reducing the formed oxide in an electrochemical cell and comparing the results with in-situ IRRAS measurements they quantified less than a nanometer as the upper limit for the thickness of the oxide layer [115]. In Figure 20 the gradual change in the shape of VSF spectra upon oxidation of ODT coated copper during exposure to dry air is depicted. This provides an indirect study of the formation of $Cu_2O$, where the spectral changes are due to the fact that Cu and $Cu_2O$ have a different non-resonant signal which thus interferes differently with the $CH_2$ and $CH_3$ signal from the ODT chains.

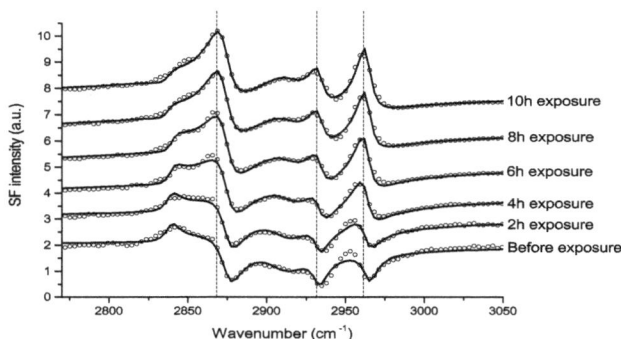

**Figure 20.** Time evolution of VSF spectra of the octadecanethiol (ODT) covered copper during exposure to dry air. Experimental data and fitted spectra are presented as empty circles and lines, respectively. The $CH_3$ symmetric, Fermi resonance, and antisymmetric stretches from ODT molecules are marked with the dash lines. The spectra are offset for clarity. Reprinted with permission from reference [115], Copyright 2011, American Chemical Society.

Integration of VSFS with a mass sensitive technique (QCM) allowed an in-situ quantification of the oxide layer thickness [116]. Hosseinpour et al., further concluded that the initial oxidation of ODT covered copper under dry air atmosphere occurs in two steps. The initial fast oxidation step was attributed to the penetration of oxygen through the ODT layer from easily accessible pathways, probably related to the gauche defects in the SAM layer. This initial fast step was followed by a slower oxidation rate with the rate limiting factor being the oxygen penetration through the SAM layer. Overall the VSFS-QCM combination allowed a quantification of oxide layer formation on copper with the resolution of 5% and 2% of an ideal cuprite layer for VSFS and QCM, respectively. Schwind et al., used an in-situ combination of VSFS-QCM and INPS to study the atmospheric corrosion and oxidation of ODT covered copper in both dry air and humidified air [117]. They found that introduction of humidity increases the oxidation rate by a factor of 4–5. Based on the VSFS results no significant change in the ordering and configuration of the protective SAM layer was observed in the course of copper oxidation. In contrast, when copper covered with alkanethiol SAMs of different chain lengths were exposed to a more corrosive atmosphere, consisting of 100 ppb formic acid and air with 85% relative humidity, an increased disordering in the SAM structure was observed during the corrosion process [118]. In Figure 21 the integration scheme for VSFS-QCM and INPS is depicted. This figure shows the advantage of integration of multiple techniques for in-situ measurements. An enhanced corrosion protection efficiency was observed for SAMs with longer chain lengths. It was also observed that the deposition of SAMs on the copper surface within experimental error completely stops the formation of copper (I) oxide, which is in stark contrast in comparison with bare copper. Such a retardation was less efficient against the formation of copper formate and copper hydroxide. VSFS studies showed that unlike alkanethiols which could be deposited on copper surfaces without any copper oxide being formed, SAMs of alkaneselenols (selenol is located below sulfur in the periodic system) always co-existed with copper (I) oxide at the surface [119]. This was attributed to the less efficient ability of alkaneselenols in replacing the very thin initial oxide on the copper surface during their adsorption. Nevertheless, both alkanethiols and alkanselenols initially formed order structures on the copper/copper oxide surface, as revealed by VSFS. However, upon prolonged exposure of the SAM covered samples to humidified air containing 100 ppb formic acid, the alkaneselenols were partially removed from the copper surface, while alkanethiols, though less ordered, remained on the surface as a protective layer. Consequently, no copper (I) oxide was observed on alkanethiol covered copper while a substantial amount of copper (I) oxide was identified on alkaneselenol covered copper after partial removal of the protective molecules from the surface due to the formation of localized galvanic couples (i.e., areas with and without alkaneselenols). VSFS studies on brass ($Cu_{20}Zn$ and $Cu_{40}Zn$) also showed that ODT can form ordered protective monolayer on both single and double phase Cu-Zn alloys [120]. The conformation of the protective layer on these samples did not change dramatically upon the sample exposure to humidified air containing formic acid. It was also showed that local galvanic effects on double phase brass resulted in less efficient corrosion inhibition [120]. However, due to the lack of lateral resolution in these VSFS measurements a direct comparison between the local structure of SAM and the induced inhibition was not feasible. Santos et al., utilized the lateral resolution offered by VSFS imaging microscopy to analyze the distribution of the copper oxide growth underneath a SAM of ODT upon spontaneous atmospheric oxidation [121]. Their findings also indicated that as the result of copper oxidation, the overall mean tilt angle of the ODT molecules at the surface decreases and the amount of gauche defects in the SAM structure increase. Furthermore, they observed a heterogeneous distribution of ODT molecules with different degrees of ordering, suggesting that oxidation of ODT covered copper is locally heterogeneous initiated from the domain boundaries moving inward.

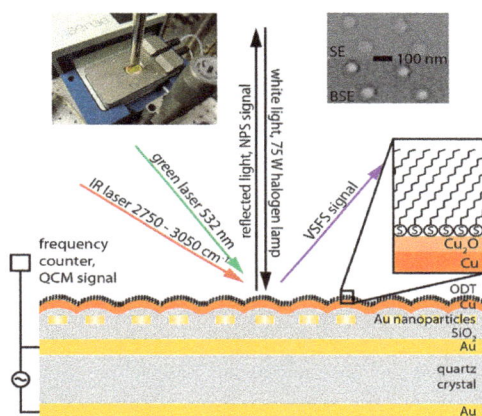

**Figure 21.** A schematic of the combined QCM-D and indirect nano plasmonic sensor, serving also as the VSFS sample. On the upper electrode of the QCM sensor, the Au nanoparticles for INPS are embedded in a $SiO_2$ film. On top of this film the actual sample is deposited, a Cu film with a corrosion protective ODT layer. The magnified picture of the latter shows the surface of the Cu film with its layer of $Cu_2O$ and with the ODT layer on top. The measurement setup is shown on the upper left including an image of the QCM-D window module with the copper coated quartz crystal-Au-nanoparticle-sensor and the reflection probe for INPS. The two SEM images (upper right), one taken with a secondary electron (SE) detector and the other with a backscattered electron (BSE) detector, show the sample surface and the INPS sensing Au nanoparticles. The low contrast in SE points to the fact that the Au nanoparticles are well covered by the $SiO_2$ layer and have no major influence on the surface roughness. Reprinted with permission from reference [117], Copyright 2013, American Chemical Society.

Hedberg et al., also used in-situ VSFS combined with IR studies and density functional theory (DFT) calculations to understand the initial stages of atmospheric corrosion of zinc initiated by formic acid in dry as well as humid conditions [122]. They found that adsorption of formic acid onto Zn/ZnO surface occurs independent of the presence or absence of humidity through a ligand exchange process. However, the presence of humidity accelerated the growth of the corrosion products layer, including formate. Their DFT calculations supported the idea of coordination of the formate to zinc ions without the participation of water molecules. They also found that formic acid is partly dissociated to fomate ions during adsorption to a ZnO surface with different hydration states depending on the relative humidity. The orientation analysis of zinc formate revealed that in humid conditions formate is ordered with the oxygen atoms toward the ZnO surface and the C-H moiety away from the surface [123]. Furthermore, VSFS studies provided evidence of zinc hydroxylation and gradual replacement of the hydroxyl groups by formate species, representing the initial stages of zinc dissolution induced by formic acid [124].

Very recently Gretic et al., applied self-assembled monolayer and multilayers of stearic acid as a corrosion protection system for copper-nickel alloy in a simulated marine exposure condition. They combined the results from electrochemical studies, AFM measurements, and contact angle measurements with those obtained from ellipsometry to assess the efficiency of the protective layers and their average thickness, respectively. In their studies, VSFS measurements were performed to assess the degree of ordering of the deposited layers on the copper-nickel surface using non-resonant suppression method developed by Lagutchev et al.[125] They found that the efficiency of the mono-/multi- layers in protecting the alloy surface from corrosion was highly dependent on the number of the layers as well as their homogeneity [126].

## 3. Conclusions

In this review article, we have summarized the main part of the work done during the last decades where vibrational spectroscopy has been employed to investigate atmospheric corrosion phenomena. A wide range of samples and exposure conditions, from outdoor atmospheric corrosion studies on exposed sculptures and historical objects to indoor and controlled exposure in-situ studies have been covered. In many cases, reference compounds are synthesized in laboratories and their spectra are compared with the sample spectra to characterize the corrosion products formed on objects with more extended and complicated corrosion. In addition, other kinds of techniques able to provide complementary information, such as the morphology or the mass of the corrosion products have been mentioned. Most of the studies presented here rely on vibrational spectroscopic techniques with a low spatial resolution (mm–cm), and hence the information obtained is an average over a large area. However, in order to further our knowledge in the field, it is desirable that a corrosion process can be studied and chemical information can be obtained at the nano level. This can now be achieved by novel IR and Raman spectroscopy techniques, with a spatial resolution as low as 10–20 nm, and has the possibility to open up new doors in the field of atmospheric corrosion.

**Author Contributions:** The authors have contributed equally to the work.

**Conflicts of Interest:** The authors declare no conflict of interest. The founding sponsors had no role in the design of the study; in the collection, analyses, or interpretation of data; in the writing of the manuscript, and in the decision to publish the results.

## References

1. Graedel, T.E. Corrosion mechanisms for silver exposed to the atmosphere. *J. Electrochem. Soc.* **1992**, *139*, 1963–1970. [CrossRef]
2. Persson, D.; Leygraf, C. Metal carboxylate formation during indoor atmospheric corrosion of Cu, Zn, and Ni. *J. Electrochem. Soc.* **1995**, *142*, 1468–1477. [CrossRef]
3. Daly, L.H.; Colthup, N.B.; Wiberley, S.E. *Introduction to Infrared and Raman Spectroscopy*, 3rd ed.; Academic Press Ltd.: London, UK, 1990.
4. Nakamoto, K. *Infrared and Raman Spectra of Inorganic and Coordination Compounds*, 4th ed.; John Wiley & Sons: New York, NY, USA, 1986.
5. Marcus, P.; Mansfeld, F. *Analytical Methods in Corrosion Science and Technology*; CRC Press, Taylor & Francis: Boca Raton, FL, USA, 2005.
6. Greenler, R.G.; Rahn, R.R.; Schwartz, J.P. The effect of index of refraction on the position, shape, and intensity of infrared bands in reflection-absorption spectra. *J. Catal.* **1971**, *23*, 42–48. [CrossRef]
7. Li, Q.X.; Wang, Z.Y.; Han, W.; Han, E.H. Characterization of the corrosion products formed on carbon steel in qinghai salt lake atmosphere. *Corrosion* **2007**, *63*, 640–647. [CrossRef]

8.  Wang, J.; Wang, Z.Y.; Ke, W. Characterisation of rust formed on carbon steel after exposure to open atmosphere in qinghai salt lake region. *Corros. Eng. Sci. Technol.* **2012**, *47*, 125–130. [CrossRef]

9.  Allam, I.M.; Arlow, J.S.; Saricimen, H. Initial stages of atmospheric corrosion of steel in the arabian gulf. *Corros. Sci.* **1991**, *32*, 417–432. [CrossRef]

10. Raman, A.; Kuban, B.; Razvan, A. The application of infrared spectroscopy to the study of atmospheric rust systems—I. Standard spectra and illustrative applications to identify rust phases in natural atmospheric corrosion products. *Corros. Sci.* **1991**, *32*, 1295–1306. [CrossRef]

11. Pacheco, A.M.G.; Teixeira, M.G.I.B.; Ferreira, M.G.S. Initial stages of chloride induced atmospheric corrosion of iron: An infrared spectroscopic study. *Br. Corros. J.* **1990**, *25*, 57–59. [CrossRef]

12. Morcillo, M.; Almeida, E.; Marrocos, M.; Rosales, B. Atmospheric corrosion of copper in ibero-america. *Corrosion* **2001**, *57*, 967–980. [CrossRef]

13. Almeida, E.; Morcillo, M.; Rosales, B. Atmospheric corrosion of zinc part 1: Rural and urban atmospheres. *Br. Corros. J.* **2000**, *35*, 284–288. [CrossRef]

14. Saxena, A.; Balasubramaniam, R.; Raman, S.; Raman, K. The forge-welded iron cannon at tanjore. In Proceedings of the Corrosion & Prevention and NDT, Melbourne, Australia, 23–26 November 2003; pp. 75/1–75/12.

15. Zhu, F.; Zhang, X.; Persson, D.; Thierry, D. In situ infrared reflection absorption spectroscopy studies of confined zinc surfaces exposed under periodic wet-dry conditions. *Electrochem. Solid State Lett.* **2001**, *4*, B19–B22. [CrossRef]

16. Wang, B.-B.; Wang, Z.-Y.; Han, W.; Wang, C.; Ke, W. Effects of magnesium chloride-based multicomponent salts on atmospheric corrosion of aluminum alloy 2024. *Trans. Nonferr. Met. Soc. China* **2013**, *23*, 1199–1208. [CrossRef]

17. Yang, L.-J.; Li, Y.-F.; Wei, Y.-H.; Hou, L.-F.; Li, Y.-G.; Tian, Y. Atmospheric corrosion of field-exposed az91d mg alloys in a polluted environment. *Corros. Sci.* **2010**, *52*, 2188–2196. [CrossRef]

18. Focke, W.W.; Nhlapo, N.S.; Vuorinen, E. Thermal analysis and ftir studies of volatile corrosion inhibitor model systems. *Corros. Sci.* **2013**, *77*, 88–96. [CrossRef]

19. Persson, D.; Leygraf, C. Analysis of atmospheric corrosion products of field exposed nickel. *J. Electrochem. Soc.* **1992**, *139*, 2243–2249. [CrossRef]

20. Fuller, M.P.; Griffiths, P.R. Diffuse reflectance measurements by infrared fourier transform spectrometry. *Anal. Chem.* **1978**, *50*, 1906–1910. [CrossRef]

21. Weissenrieder, J.; Kleber, C.; Schreiner, M.; Leygraf, C. In situ studies of sulfate nest formation on iron. *J. Electrochem. Soc.* **2004**, *151*, B497–B504. [CrossRef]

22. Kotenev, V.A.; Petrunin, M.A.; Maksaeva, L.B.; Sokolova, N.P.; Gorbunov, A.M.; Kablov, E.N.; Tsivadze, A.Y. Gravimetry, resistometry, and fourier-transform infrared spectroscopy for monitoring the corrosivity of the atmosphere with the use of an iron-oxide nanocomposite sensor layer. *Prot. Met. Phys. Chem. Surf.* **2013**, *49*, 597–603. [CrossRef]

23. Aramaki, K. Protection of iron corrosion by ultrathin two-dimensional polymer films of an alkanethiol monolayer modified with alkylethoxysilanes. *Corros. Sci.* **1999**, *41*, 1715–1730. [CrossRef]

24. Grundmeier, G.; Matheisen, E.; Stratmann, M. Formation and stability of ultrathin organosilane polymers on iron. *J. Adhes. Sci. Technol.* **1996**, *10*, 573–588. [CrossRef]

25. Persson, D.; Thierry, D.; LeBozec, N.; Prosek, T. In situ infrared reflection spectroscopy studies of the initial atmospheric corrosion of Zn–Al–Mg coated steel. *Corros. Sci.* **2013**, *72*, 54–63. [CrossRef]

26. Jang, J.; Kim, E.K. Corrosion protection of epoxy-coated steel using different silane coupling agents. *J. Appl. Polym. Sci.* **1999**, *71*, 585–593. [CrossRef]

27. Persson, D.; Leygraf, C. Vibrational spectroscopy and xps for atmospheric corrosion studies on copper. *J. Electrochem. Soc.* **1990**, *137*, 3163–3169. [CrossRef]

28. Persson, D.; Leygraf, C. In situ infrared reflection absorption spectroscopy for studies of atmospheric corrosion. *J. Electrochem. Soc.* **1993**, *140*, 1256–1260. [CrossRef]

29. Kleber, C.; Kattner, J.; Frank, J.; Hoffmann, H.; Kraft, M.; Schreiner, M. Design and application of a new cell for in situ infrared reflection–absorption spectroscopy investigations of metal–atmosphere interfaces. *Appl. Spectrosc.* **2003**, *57*, 88–92. [CrossRef] [PubMed]

30. Persson, D.; Leygraf, C. Initial interaction of sulfur dioxide with water covered metal surfaces: An in situ iras study. *J. Electrochem. Soc.* **1995**, *142*, 1459–1468. [CrossRef]

31. Aastrup, T.; Leygraf, C. Simultaneous infrared reflection absorption spectroscopy and quartz crystal microbalance measurements for in situ studies of the metal/atmosphere interface. *J. Electrochem. Soc.* **1997**, *144*, 2986–2990. [CrossRef]

32. Itoh, J.; Sasaki, T.; Seo, M.; Ishikawa, T. In situ simultaneous measurement with ir-ras and qcm for investigation of corrosion of copper in a gaseous environment. *Corros. Sci.* **1997**, *39*, 193–197. [CrossRef]

33. Wadsak, M.; Schreiner, M.; Aastrup, T.; Leygraf, C. Combined in-situ investigations of atmospheric corrosion of copper with sfm and iras coupled with qcm. *Surf. Sci.* **2000**, *454–456*, 246–250. [CrossRef]

34. Gil, H.; Leygraf, C. Quantitative in situ analysis of initial atmospheric corrosion of copper induced by acetic acid. *J. Electrochem. Soc.* **2007**, *154*, C272–C278. [CrossRef]

35. Gil, H.; Leygraf, C. Initial atmospheric corrosion of copper induced by carboxylic acids: A comparative in situ study. *J. Electrochem. Soc.* **2007**, *154*, C611–C617. [CrossRef]

36. Gil, H.; Leygraf, C.; Tidblad, J. Gildes model simulations of the atmospheric corrosion of zinc induced by low concentrations of carboxylic acids. *J. Electrochem. Soc.* **2012**, *159*, C123–C128. [CrossRef]

37. Kleber, C.; Schreiner, M. Multianalytical in-situ investigations of the early stages of corrosion of copper, zinc and binary copper/zinc alloys. *Corros. Sci.* **2003**, *45*, 2851–2866. [CrossRef]

38. Faguy, P.W.; Richmond, W.N.; Jackson, R.S.; Weibel, S.C.; Ball, G.; Payer, J.H. Real-time polarization modulation in situ infrared spectroscopy applied to the study of atmospheric corrosion. *Appl. Spectrosc.* **1998**, *52*, 557–564. [CrossRef]

39. Malvault, J.Y.; Lopitaux, J.; Delahaye, D.; Lenglet, M. Cathodic reduction and infrared reflectance spectroscopy of basic copper(ii) salts on copper substrate. *J. Appl. Electrochem.* **1995**, *25*, 841–845. [CrossRef]

40. Chen, Z.Y.; Persson, D.; Samie, F.; Zakipour, S.; Leygraf, C. Effect of carbon dioxide on sodium chloride-induced atmospheric corrosion of copper. *J. Electrochem. Soc.* **2005**, *152*, B502–B511. [CrossRef]

41. Chen, Z.Y.; Persson, D.; Leygraf, C. In situ studies of the effect of SO$_2$ on the initial nacl-induced atmospheric corrosion of copper. *J. Electrochem. Soc.* **2005**, *152*, B526–B533. [CrossRef]

42. Itoh, J.; Sasaki, T.; Ohtsuka, T.; Osawa, M. Surface layers formed initially on copper in air containing water vapor and so2 as determined by ir-ras and 2d-ir. *J. Electroanal. Chem.* **1999**, *473*, 256–264. [CrossRef]

43. Patois, T.; Et Taouil, A.; Lallemand, F.; Carpentier, L.; Roizard, X.; Hihn, J.-Y.; Bondeau-Patissier, V.; Mekhalif, Z. Microtribological and corrosion behaviors of 1H,1H,2H,2H-perfluorodecanethiol self-assembled films on copper surfaces. *Surf. Coat. Technol.* **2010**, *205*, 2511–2517. [CrossRef]

44. Nilsson, J.O.; Törnkvist, C.; Liedberg, B. Photoelectron and infrared reflection absorption spectroscopy of benzotriazole adsorbed on copper and cuprous oxide surfaces. *Appl. Surf. Sci.* **1989**, *37*, 306–326. [CrossRef]

45. Yoshida, S.; Ishida, H. A study on the orientation of imidazoles on copper as corrosion inhibitor and possible adhesion promoter for electric devices. *J. Chem. Phys.* **1983**, *78*, 6960–6969. [CrossRef]

46. Yoshida, S.; Ishida, H. An investigation of the thermal stability of undecylimidazole on copper by ft-ir reflection-absorption spectroscopy. *Appl. Surf. Sci.* **1995**, *89*, 39–47. [CrossRef]

47. Yoshida, S.; Ishida, H. The effect of chain length on the thermal stability of 2-alkylimidazoles on copper and 2-alkylimidazolato copper(ii) complexes. *Appl. Surf. Sci.* **1985**, *20*, 497–511. [CrossRef]

48. Kim, H.; Jang, J. Effect of copolymer composition in vinyl silane modified polyvinylimidazole on copper corrosion protection at elevated temperature. *Polymer* **1998**, *39*, 4065–4074. [CrossRef]

49. Opila, R.L.; Krautter, H.W.; Zegarski, B.R.; Duboisa, L.H.; Wenger, G. Thermal stability of azole-coated copper surfaces. *J. Electrochem. Soc.* **1995**, *142*, 4074–4077. [CrossRef]

50. Wiesinger, R.; Schade, U.; Kleber, C.; Schreiner, M. An experimental set-up to apply polarization modulation to infrared reflection absorption spectroscopy for improved in situ studies of atmospheric corrosion processes. *Rev. Sci. Instrum.* **2014**, *85*, 064102. [CrossRef] [PubMed]

51. Wiesinger, R.; Kleber, C.; Frank, J.; Schreiner, M. A new experimental setup for in situ infrared reflection absorption spectroscopy studies of atmospheric corrosion on metal surfaces considering the influence of ultraviolet light. *Appl. Spectrosc.* **2009**, *63*, 465–470. [CrossRef] [PubMed]

52. Wiesinger, R.S.J.; Hutter, H.; Schreiner, M.; Kleber, C. About the Formation of Basic Silver Carbonate on Silver Surfaces—An In Situ IRRAS Study. *Open Corros. J.* **2009**, *2*, 96–104. [CrossRef]

53. Kleber, C.; Hilfrich, U.; Schreiner, M. In situ investigations of the interaction of small inorganic acidifying molecules in humidified air with polycrystalline metal surfaces by means of tm-afm, irras, and qcm. *Surf. Interface Anal.* **2007**, *39*, 702–710. [CrossRef]

54. Johnson, C.M.; Leygraf, C. Atmospheric corrosion of zinc by organic constituents: Iii. An infrared reflection-absorption spectroscopy study of the influence of formic acid. *J. Electrochem. Soc.* **2006**, *153*, B547–B550. [CrossRef]

55. Johnson, C.M.; Leygraf, C. Atmospheric corrosion of zinc by organic constituents: Ii. Reaction routes for zinc-acetate formation. *J. Electrochem. Soc.* **2006**, *153*, B542–B546. [CrossRef]

56. Johnson, C.M.; Tyrode, E.; Leygraf, C. Atmospheric corrosion of zinc by organic constituents: I. The role of the zinc/water and water/air interfaces studied by infrared reflection/absorption spectroscopy and vibrational sum frequency spectroscopy. *J. Electrochem. Soc.* **2006**, *153*, B113–B120. [CrossRef]

57. Qiu, P.; Persson, D.; Leygraf, C. Initial atmospheric corrosion of zinc induced by carboxylic acids: A quantitative in situ study. *J. Electrochem. Soc.* **2009**, *156*, C441–C447. [CrossRef]

58. Petrie, W.T.; Vohs, J.M. An hreels investigation of the adsorption and reaction of formic acid on the (0001)-zinc surface of zinc oxide. *Surf. Sci.* **1991**, *245*, 315–323. [CrossRef]

59. Persson, D.; Axelsen, S.; Zou, F.; Thierry, D. Simultaneous in situ infrared reflection absorption spectroscopy and kelvin probe measurements during atmospheric corrosion. *Electrochem. Solid State Lett.* **2001**, *4*, B7–B10. [CrossRef]

60. Qiu, P.; Leygraf, C. Initial oxidation of brass induced by humidified air. *Appl. Surf. Sci.* **2011**, *258*, 1235–1241. [CrossRef] [PubMed]

61. Qiu, P.; Leygraf, C. Multi-analysis of initial atmospheric corrosion of brass induced by carboxylic acids. *J. Electrochem. Soc.* **2011**, *158*, C172–C177. [CrossRef]

62. Takeshi, S.; Ryoji, K.; Toshiaki, O. In situ ir-ras investigation of corrosion of tin in air containing $H_2O$, $NO_2$ and $SO_2$ at room temperature. *J. Univ. Sci. Technol. Beijing* **2002**, *10*, 35–38.

63. Wadsak, M.; Aastrup, T.; Odnevall Wallinder, I.; Leygraf, C.; Schreiner, M. Multianalytical in situ investigation of the initial atmospheric corrosion of bronze. *Corros. Sci.* **2002**, *44*, 791–802. [CrossRef]

64. Dai, Q.; Freedman, A.; Robinson, G.N. Sulfuric acid-induced corrosion of aluminum surfaces. *J. Electrochem. Soc.* **1995**, *142*, 4063–4069. [CrossRef]

65. Matheisen, E.; Nazarov, A.P.; Stratmann, M. In situ investigation of the adsorption of alkyltrimethoxysilanes on iron surfaces. *Fresenius' J. Anal. Chem.* **1993**, *346*, 294–296. [CrossRef]

66. Chen, Z.Y.; Persson, D.; Nazarov, A.; Zakipour, S.; Thierry, D.; Leygraf, C. In situ studies of the effect of $CO_2$ on the initial nacl-induced atmospheric corrosion of copper. *J. Electrochem. Soc.* **2005**, *152*, B342–B351. [CrossRef]

67. Chen, Z.Y.; Persson, D.; Leygraf, C. Initial nacl-particle induced atmospheric corrosion of zinc—Effect of $CO_2$ and $SO_2$. *Corros. Sci.* **2008**, *50*, 111–123. [CrossRef]

68. Le Bozec, N.; Persson, D.; Nazarov, A.; Thierry, D. Investigation of filiform corrosion on coated aluminum alloys by ftir microspectroscopy and scanning kelvin probe. *J. Electrochem. Soc.* **2002**, *149*, B403–B408. [CrossRef]

69. LeBozec, N.; Persson, D.; Thierry, D. In situ studies of the initiation and propagation of filiform corrosion on aluminum. *J. Electrochem. Soc.* **2004**, *151*, B440–B445. [CrossRef]

70. LeBozec, N.; Persson, D.; Thierry, D.; Axelsen, S.B. Effect of climatic parameters on filiform corrosion of coated aluminum alloys. *Corrosion* **2004**, *60*, 584–593. [CrossRef]

71. Nazarov, A.; Olivier, M.G.; Thierry, D. Skp and ft-ir microscopy study of the paint corrosion de-adhesion from the surface of galvanized steel. *Prog. Org. Coat.* **2012**, *74*, 356–364. [CrossRef]

72. Nazarov, A.P.; Thierry, D. Probing of atmospheric corrosion of metals: Carbon steel. *Prot. Met.* **2004**, *40*, 377–388. [CrossRef]

73. Lefez, B.; Jouen, S.; Kasperek, J.; Hannoyer, B. Ft-ir microscopic base imaging system: Applications for chemical analysis of zn and ni atmospheric corrosion. *Appl. Spectrosc.* **2001**, *55*, 935–938. [CrossRef]

74. Johnson, C.M.; Böhmler, M. Nano-ftir microscopy and spectroscopy studies of atmospheric corrosion with a spatial resolution of 20 nm. *Corros. Sci.* **2016**, *108*, 60–65. [CrossRef]

75. Morsch, S.; Liu, Y.; Lyon, S.B.; Gibbon, S.R. Insights into epoxy network nanostructural heterogeneity using afm-ir. *ACS Appl. Mater. Interfaces* **2016**, *8*, 959–966. [CrossRef] [PubMed]

76. Morsch, S.; Lyon, S.; Smith, S.D.; Gibbon, S.R. Mapping water uptake in an epoxy-phenolic coating. *Prog. Org. Coat.* **2015**, *86*, 173–180. [CrossRef]

77. Morsch, S.; Lyon, S.; Greensmith, P.; Smith, S.D.; Gibbon, S.R. Mapping water uptake in organic coatings using afm-ir. *Faraday Discuss.* **2015**, *180*, 527–542. [CrossRef] [PubMed]

78. Faubel, W.; Heissler, S.; Palmer, R.A. Quantitative analysis of corroded copper patina by step scan and rapid scan photoacoustic fourier transform infrared spectroscopy. *Rev. Sci. Instrum.* **2003**, *74*, 331–333. [CrossRef]
79. Raman, C.V.; Krishnan, K.S. A new type of secondary radiation. *Nature* **1928**, *121*, 501–502. [CrossRef]
80. Li, S.; Hihara, L.H. A micro-raman spectroscopic study of marine atmospheric corrosion of carbon steel: The effect of akaganeite. *J. Electrochem. Soc.* **2015**, *162*, C495–C502. [CrossRef]
81. Dhaiveegan, P.; Elangovan, N.; Nishimura, T.; Rajendran, N. Corrosion behavior of 316l and 304 stainless steels exposed to industrial-marine-urban environment: Field study. *RSC Adv.* **2016**, *6*, 47314–47324. [CrossRef]
82. Morcillo, M.; Chico, B.; Alcantara, J.; Daiaz, I.; Wolthuis, R.; de la Fuente, D. Sem/micro-raman characterization of the morphologies of marine atmospheric corrosion products formed on mild steel. *J. Electrochem. Soc.* **2016**, *163*, C426–C439. [CrossRef]
83. Li, S.; Hihara, L.H. In situ raman spectroscopic study of nacl particle-induced marine atmospheric corrosion of carbon steel. *J. Electrochem. Soc.* **2012**, *159*, C147–C154. [CrossRef]
84. Cano, H.; Neff, D.; Morcillo, M.; Dillmann, P.; Diaz, I.; de la Fuente, D. Characterization of corrosion products formed on ni 2.4 wt %-Cu 0.5 wt %-Cr 0.5 wt % weathering steel exposed in marine atmospheres. *Corros. Sci.* **2014**, *87*, 438–451. [CrossRef]
85. Cook, D.C.; Oh, S.J.; Balasubramanian, R.; Yamashita, M. The role of goethite in the formation of the protective corrosion layer on steels. *Hyperfine Interact.* **1999**, *122*, 59–70. [CrossRef]
86. Zhang, Q.C.; Wu, J.S.; Wang, J.J.; Zheng, W.L.; Chen, J.G.; Li, A.B. Corrosion behavior of weathering steel in marine atmosphere. *Mater. Chem. Phys.* **2003**, *77*, 603–608. [CrossRef]
87. Oh, S.J.; Cook, D.C.; Townsend, H.E. Atmospheric corrosion of different steels in marine, rural and industrial environments. *Corros. Sci.* **1999**, *41*, 1687–1702. [CrossRef]
88. de la Fuente, D.; Alcantara, J.; Chico, B.; Diaz, I.; Jimenez, J.A.; Morcillo, M. Characterization of rust surfaces formed on mild steel exposed to marine atmospheres using xrd and sem/micro-raman techniques. *Corros. Sci.* **2016**, *110*, 253–264. [CrossRef]
89. Diaz, I.; Cano, H.; de la Fuente, D.; Chico, B.; Vega, J.M.; Morcillo, M. Atmospheric corrosion of ni-advanced weathering steels in marine atmospheres of moderate salinity. *Corros. Sci.* **2013**, *76*, 348–360. [CrossRef]
90. Hazan, E.; Sadia, Y.; Gelbstein, Y. Characterization of aisi 4340 corrosion products using raman spectroscopy. *Corros. Sci.* **2013**, *74*, 414–418. [CrossRef]
91. Duennwald, J.; Otto, A. An investigation of phase transitions in rust layers using raman spectroscopy. *Corros. Sci.* **1989**, *29*, 1167–1176. [CrossRef]
92. Evans, U.R. Electrochemical mechanism of atmospheric rusting. *Nature* **1965**, *206*, 980–982. [CrossRef]
93. Aramendia, J.; Gomez-Nubla, L.; Castro, K.; Martinez-Arkarazo, I.; Vega, D.; Sanz Lopez de Heredia, A.; Garcia Ibanez de Opakua, A.; Madariaga, J.M. Portable raman study on the conservation state of four corten steel-based sculptures by eduardo chillida impacted by urban atmospheres. *J. Raman Spectrosc.* **2012**, *43*, 1111–1117. [CrossRef]
94. Veneranda, M.; Aramendia, J.; Gomez, O.; Fdez-Ortiz de Vallejuelo, S.; Garcia, L.; Garcia-Camino, I.; Castro, K.; Azkarate, A.; Madariaga, J.M. Characterization of archaeometallurgical artefacts by means of portable raman systems: Corrosion mechanisms influenced by marine aerosol. *J. Raman Spectrosc.* **2016**, *48*, 258–266. [CrossRef]
95. Yucel, N.; Kalkanli, A.; Caner-Saltik, E.N. Investigation of atmospheric corrosion layers on historic iron nails by micro-raman spectroscopy. *J. Raman Spectrosc.* **2016**, *47*, 1486–1493. [CrossRef]
96. Monnier, J.; Bellot-Gurlet, L.; Baron, D.; Neff, D.; Guillot, I.; Dillmann, P. A methodology for raman structural quantification imaging and its application to iron indoor atmospheric corrosion products. *J. Raman Spectrosc.* **2011**, *42*, 773–781. [CrossRef]
97. Neff, D.; Bellot-Gurlet, L.; Dillmann, P.; Reguer, S.; Legrand, L. Raman imaging of ancient rust scales on archaeological iron artefacts for long-term atmospheric corrosion mechanisms study. *J. Raman Spectrosc.* **2006**, *37*, 1228–1237. [CrossRef]
98. Hayez, V.; Guillaume, J.; Hubin, A.; Terryn, H. Micro-raman spectroscopy for the study of corrosion products on copper alloys: Setting up of a reference database and studying works of art. *J. Raman Spectrosc.* **2004**, *35*, 732–738. [CrossRef]

99.  Hayez, V.; Costa, V.; Guillaume, J.; Terryn, H.; Hubin, A. Micro raman spectroscopy used for the study of corrosion products on copper alloys: Study of the chemical composition of artificial patinas used for restoration purposes. *Analyst (Camb. UK)* **2005**, *130*, 550–556. [CrossRef] [PubMed]

100. Bernardi, E.; Chiavari, C.; Martini, C.; Morselli, L. The atmospheric corrosion of quaternary bronzes: An evaluation of the dissolution rate of the alloying elements. *Appl. Phys. A Mater. Sci. Process.* **2008**, *92*, 83–89. [CrossRef]

101. Martina, I.; Wiesinger, R.; Jembrih-Simbuerger, D.; Schreiner, M. Micro-raman characterisation of silver corrosion products: Instrumental set up and reference database. *e-Preserv. Sci.* **2012**, *9*, 1–8.

102. Martina, I.; Wiesinger, R.; Schreiner, M. Micro-raman investigations of early stage silver corrosion products occurring in sulfur containing atmospheres. *J. Raman Spectrosc.* **2013**, *44*, 770–775. [CrossRef]

103. Ohtsuka, T.; Matsuda, M. In situ raman spectroscopy for corrosion products of zinc in humidified atmosphere in the presence of sodium chloride precipitate. *Corrosion (Houston, TX, USA)* **2003**, *59*, 407–413. [CrossRef]

104. Jayasree, R.S.; Mahadevan Pillai, V.P.; Nayar, V.U.; Odnevall, I.; Keresztury, G. Raman and infrared spectral analysis of corrosion products on zinc $NaZn_4Cl(OH)_6SO_4 \cdot 6H_2O$ and $Zn_4Cl_2(OH)_4SO_4 \cdot 5H_2O$. *Mater. Chem. Phys.* **2006**, *99*, 474–478. [CrossRef]

105. Hedberg, J.; Baldelli, S.; Leygraf, C.; Tyrode, E. Molecular structural information of the atmospheric corrosion of zinc studied by vibrational spectroscopy techniques: I. Experimental approach. *J. Electrochem. Soc.* **2010**, *157*, C357–C362. [CrossRef]

106. Hedberg, J.; Baldelli, S.; Leygraf, C. Molecular structural information of the atmospheric corrosion of zinc studied by vibrational spectroscopy techniques. Ii. Two and 3-dimensional growth of reaction products induced by formic and acetic acid. *J. Electrochem. Soc.* **2010**, *157*, C363–C373. [CrossRef]

107. Forslund, M.; Leygraf, C.; Claesson, P.M.; Lin, C.; Pan, J. Micro-galvanic corrosion effects on patterned copper-zinc samples during exposure in humidified air containing formic acid. *J. Electrochem. Soc.* **2013**, *160*, C423–C431. [CrossRef]

108. Forslund, M.; Leygraf, C.; Claesson, P.M.; Pan, J. Octadecanethiol as corrosion inhibitor for zinc and patterned zinc-copper in humidified air with formic acid. *J. Electrochem. Soc.* **2014**, *161*, C330–C338. [CrossRef]

109. De Faria, D.L.A.; Cavicchioli, A.; Puglieri, T.S. Indoors lead corrosion: Reassessing the role of formaldehyde. *Vib. Spectrosc.* **2010**, *54*, 159–163. [CrossRef]

110. Matikainen, A.; Nuutinen, T.; Itkonen, T.; Heinilehto, S.; Puustinen, J.; Hiltunen, J.; Lappalainen, J.; Karioja, P.; Vahimaa, P. Atmospheric oxidation and carbon contamination of silver and its effect on surface-enhanced raman spectroscopy (sers). *Sci. Rep.* **2016**, *6*, 37192. [CrossRef] [PubMed]

111. Kumar, N.; Spencer, S.J.; Imbraguglio, D.; Rossi, A.M.; Wain, A.J.; Weckhuysen, B.M.; Roy, D. Extending the plasmonic lifetime of tip-enhanced raman spectroscopy probes. *PCCP* **2016**, *18*, 13710–13716. [CrossRef] [PubMed]

112. Wang, K.; Li, Y.-S.; He, P. In situ identification of surface species on molybdenum in different media. *Electrochim. Acta* **1998**, *43*, 2459–2467. [CrossRef]

113. Tormoen, G.; Burket, J.; Dante, J.F.; Sridhar, N. Monitoring the adsorption of volatile corrosion inhibitors in real time with surface-enhanced raman spectroscopy. *Corrosion* **2006**, *62*, 1082–1091. [CrossRef]

114. Miranda, P.B.; Shen, Y.R. Liquid interfaces: A study by sum-frequency vibrational spectroscopy. *J. Phys. Chem. B* **1999**, *103*, 3292–3307. [CrossRef]

115. Hosseinpour, S.; Hedberg, J.; Baldelli, S.; Leygraf, C.; Johnson, M. Initial oxidation of alkanethiol-covered copper studied by vibrational sum frequency spectroscopy. *J. Phys. Chem. C* **2011**, *115*, 23871–23879. [CrossRef]

116. Hosseinpour, S.; Schwind, M.; Kasemo, B.; Leygraf, C.; Johnson, C.M. Integration of quartz crystal microbalance with vibrational sum frequency spectroscopy–quantification of the initial oxidation of alkanethiol-covered copper. *J. Phys. Chem. C* **2012**, *116*, 24549–24557. [CrossRef]

117. Schwind, M.; Hosseinpour, S.; Johnson, C.M.; Langhammer, C.; Zorić, I.; Leygraf, C.; Kasemo, B. Combined in situ quartz crystal microbalance with dissipation monitoring, indirect nanoplasmonic sensing, and vibrational sum frequency spectroscopic monitoring of alkanethiol-protected copper corrosion. *Langmuir* **2013**, *29*, 7151–7161. [CrossRef] [PubMed]

118. Hosseinpour, S.; Johnson, C.M.; Leygraf, C. Alkanethiols as inhibitors for the atmospheric corrosion of copper induced by formic acid: Effect of chain length. *J. Electrochem. Soc.* **2013**, *160*, C270–C276. [CrossRef]

119. Hosseinpour, S.; Göthelid, M.; Leygraf, C.; Johnson, C.M. Self-assembled monolayers as inhibitors for the atmospheric corrosion of copper induced by formic acid: A comparison between hexanethiol and hexaneselenol. *J. Electrochem. Soc.* **2014**, *161*, C50–C56. [CrossRef]

120. Forslund, M.; Pan, J.; Hosseinpour, S.; Zhang, F.; Johnson, M.; Claesson, P.; Leygraf, C. Corrosion inhibition of two brass alloys by octadecanethiol in humidified air with formic acid. *Corrosion* **2015**, *71*, 908–917. [CrossRef]

121. Santos, G.M.; Baldelli, S. Monitoring localized initial atmospheric corrosion of alkanethiol-covered copper using sum frequency generation imaging microscopy: Relation between monolayer properties and $Cu_2O$ formation. *J. Phys. Chem. C* **2013**, *117*, 17591–17602. [CrossRef]

122. Hedberg, J.; Henriquez, J.; Baldelli, S.; Johnson, C.M.; Leygraf, C. Initial atmospheric corrosion of zinc exposed to formic acid, investigated by in situ vibrational sum frequency spectroscopy and density functional theory calculations. *J. Phys. Chem. C* **2009**, *113*, 2088–2095. [CrossRef]

123. Hedberg, J.; Baldelli, S.; Leygraf, C. Initial atmospheric corrosion of zn: Influence of humidity on the adsorption of formic acid studied by vibrational sum frequency spectroscopy. *J. Phys. Chem. C* **2009**, *113*, 6169–6173. [CrossRef]

124. Hedberg, J.; Baldelli, S.; Leygraf, C. Evidence for the molecular basis of corrosion of zinc induced by formic acid using sum frequency generation spectroscopy. *J. Phys. Chem. Lett.* **2010**, *1*, 1679–1682. [CrossRef]

125. Lagutchev, A.; Hambir, A.; Dlott, D. Nonresonant Background Suppression in Broadband Vibrational Sum-Frequency Generation Spectroscopy. *J. Phys. Chem. C* **2007**, *111*, 13645–13647. [CrossRef]

126. Hajdari Gretić, Z.; Kristan Mioč, E.; Čadež, V.; Šegota, S.; Otmačić Ćurković, H.; Hosseinpour, S. The influence of thickness of stearic acid self-assembled film on its protective properties. *J. Electrochem. Soc.* **2016**, *163*, C937–C944. [CrossRef]

MDPI AG

St. Alban-Anlage 66

4052 Basel, Switzerland

Tel. +41 61 683 77 34

Fax +41 61 302 89 18

http://www.mdpi.com

*Materials* Editorial Office

E-mail: materials@mdpi.com

http://www.mdpi.com/journal/materials